CHANGING CHINA

CHANGING CHINA

A Geographic Appraisal

Chiao-min Hsieh and Max Lu

Editors

Routledge
Taylor & Francis Group
New York London

First published 2004 by Westview Press

Published 2018 by Routledge
605 Third Avenue, New York, NY 10017
2 Park Square, Milton Park, Abingdon, Oxon OX14 4RN

Routledge is an imprint of the Taylor & Francis Group, an informa business

Library of Congress Cataloging-in-Publication Data

Changing China : a geographic appraisal / Chiao-min Hsieh and Max Lu, editors.
     p.   cm.
   Includes bibliographical references and index.
   ISBN 0-8133-3473-X (alk. paper) — ISBN 0-8133-3474-8 (pbk. : alk. paper)
   1. China—Economic policy—1976-2000.   2. China—Social conditions—1976-2000.
I. Hsieh, Chiao-min, 1921-   II. Lu, Max.
HC427.92.C343   2003
320.6'0951—dc22

                                                                                                2003014826

ISBN 13: 978-0-8133-3474-5 (pbk)

Typeface used in this text: AGaramond

Contents

Part Two   Social Changes

Part Three   Changes Along China's Periphery

Editors

Chiao-min Hsieh is Professor Emeritus of Geography at the University of Pittsburgh. He has been awarded the Fulbright Research Professorship three times and has been a Senior Fellow of the National Endowment for the Humanities.

Max Lu is Associate Professor of Geography at Kansas State University. He received his Ph.D. from Indiana University. His research concerns population, regional development and the interface between population, resources, environment, and development.

Contributors

Carolyn L. Cartier is an Associate Professor in the Department of Geography, University of Southern California, Los Angeles, U.S.A.

Kam Wing Chan is a Professor in the Department of Geography, University of Washington, Seattle, U.S.A.

Liping Di is a Research Professor and Director of the Laboratory for Advanced Information Technology and Standards (LAITS), George Mason University, Fairfax, Virginia, U.S.A.

C. Cindy Fan is a Professor of geography at the University of California, Los Angeles, U.S.A.

Charles Greer is an Associate Professor in the Department of Geography, Indiana University, Bloomington, U.S.A.

Chiao-min Hsieh is a Professor Emeritus of geography at the University of Pittsburgh, U.S.A.

Sun Sheng Han is an Associate Professor in the Department of Real Estate, School of Design and Environment, National University of Singapore.

Ronald G. Knapp is a Distinguished Professor Emeritus in the Department of Geography, State University of New York at New Paltz, U.S.A.

Chi Kin Leung is a Professor in the Department of Geography, California State University, Fresno, U.S.A.

C. P. Lo is a Professor of geography in the Department of Geography, University of Georgia, Athens, Georgia, U.S.A.

Dadao Lu is a Research Professor and former Director of the Institute of Geographical Sciences and Natural Resources Research, Chinese Academy of Sciences, Beijing, China.

Max Lu is an Associate Professor in the Department of Geography at Kansas State University, Manhattan, Kansas, U.S.A.

Kevin Matthews is a graduate student in the Department of Geography and Earth Science at George Mason University, Fairfax, Virginia, U.S.A.

Robert W. McColl is a Professor in the Department of Geography, University of Kansas, Lawrence, Kansas, U.S.A.

Clifton W. Pannell is a Professor of geography and Associate Dean of the Franklin College of Arts and Sciences at the University of Georgia, Athens, Georgia, U.S.A.

Mei-e Ren is a Professor and Honorary Chair in the Department of Geography, Nanjing University, Nanjing, China, and an academician of the Chinese Academy of Sciences.

Jianfa Shen is an Associate Professor in the Department of Geography and research fellow at the Center for Environmental Studies, Chinese University of Hong Kong, Shatin, Hong Kong, China.

Stanley W. Toops is an Associate Professor of geography and the Director of International Studies at Miami University, Oxford, Ohio, U.S.A.

Fahui Wang is an Associate Professor in the Department of Geography, Northern Illinois University, DeKalb, Illinois, U.S.A.

Shuguang Wang is an Associate Professor of geography in the School of Applied Geography, Ryerson Polytechnic University in Toronto, Canada.

Yehua Dennis Wei is an Associate Professor in the Department of Geography, University of Wisconsin, Milwaukee, U.S.A.

David Wong is an Associate Professor in the Department of Earth Systems and GeoInformation Sciences, School of Computational Sciences and Geography, George Mason University, Fairfax, Virginia, U.S.A.

Gang Xu is a Visiting Assistant Professor in the Department of Geography, University of Vermont, Burlington, U.S.A.

Runsheng Yin is an Assistant Professor in the Department of Forestry at Michigan State University, U.S.A.

Yi-Xing Zhou is a Professor of geography at Peking University, Beijing, China.

Preface

For thousands of years, a great human drama has been unfolding among the broad coastal plains, deep river valleys, and soaring mountains of a land called China. Today, at the dawn of the twenty-first century, the world is witnessing a new era of change in the lives of the people of this enduring civilization. China is undergoing a profound metamorphosis that is unmatched in its long history—simultaneous transitions from a command economy to a market-based one, and from a rural, agricultural society to an urban industrial power. Its double-digit growth rate, emerging market economy, expanding private enterprises, urban construction boom, rural industrialization, internal and external migration, and increasing links to the world economy are all reshaping the landscape of this fascinating realm. The whole world is watching the rapid rise of China and wondering about its intentions and aspirations as a superpower.

These concerns arise from the increasing scope and speed of China's impact on the rest of the world, both politically and economically, with one-fifth of the world's population and a territory rich with resources. China's potential influence on world affairs could be matched only by an imaginary megastate combining the nations of the United States of America and Western Europe.

We felt there was a need to examine more systematically and from a geographic perspective the dramatic changes that have taken place in China since the inception of its economic reform in 1978. Geographers have talents that allow them to describe the phenomena of changing China, such as migration, urbanization, rural industrialization, and the disparity between coastal and interior regions. Their sense of place, sympathy for synthesis, emphasis on man–land relations, comparative viewpoints of different regions and systems, ability to work and interview in the field, and talent in creating maps and charts—all make geographers well-qualified for such an important task. We are very fortunate to have assembled in *Changing China* a distinguished team of geographers from around the world, both inside and outside China, to provide a much-needed appraisal of post-reform China and its unprecedented social and demographic changes.

The book is divided into three parts that focus on different aspects of the changing geographies of post-reform China. The first part, with ten chapters, explains and
documents economic changes that have occurred since the reform and, in some
cases, prospects for future change. The issues discussed range from land-use patterns,
agricultural and industrial development, foreign direct investment, and high-tech
zones to China's trade links with other countries. The eight chapters in the second
part profile the contours of recent social changes in the country, including demographic changes, trends in internal migration, gender issues in industrialization,
management of large cities, suburbanization, and village transformation. Six chapters constitute the third part, which is titled "Changes Along China's Periphery."
The contributors of this part examine the future of Hong Kong, Taiwan, Inner
Mongolia, and Xinjiang as well as China's changing boundaries and the Chinese
mega-state. An introductory chapter provides an overview of the major themes and
conclusions of the chapters included in this volume, and in the afterword, one of us
speculates about China's future in the twenty-first century. To facilitate reading, this
book also contains an administrative map of China, a chronology of major events in
modern Chinese history and a glossary of the Chinese terms that appear in various
chapters. Most chapters were peer-reviewed and subsequently revised by their
authors. The chapters can be read in sequence, but they can also be treated as stand-
alone essays.

With a wide range of topics, this book provides the most comprehensive and
detailed account of a China that is undergoing a historical transition from a centralized command economy to a market-based one, and from a rural, agricultural
society to an urban, industrial power. Although this book has been designed as a
core text for upper-division undergraduate and graduate courses on China, or as a
supplemental text for courses on East Asia and international relations, anyone who
wants to gain a knowledge of contemporary China will find this book invaluable.

Chiao-min Hsieh and Max Lu

Acknowledgments

It is never easy to undertake a project as ambitious and complex as this one. We would like to acknowledge the people who helped us along the way. Our sincere thanks go to all the contributors whose work made this collection possible and whose insights and analyses constitute the virtues of this volume. A number of the authors and Dr. Hongmian Gong of Hunter College served as reviewers of the manuscripts, sometimes on short notice. We thank them for being so responsive and helpful. The complete set of manuscripts was later reviewed by two anonymous reviewers. Their constructive comments are much appreciated.

We would like to thank Drs. Richard Seckinger, Mark Periman, and Charles Coachman of the University of Pittsburgh and Dr. Martin Glassner of the University of Central Connecticut, who provided valuable suggestions and editorial assistance. We would like to express our gratitude to Karl Yambert, senior editor for geography at Westview, and Jennifer Chen, formerly assistant editor for geography at Westview, for their suggestions for improving the manuscripts and also for their patience. Thanks also go to Nanyan Weng and Stacia Wood of Kansas State University, who provided the much-appreciated editorial assistance while we worked on the final editing of the book. Without their help, the editing would have taken longer. Max would also like to thank his daughter, Andrea, who was a toddler while he worked on the demanding task of editing the book, for bringing him an enormous amount of joy and just the right amount of distraction. Finally, we would be remiss without acknowledging our respective universities and departments for providing an environment that facilitated our work.

Chiao-min Hsieh and Max Lu

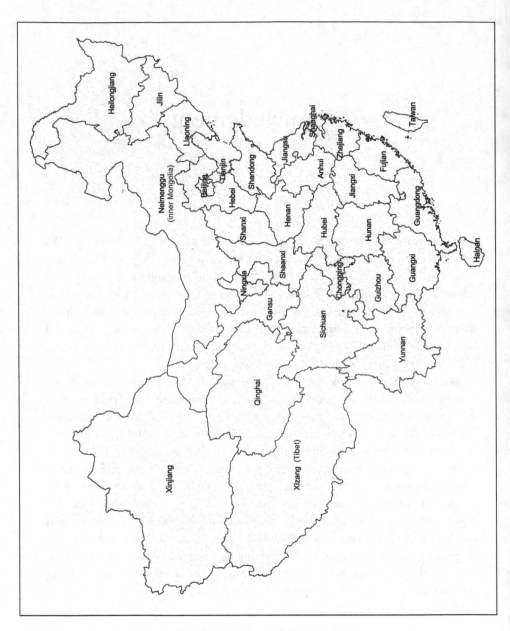

Administrative Map of China. Created by Max Lu

1

Introduction: The Changing Geographies of China

Max Lu

In 1978, the Chinese government launched its ambitious economic reform and opened the country to the outside world after nearly thirty years of rigid, Soviet-style development. The reform started in the countryside as a way of tackling the chronic low productivity and incentive problems that plagued Chinese agriculture. Following a remarkable success in rural reforms, the government began to introduce market principles into the industrial sector. It encouraged the growth of private enterprises and established special economic zones (SEZs) and open cities along the coast, in which tax and other incentives are provided to attract foreign investment and technological know-how. Although punctuated by such political events as the military crackdown of the 1989 pro-democracy movement, the economic reform and open-door policy have been a great success.

A quarter century has passed since the reform started. China has undergone a profound metamorphosis that is unmatched in its long history. Its sustained fast economic growth, burgeoning market economy, expanding private sector, rural industrialization, unprecedented interregional migration, and increasing links between the Chinese economy and the world economy are all reshaping the landscape of this fascinating and complex giant. For the first time in the People's Republic, the private sector has become the most vital part of the economy, producing a significant portion of the country's GDP and creating new jobs. The sizable middle class that emerged after the reform is growing. The vast majority of the Chinese are getting a chance to savor freedom in their lives and work. With their newly found wealth, many are purchasing color televisions, washers, computers,

1

cellular phones, even cars and homes—things they would not have even dreamed of not long ago. Millions of Chinese are also surfing on the Internet. The open-door policy has brought the Chinese people closer to the world. They pack the McDonald's, the Pizza Huts, the Ikea warehouse stores, the Putt Putt miniature golf courses, and Western-style nightclubs. They wear stylish clothes just as people in Paris or New York do. The glitzy shopping centers in many cities sell everything from the latest Celine Dion albums to expensive designer clothes and other luxuries. For better or worse, the youth of China are as familiar with Janet Jackson, Mariah Carey, Eminem, and Britney Spears as their counterparts in the West. New skylines have emerged in many Chinese cities as old, crumbling buildings give way to skyscrapers. Even as Japan sinks further into recession and Southeast Asian countries continue to struggle following the July 1997 currency crisis in Thailand, China remains an oasis of prosperity, with an annual growth rate of around 8 percent. As the twenty-first century unfolds, China's emergence as an economic superpower has implications not only for the Asia-Pacific realm but also for the world.

The reform and the open-door policy have also significantly altered the spatial pattern of China's economy. The coastal area has enjoyed a disproportionate share of domestic and foreign investment and rapid economic growth; the special economic zones and other types of open areas located mainly along the coast have become new centers of growth and trade. Coastal cities like Shenzhen, Shanghai, and Guangdong set the pace for the rest of the country. With the pre-reform balanced development strategy giving way to one that emphasizes efficiency, the coastal area of the country has benefited from the reform far more than the interior has. The regional disparity between the coast and the inland has widened substantially.

China has changed, and continues to change, rapidly. Understanding the changes that are occurring in China is important not only because of the sheer size of the country but also because of the expanding role China plays in various world affairs. This book documents, examines, and explains the changes that have taken place in China since the reform from a geographic perspective. As a science of space and place, geography provides a unique perspective for unraveling the forces behind the changes. Such knowledge helps to shed light on the future course China may take. The book focuses particularly on the themes and issues that have exerted and will likely continue to exert a significant impact on China's economy and society. The topics include, for example, land-use patterns and land-use change, agricultural growth and food security, rural surplus labor transfer, foreign investment distribution, industrialization, changing regional development patterns, migration, village transformation, large city expansion, ethnic diversity, and the future of Hong Kong, Taiwan, and China's minority areas. Although there are many other books on China's reform, economy, and politics, this book represents the first attempt to discuss the wide range of issues concerning contemporary China in a single volume.

In addition to this introductory chapter and an epilogue, the main body of the book consists of twenty-four chapters, which are organized into three sections. The first section, with ten chapters, focuses on various aspects of economic changes. The second section contains eight chapters about social changes. Six chapters form the third and final section, which discusses changes along China's periphery. The rest of this introduction provides an overview of the chapters in this volume. The overview is organized around the section themes.

Economic Changes

Part I addresses a number of very important economic issues in post-reform China. The first four chapters complement each other very well because they deal with different aspects of agricultural and rural development. Liping Di (Chapter 2) focuses his study on the types, determinants, patterns, and implications of land-use change in China. He states that although China's land-use patterns are the result of long-term actions of both natural and human forces on the land surface, human forces dominate current land-use change in China. He discusses the major types of land-use change since the reform, such as the conversion of farmland to urban and industrial use, the shift from growing grain to growing cash crops, farmland abandonment due to both the high opportunity cost of farming and industrialization, land reclamation, deforestation and reforestation, and desertification. The conversion of farmland to nonfarm uses is a key land-use change. The loss of prime farmland is particularly substantial in China's coastal provinces. Di also discusses the implications of farmland depletion and other land-use changes for China's food production and environment.

Clifton Pannell and Runsheng Yin (Chapter 3) address more specifically the problem of diminishing cropland. They argue that rapid farmland conversion to urban and industrial uses is putting a strain on China's grain production capability. The problem is particularly serious in densely populated coastal areas. However, the losses there are somewhat offset by gains in interior provinces, where limited economic opportunities offer more incentives for farming. Location and environmental characteristics, such as erosion and water supply, are significant determinants of cropland loss. Pannell and Yin point out that while conflicting data and unreliable statistics make an accurate assessment of farmland supply difficult, the situation may not be as serious as previously claimed, since recent studies indicate China's overall stock of cultivated land is greater than officially reported.

Although Di, Pannell, and Yin are all concerned with China's agricultural growth and food production in their studies, Jianfa Shen (Chapter 4) tackles the issue head-on. The debate over whether China will be able to feed itself unfolded soon after the publication of Lester Brown's 1995 book *Who Will Feed China?* Shen contributes to this ongoing debate by assessing China's future food security and the need for grain imports based on a demand and supply framework. He analyzes changes in national and regional grain production during 1990–1995 by examining four production factors: changes in the supply of arable land, the cropping index, the proportion of cropland for grain production, and grain yield per hectare. His analysis shows that major gains in grain output have come from increases in grain yield per capita and that major losses have resulted from the conversion of cropland to nongrain crops rather than from a decline in the amount of arable land. Changes in grain production differ among regions. Grain production dropped in eleven provinces between 1990 and 1995 because of diminished arable land, smaller proportion of cropland devoted to grain production, or decline in grain yield. Shen's simulations indicate that estimates of the amount of grain China will need to import from the world market in the future are very much dependent upon what assumptions one uses about per-capita grain consumption, yield, and proportion of cropland used for grain production. However, further boosts in yield are considered essential to securing sufficient grain supply for China's ever expanding population.

Sun Sheng Han (Chapter 5) deals with another very important issue in China: agricultural surplus labor transfer. Although rural labor surplus has always existed, economic reform brought the issue to the fore. By some estimates, the surplus amounts to about 250 million people. Han examines the concepts, characteristics, and impacts of agricultural surplus labor. He argues that the surplus labor has generated an active force toward the formation of innovative development policies. Whereas previous policy attempted to prevent the exodus of rural surplus labor, farmers are now allowed to settle in small towns and cities. Innovative policies have been devised to turn the surplus labor into a positive force that promotes rural industrialization and regional economic growth. The skills and remittances that rural migrants bring back to their hometowns are giving many poor areas a leg up in their attempt to spur economic development. Han also examines the problems related to rural labor transfer, such as the increased crime rates in cities, the feminization of the agricultural labor force due to the gender selectivity of rural out-migration, and the widespread environmental pollution caused by geographically dispersed rural industries. Appropriate government policies are therefore critically important in mitigating the negative effects while maximizing the benefits of rural surplus labor transfer.

The next two chapters deal with the changes in China's space economy since the initiation of economic reform. In Chapter 6, Dadao Lu documents the changes both in China's industrial development policies and in spatial patterns of development since the founding of the People's Republic in 1949. His discussion is organized around China's Five-Year Plans (FYPs) and based on a framework that includes three major points. First, industrial development was inward-looking and domestic resource-based in the pre-reform years but has moved toward globalization since the early 1980s. Second, the emphasis in industrial development has shifted from heavy industry to production of consumer goods and, more recently, to high-tech industry. Third, the planned economy has been gradually replaced by a market-oriented system. After the reform, industrial structure and spatial patterns have been shaped mainly by market forces. The regional disparity of industrial development, though mitigated to some extent in the three decades prior to the reform, remains substantial.

Max Lu (Chapter 7) further examines the recent changes in China's regional development patterns. He posits that two opposing forces are shaping the changes. One is the search for faster economic growth by local governments through developing industries with comparative advantages, which leads to regional specialization. The other is the regional imitation behavior prevalent in local governments' decision-making, which breeds homogenization of regional industrial structures. Empirical analysis of the employment data based on dissimilarity measures indicates that although the economic structure of most provinces is still very similar to that of the rest of the country, specialization has grown during the post-reform period. The author also argues that China's space economy may be experiencing an axial shift. The dominant role of the Yangtze River Delta as a traditional hinge region of China has been eroded to some extent by fast-growing South China, particularly Guangdong province. With the rise of South China, the return of Hong Kong to Chinese rule in July 1997, and the completion of the Beijing-Kowloon Railway, a north-south axis is emerging as the dominant economic flow, surpassing the traditional east-west one led by Shanghai. Nonetheless, the area surrounding Shanghai and the Yangtze River Delta, is likely to remain an important economic region of China.

The future role of the Yangtze River Delta area in China's space economy is explored by Mei-e Ren (Chapter 8). Ren argues that to achieve sustainable development in the region, four important issues must be addressed. They include the establishment of the Shanghai International Shipping Center, the readjustment of agriculture in response to increasing population and diminishing farmland, industrial restructuring, and environmental problems. He discusses the challenges facing the region's future.

Establishing various kinds of special zones seems to have become the method by which China furthers its reforms. Following the establishment of the special economic zones (SEZs) shortly after the inception of the reform, numerous economic and technological development zones have sprouted up all over China. These zones have already generated and will continue to generate profound impacts on China's regional economic development. In Chapter 9, Shuguang Wang examines the Chinese experience with the relatively recent incarnation of the SEZs—the New and High Technology Industrial Development Zones (NHTIDZs). These zones are designed to promote development of new products and industries supported by China's own scientific and high-tech capabilities and are modeled after technology parks or science cities in various Western countries such as the United States, Britain, and Japan. The influence of Northern California's Silicon Valley and Boston's Route 128 in the United States and the M4 Corridor in Britain has been especially significant. The NHTIDZs were designed to serve as new growth poles, spurring regional economic growth. Wang's analysis shows that remarkable achievements have indeed been made in those zones.

One objective of China's new development strategy is to attract foreign invest-ment (including capital from Hong Kong and Taiwan) and technological know-how to support its modernization drive. To that end, China set up special economic zones and devised many preferential policies. This strategy has been vastly successful. China has become the world's second largest recipient country of foreign direct investment, just after the United States. In 2002, China received $50 billion in foreign investment, more than any other country. Chi Kin Leung (Chapter 10) examines the factors that determine the locational choices of foreign investors and how the foreign capital inflow has affected regional economic devel-opment. Using North China's coastal region as a case study, Leung has analyzed both the provincial investment data and the information collected from 2,417 for-eign ventures in the region. He first details the locational and sectoral composi-tion of the foreign direct investment (FDI) with respect to three types of regions: city proper and inner suburb, outer suburb, and county-level units. He then dis-cusses the factors underlying the spatial and sectoral patterns of FDI and its local economic impact.

China's economic reform and open-door policy have not only attracted an enor-mous amount of foreign (including Hong Kong and Taiwan) investment but also significantly increased China's foreign trade. China is now the world's tenth largest exporter, with its exports in 2002 reaching $294 billion. Trade has transformed China from a virtually closed economy to a major participant in the international market. China's emergence as a major trading nation will have a tremendous impact on the rest of the world. In Chapter 11, Gang Xu discusses the characteristics of China's foreign trade and exchange regimes and the changes in trade balance and commodity composition since the reform, with special attention to China's trade links with countries in the Pacific Rim region. He argues that China's rapid rise to the ranks of the world's major trading powers is attributed to its export-oriented coastal development strategy. Xu analyzes the circumstances under which China launched the new development strategy and its success in drawing foreign direct investment to spur industrialization. The rapid economic growth in the coastal area strengthened China's trade links with Hong Kong and Taiwan, so much so that Xu believes a new Chinese Economic Zone is emerging.

By 1997, China had enjoyed two decades of almost uninterrupted double-digit economic growth. This remarkable economic performance alone made China the envy of the world. However, the lingering Asian financial crisis in the late 1990s was the true test of Chinese economy's capability to absorb external shocks and move ahead. The challenge to China's economy has been complicated by Japan's major economic slump in the 1990s, its first since World War II. Up until the end of 2002, China seemed immune to the severe "Asian financial flu." Its economy appeared to be growing at a healthy annual rate of 7–8 percent. Xu attributes China's resilience in the midst of the widespread Asian financial crisis to its persistent trade surpluses in recent years, substantial growth in FDI inflows, huge buildup of its foreign exchange reserve, which, at the end of 2002 had grown to $286 billion, or the second largest in the world, and last but not the least, the small proportion of foreign capital inflows devoted to portfolio investment in China. Portfolio investment, whose investors tend to take flight at the first sign of economic trouble, is widely believed to be the major cause of the recent Asian crisis.

Social Changes

The various social changes taking place in China are the focus of Part II of this book. The topics explored include population growth and characteristics, ethnic diversity, migration, gender, growth and management of large cities, suburbanization, and village transformation. In Chapter 12, Chiao-min Hsieh examines the shifts in China's population and composition over the past fifty years based on data from the five Chinese censuses conducted in 1953, 1964, 1982, 1990, and 2000. After comparing the five censuses and investigating patterns of population density and redistribution revealed in the 1990 census, he discusses China's population problem and population policies, which have changed from advocating large families to encouraging family planning. After 1979, the government began to enforce a policy of one-child families, a practice that has caused several social problems such as aging. Hsieh maintains that population control will be difficult, but China has no choice but to attempt to curb its rapid population growth.

Also using the Chinese census data, David Wong and Kevin Matthews (Chapter 13) examine the spatial dimensions of minority groups in China between 1980 and 1990. Several segregation and diversity measures are used to study the changing geographical dimensions of ethnic mixes at the national and provincial levels. Their study confirms earlier evidence of a massive reclassification of ethnic identity among minority populations. Although reclassification created dramatic increases in population size of some minority groups, they are still very small compared to the Han majority. The rising growth rates among minority groups have tended to alter the regional ethnic mixes, but the degree of change is dependent upon the scale of analysis. The population increases registered by many minorities have only slightly changed the national levels of segregation and diversity due to the relatively small size of most minority groups, but moderate changes were detected at the local level. In provinces such as Inner Mongolia (Neimenggu), Liaoning, and Sichuan, where the control over the growth of minority population had previously been rather tight, the minority's share in the total population has been rising.

The rapid structural changes in the Chinese economy, particularly its transition to a market economy, have resulted in a dramatic rise in population mobility. China's State Statistics Bureau estimated the number of internal migrants in 2002 at more than 120 million. With 73 percent of the migrants originating in rural areas, there is no doubt that China is experiencing a mass rural population exodus, much like many other developing countries. Kam Wing Chan (Chapter 14) provides an overview of China's recent internal migration. He first discusses the institutional factors that influence migration, particularly the role of *hukou,* or the household registration system, in controlling population movement. In his examination of the characteristics and patterns of recent migration, Chan demonstrates that at the macrosocietal level there are two different migration streams from different socioeconomic strata. They operate within fairly distinct "circuits" that are fixed by social and economic institutions based on the *hukou* system. While sharing some general demographic characteristics, they exhibit dissimilar socioeconomic characteristics and geographies due to the different opportunities and constraints they face. He also discusses major policy issues regarding China's internal migration and asserts that migration is crucial to China's current reform and future development. The "peasant invasion" of cities has not only challenged established attitudes and norms but also opened up possibilities for new social change and for the formation of a more diverse, tolerant, and pluralistic society. Because migration impacts both rural and urban sectors, as well as regional development, it has a strategic role to play in China's future development.

Chan's overview is followed by Cindy Fan's study (Chapter 15) on gender differences in China's internal migration. Fan points out that research on this topic has been meager, which reflects a prevailing view that female migration is problematic. The widely held notions that men's migration propensity is higher than women's, that men travel longer distances, and that men move primarily for economic reasons whereas women move for social reasons are mostly based on cursory observations and attributed to China's sociocultural traditions. Her argument is that gender differences in migration can no longer be explained only as an outcome of sociocultural factors but must also be interpreted in relation to other structural factors pertinent to China's transitional economy, such as spatial-economic changes and the *hukou* system. Her empirical analysis, based on the one-percent sample of the 1990 census, compares systematically the migration patterns and processes of Chinese men and women, evaluates the gender differences, documents the spatial patterns of male and female migration, and examines migrants' participation in the labor market. She concludes that although men's migration propensity is higher than women's, the latter is more highly represented in interprovincial long-distance moves and that female migration is characterized by strong economic rationale. In addition, occupational data suggest that migration has further reinforced gendered segmentation of the labor market. Her findings underscore the importance of examining male and female migration separately and of including the study of female migration in mainstream migration research.

Carolyn Cartier (Chapter 16) addresses a thus far largely ignored topic: gender issues in China's reform process. She echoes Fan's call for including gender analysis as a regular component of geographical research on China. She finds that changes in social and economic conditions since the reform have affected men and women differently. At the aggregate scale, the macroeconomic reform policies restructuring the organization of production have promoted new patterns of gendered labor as well as heightened awareness of gender issues. More specifically, gendered geographies have been structured particularly by trends produced by the complex interactions among five initiatives: 1) the open-door policy and the emphasis on export-oriented industrialization, which relies on low-wage women's labor for manufacturing; 2) the family planning policy emphasizing one child, which has tipped the birth rate in favor of males; 3) the rural household responsibility system, which has anchored many rural women's productive and reproductive activities in the household economy; 4) the loosening up of the *hukou* system, which has given rise to gendered migration patterns; and 5) the restructuring of state-owned enterprises, which has resulted in significant layoffs and has particularly limited women's employment options in a climate of high unemployment. Cartier assesses and critiques the gendered conditions in the Chinese society and economy since the reform.

China's urban policy has attempted to control the growth of large cities, but Chinese cities have nonetheless recorded rapid growth. Yehua Dennis Wei (Chapter 17) argues that the economic reforms, which emphasize decentralization, marketization, and globalization, have had a great impact on the growth of large cities. Local governments and foreign investment have also greatly shaped urban spaces in China. Wei discusses the growth and management of large cities following the reform. He concludes that the growth of large cities has had both positive and negative consequences and has seriously challenged China's approaches to urban planning and management. Following Wei's chapter, Yi-xing Zhou and Fahui Wang (Chapter 18) specifically discuss the trend toward suburbanization in Beijing based on China's third and fourth censuses. They found that most central city subdistricts *(jiedao)* have experienced significant losses of population, while suburban subdistricts have registered growth. However, suburbanization in China has mainly been caused by urban land-use reform, construction of thoroughfares in cities, suburban housing development, and overall improvement in suburban infrastructure. This is very different from the case in Western countries such as the United States.

Ronald Knapp (Chapter 19) examines rural habitat changes across space and through time as well as the role of the state in reshaping villages in Fujian and Taiwan. He argues that Chinese rural settlements traditionally took shape over a lengthy period of time as a result of population growth, small-scale agricultural development, and popular beliefs concerning *fengshui*. But over the past half century, more formal planning, top-down political decisions, and the broad economic policies of the state have significantly reshaped the Chinese countryside. Although there were similarities in traditional dwellings and village forms between Fujian and Taiwan, quite different political and social systems since 1949 have acted to alter the geometry and morphology of villages. The appearance and function of the hybrid rural settlements echo traditional village forms but are often neither rural, nor urban, nor suburban settlements. He documents the recent trend toward convergence in the appearance of Taiwan and mainland villages and also notes problems that exist in rural villages, from wasteful use of building materials to environmental deterioration.

Changes Along China's Periphery

Six chapters in Part III of this book deal with issues related to China's peripheral regions (Xinjiang and Inner Mongolia) and what has been sometimes called Greater China (Hong Kong, Taiwan, etc.). In Chapter 20, Chiao-min Hsieh reviews the past and present conditions of China's boundaries. He states that the approach to different boundary situations varies, depending on the historical, ethnic, physical, and political overtones that form the basis for conflict and change. China's territorial claims seem to be predicated upon the boundaries of the Manchu Empire of 1840. The Chinese regard their territorial claims as political and ideological, rather than judicial, issues. To China, territorial claims are "a problem left from imperialism." There are boundary disputes between China and most of its fourteen neighbors. Hsieh argues that the two most critical boundary disputes are with Russia and India. The disputes between Russia and China in inner Asia stem from ethnic problems rather than boundary problems per se. Outer Mongolia has been set up as a buffer zone between the two countries, but that boundary has never been a rigid one and remains in dispute. The Sino-India Agreement of Trade and Intercourse contained five principles of peaceful coexistence but was abandoned in 1962. Since then, the two nations have been engaged in a bitter dispute. Much of the trouble stems from the creation of buffer zones and states by Russia and Britain. The geography of the region does not allow for simple delineation along physical lines, and the nomadic local people are characterized by their shifting allegiances. To deal with boundary problems of the small Himalayan states, China has tried to isolate India from other Asian states that have policies unfavorable to China. Although disputes over the China-North Korea border and the China-Vietnam border exist, these will probably not become serious because of close political ties between those nations. It is worth pointing out, however, that since the early 1990s China has signed agreements with Russia, Kyrgyzstan, Tajikistan, and Vietnam to settle their border disputes. China has also been engaged in negotiations with India on the demarcation of the 2,486-mile-long boundary between them.

In Chapter 21, C. P. Lo evaluates the political and economic significance of Hong Kong's return to China on July 1, 1997, by examining the factors that affected Hong Kong's economic development before and after the changeover. He states that the decade preceding the reversion witnessed increasing economic integration between Hong Kong and mainland China, which diminished the importance of the British protective umbrella. The restructured Hong Kong economy complements China's by way of the "front shops, rear factories" model. Politically, the "one country, two systems" principle has given Hong Kong Special Administrative Region (SAR) a high degree of autonomy, which will keep Hong Kong's capitalist system and way of life intact for fifty years. Hong Kong's more open and democratic government system provides its residents a higher degree of freedom in speech and press than exists in mainland China, although recent security legislation by the Hong Kong government may sharply curtail that freedom. The author claims that Hong Kong may provide a model of democratization for China to emulate in the future. Hong Kong's Chief Executive Tung Chee-hwa has so far managed to meet all the political and economic challenges. Lo argues that if Hong Kong is to maintain its status as a world city and finance center, it must have strong leadership and project to the world an image of an open, pollution-free environment where law and justice prevail.

In Chapter 22, Chiao-min Hsieh traces the background of the Taiwan Strait conflict. He notes that in the late 1970s, the communists abandoned their plan to "liberate Taiwan" and began a program of "peaceful unification." The nationalists in Taiwan responded by shifting their focus to the "principles and doctrine to unify China." Taiwan's former president, Chiang Ching-kuo, lifted a travel ban across the strait, revoked martial law, and tolerated the formation of the Democratic Progressive Party, which radically transformed the relations between the two adversaries. But when Lee Teng-hui succeeded Chiang Ching-kuo in 1988, relations across the strait became uncertain. Without attachments to the mainland, Lee worked toward Taiwanese independence. Hsieh argues that as the civil war generation leaves the political scene, the leadership of both governments needs to become more pragmatic, rational, and willing to coexist peacefully. Whether the outcome is along the lines of "two nations, two systems," a loose confederation, or some kind of commonwealth under the name of China is uncertain. If history provides any insight, Taiwan and the mainland, with their shared race, language, and customs, will eventually reunite. A reunified China, with its vast resources, would contribute to the world's prosperity and significantly reduce the likelihood of hostilities in the Pacific region.

Robert McColl (Chapter 23) focuses his discussion on the Inner Mongolia Autonomous Region (IMAR), which he considers a relict of China's frontier history. Though the IMAR's current size and shape reflect its geographic location contiguous to the country of Mongolia, there is little formal or functional contact between the two areas. The vast east-west extent makes it virtually impossible for any real homogeneity, other than a general linguistic base of Mongolian. Since there are no significant Mongolian cities or economies along the international boundary, there is no border development to the north. Virtually all of the IMAR's economic and human geography is related to areas contiguous to the rest of the People's Republic of China. The IMAR is like an apex of a crude triangle. The greatest distance and geographic variety occurs along the base with China proper. The IMAR touches such diverse economic, political, and cultural areas as the Northeast (Manchuria), North China and Beijing, the *Guannei* or loess regions, and the Gobi Desert contiguous to Gansu and China's Northwest provinces. This means that transport distances between the Northeast and Xinjiang (Far East to Far West) are much less via IMAR than via the current transport systems. The natural geographic comparative advantages of the IMAR include its mixed animal and crop agriculture, its rich mineral resources, its still largely undeveloped grasslands, and not least, its proximity to so many distinct geographies of China proper. The economic future of the IMAR clearly is linked to each of the distinct regions it touches in China. In fact, it is logical to imagine a future IMAR divided into distinct provinces or administrative units like those in later republican times. Only a desire to maintain a sense of minority homogeneity and privilege would mitigate this otherwise logical development.

Stanley Toops (Chapter 24) provides an interesting perspective on Xinjiang, the largest autonomous region in China. He does so using three organizing themes: the names of places and people, the three subregions, and the different dimensions of the landscape in Xinjiang. For him, the names of places and people show the identity of Xinjiang as a central Asian area, which may be divided into South, Central, and North subregions. The main external forces that affect Xinjiang are China's economic reform and Xinjiang's Central Asian linkages. Like everywhere else in China, Xinjiang has changed in the past two decades. However, ethnic conflict has always been just below the surface, and the tension may have increased following the breakup of the former Soviet Union in the early 1990s. Toops believes that most Uyghurs in Xinjiang are not in favor of separation, given the experience of the former Soviet Union's central Asian republics just across the border, but the future for the region is uncertain.

Finally, Charles Greer (Chapter 25) deals with an intriguing issue regarding China's administrative geography, that is, the importance of China's provinces as units of political economy, nationality, and administration. He draws some parallels between China's provinces and Europe's nation-states. China's population is more than half again the size of Europe's, including Russia's, and its provinces are units of territory, population, and regional culture comparable to European countries. Similar regional entities evolved over two thousand years of history into nation-states in Europe and province units of the Chinese empire. The apparatus that linked these units into China's historical empire slowed the transformation toward modern society, but now that modernization is accelerating dramatically and some provinces are gaining the weight of countries in other parts of the world, the same advantages recognized in Europe's trend toward greater union may be anticipated to counter-balance the problems of provincialism in China. He further argues that the importance of China's provinces as geographic entities in the configuration of the country is worthy of much more attention than it now receives.

Concluding Remarks

China is a country in transition. Its economic reform and opening up have had a very profound impact on virtually every aspect of Chinese society. The tumultuous first three decades following the communist takeover have been succeeded by a pervasive sense of hope and optimism that is fueled by rapid economic growth, dramatic improvement in living standards, more choices in the way of consumer goods, and an increased awareness of the outside world. Few countries in recent history have experienced more drastic social and economic changes within a two-decade time span than has China. As China vigorously transforms itself from a centralized command economy to a market-oriented one and from an agricultural society to an urban, industrial giant, its fast-growing economy and ever-changing society make China one of the most exciting countries to study. The chapters that follow provide a much needed examination of the changing geographies of this giant country in transition.

As one reads about China's remarkable transformations in the reform period, it is natural to want to look into the future and speculate on what China will be like in the years ahead. In the epilogue of the book, Chiao-min Hsieh provides his thoughts on the future of China in the twenty-first century. Needless to say, social and economic predictions are difficult in any case. And in the Chinese case they may be even more so. Who would have predicted the 1989 Tiananmen incident and the subsequent turn-back-the-clock period? It took a visit by the late paramount leader Deng Xiaoping to the south in early 1992 to reorient the country in the course of reform. Since then, China seems to be moving in the right direction. The dangers of social unrest exist, particularly with massive layoffs from state-owned enterprises and the inevitable widening of income gaps between the rich and poor. China's entry into the World Trade Organization, while widely acclaimed in China as a major accomplishment, may prove to be a double-edged sword. A more open environment and heightened competition will exacerbate unemployment and inequality in income. But few in China would want to go back to the old days of equal poverty for all. One thing is clear about China's reform and opening: The genie is already out of the bottle; there can be no turning back. The contributors of this book have documented the changes that have taken place in the country since the reform, provided their analysis, explanations, and in some cases, predictions. The reader may judge for him- or herself how China's future may play out economically, politically, and geographically.

PART ONE
Economic Changes

2

Land-Use Patterns and Land-Use Change

Liping Di

In the past two decades, China has experienced unparalleled economic growth in its history. Since 1980, China's economy has more than quadrupled. Rapid industrialization and urbanization have greatly changed land use and land-management practices throughout the country. This chapter discusses the types, driving forces, patterns, and implications of the recent land-use change in China.

An Overview of Land Use

China has a total territory of over 3.7 million square miles (9.6 million square kilometers). While the western part of the country reaches into the heart of Eurasia, the largest continent on Earth, the eastern and southeastern parts of the country abut the world's largest ocean, the Pacific. China's landscape consists of mostly mountains, high plateaus, and deserts in the west; hills and basins in the middle section; and plains, deltas, and low hills in the east. Plains account for only about 10 percent of China's total land area.

China's climate belongs to the famous East Asia monsoon regime with severe impacts from tropical storms. During the winter, dry, cold air from the continent's interior in the north and northwest sweeps over north China and frequently invades south China, resulting in dry and clear weather throughout most of China. In the summer, moist air from the Pacific occupies the south and southeast and reaches the continent's interior as far as the 105th meridian. The moist air brings humidity and precipitation to a large part of China. This weather pattern results in significant differences in seasonal and regional distribution of precipitation. The general climate patterns are characterized by wet summers and dry winters, and humid east and dry west, with annual precipitation decreasing from more than 60 inches along the southeast coast to less than 10 inches in the northwest interior.

With nearly 1.3 billion people, China is the most populous country in the world. However, population distribution is highly uneven. Most people live in the eastern part of the country, especially along the coast, where the population density exceeds 2,500 persons per square mile. The coastal area is also the most developed area in China. In contrast, the west and northwest regions are sparsely populated because the physical environment cannot support intensive human socioeconomic activities. For example, the population density of the Xinjiang Autonomous Region in Northwest China is less than 25 people per square mile (Zhang et al. 1992).

A combination of physical environment and socioeconomic conditions determines land-use patterns. China's overall land use includes 10 percent for agriculture, 31 percent for meadows and pastures, 14 percent for forest and woodland, 2–3 percent for urban areas, and 2–3 percent for inland water; all other land is barren (CIA 1997). The overall current land use survey, derived from the data collected by the Advanced Very High Resolution Radiometer (AVHRR) onboard the National Oceanic and Atmospheric Administration (NOAA) polar-orbiting satellites, shows that most of the cropland and urban areas are concentrated in the eastern region, where most of the plains are located and the climate is moist. Barrens are mostly located in West China where the weather is cool and dry. Grassland is concentrated in North and West China where the climate is too cold and dry to support agriculture, or along hills and low mountains in Central, South, and Southwest China where the terrain is not suitable for agriculture, although the climate is moist. The forests and woodlands are distributed along hills and low mountains in Northeast and Southwest China.

Driving Forces of Land-Use Change

Both natural and human forces control land-use patterns and drive land-use changes. Under normal conditions, natural forces dominate the potential land-use types, while human forces decide the actual land-use type within the potential. When the driving forces surpass a limit, land use will change. Changes in natural forces include climate, environment, or terrain changes. The natural course of terrain change is normally slow and steady. In human history, climate and environment changes have been the dominant natural driving forces in land-use changes.

Human forces consist of mainly socioeconomic activities. They not only directly cause land-use changes, such as reclamation, but also induce changes in the natural forces. For example, industrial activities have both directly changed land use by urbanization and increased the greenhouse gases in the atmosphere. Human forces have become dominant in land-use change in most areas of the world.

China's current land-use patterns are the result of long-term actions of both natural and human forces on the land surface. Because of the long history of Chinese civilization, the human forces and human-induced natural forces have played key roles over a long period of time in shaping land-use patterns and driving land-use changes. For example, most of the current pastureland in central and south China used to be forests. Because of the long-term, continuous harvesting of trees for use as firewood and in construction, the resultant deforestation has transformed large areas of forest into barrens or grassland. With the rapid industrialization and urbanization in the past two decades, China's land use and land-use patterns have changed rapidly and significantly. The scale, extension, and speed of such changes during this period are unique in the 5,000-year history of China. Human forces are the absolute dominant forces in driving recent land-use changes in China.

Land-Use Change Since the Reform

Many types of land-use changes have occurred in China since the implementation of the open-door policy, but the major changes are nearly always related to agricultural land, such as conversion of farmland to urban and industrial uses, the shift from grain crops to cash crops, and reclamation of wetlands for agricultural uses. This section discusses the major types of land-use change during the past two decades.

Conversion of Farmland to Nonagricultural Uses

China's rapid economic growth in the past two decades has spurred large-scale industrialization and urbanization. The number of cities in China increased from 193 in 1978 to 666 in 1998 (Wang 1995b; Li 1998). Most of the new cities grew from smaller towns. Meanwhile, the number of towns increased from about 6,000 in the early 1980s to 16,702 in 1995, and the urban population increased from 200 million in 1978 to 334 million in 1993 (Wang 1995b). This urban population does not include the millions of people, mainly farmers, who migrate from rural areas to cities seeking work opportunities. The rapid expansion of the economy not only demands large amounts of land for urban expansion and factory construction but also requires larger, denser highway and railway transportation networks, all of which depletes farmland. Conservative estimates indicate that about 1,400,000 hectares of farmland were consumed annually by urban sprawl and transportation networks during the period of the eighth 5-year plan (1991–1995) (New China News Agency 1995a,b).

Satellite remote sensing provides direct evidence of rapid urban expansion. For example, overlaying classification results of Landsat Thematic Mapper (TM) images acquired in 1988, 1992, and 1996 for Chengdu, the capital of Sichuan province, clearly shows that from 1988 to 1996, its size more than doubled, and urban expansion has accelerated in recent years. Furthermore, the urban expansion happened not only in the suburbs of big cities but also around towns, with urban sprawl in towns occurring just as fast as or even faster than in big cities.

Although urban expansion is significant in farmland conversion, a statistically more significant conversion is actually happening in rural areas. One of the major characteristics of China's recent economic development is the growth of small family, village, or township enterprises. The workers, managers, and owners are all former farmers, who live nearby or operate the factory in their backyard. Those factories currently employ more than 80 million former farmers, produce one-third of the total national industrial output, and are one of the driving forces behind the rapid economic growth in the past two decades. In order to accommodate those small factories, almost every township in the coastal provinces has set up industrial parks on what was previously farmland. The size of such parks varies from an acre or so to a couple of square miles. Since there are many such industrial parks, the amount of farmland involved in the conversion is very large.

Besides rural industry, rural residential lots also consume considerable amounts of farmland. Since the reform, farmers' incomes have increased substantially; for example, in some coastal areas, farmers' incomes have increased more than tenfold. When farmers become rich, the first thing they tend to do is to improve their living standard by building a newer and bigger home. As a result, the national average of square footage per capita has more than doubled in the past twenty years. Because farmhouses traditionally have big yards, they occupy much more land than city apartments do for the same amount of living space. Because of China's dense population in the east (with more than 2,500 people per square mile in many rural areas), the conversion of farmland to residential use is very significant. In some rural areas, more than 15 percent of the land is occupied by housing (Dai 1996).

The conversion of farmland to recreational use is also significant in many areas. With the rapid increase in income, recreation is gradually becoming part of Chinese people's lives. Domestic tours are currently very popular in China. In addition, China is also the world's sixth most popular and fastest growing destination of international tourists. Statistics show that 23.77 million foreign tourists visited China in 1997, an increase of 16.8 percent from 1996 (International Statistics Information Center 1998). In order to accommodate this rapid increase, a substantial amount of farmland has been converted to recreational uses, such as golf courses, recreational grounds, and vacation villages. For example, more than 115 square miles of farmland was approved by the government to be converted to recreational uses in the Yangtze River Delta area alone (Yang 1995).

Shift from Grain Production to Cash Crops

Food supply has always been a problem in China. Under the planned economy before 1978, the government decided how much land was allocated for growing grains and how much for cash crops. Because of the tight food supply, most of the farmland was allocated for growing grains, leaving a small amount of land for growing cash crops such as fruits and vegetables. Since all agricultural products were purchased by the government with government-specified prices and rationed back to consumers at government-specified prices, there was no incentive for farmers to grow cash crops. All farmers worked at collective farms (People's Communes) without any say in the type of crops they grew. This situation continued in the early years of the reform.

The rural areas underwent a fundamental change between 1979 and 1982. In response to popular demand, the government abolished collective farms and distributed farmland to individual farmers under a new household responsibility system. Farmers must sell a portion of their agricultural products to the government at government-specified prices, but the rest of their products may be sold freely at the newly legalized free markets, usually at much higher prices. This greatly encouraged farmers to produce more and better agricultural products. As a result, the grain output of China increased from 304.8 million tons in 1978 to 407.3 million tons in 1984, a 4.9 percent increase per year. This period marked the fastest growth in grain products since 1949 (Press Office of the State Council 1996).

Because of the abundant supply of grain, in 1984 the grain price at the free market dropped to a level very close to the government procurement price. However, the prices of fruits, vegetables, and other cash crops were still very high because of the limited supply. Compared with growing grain crops, growing cash crops became much more profitable. Since the obligation of farmers to sell a specific amount of grain to the government could be fulfilled by purchasing the grain in the free market, many farmers switched to cash crops in the pursuit of higher profits. During 1984–1985, "To be rich is glorious" was the most popular slogan in China. The local government officials encouraged farmers to grow cash crops, although the contract did not allow them to do so freely. As a result, a significant amount of cropland was used for growing fruits, vegetables, and other cash crops or for raising fish. For example, the water area in suburban Beijing increased from 1.5 square miles in 1986 to 6 square miles in 1992, and most of the increased water area was fish ponds converted from cropland (Dai et al. 1995). Because a large amount of farmland was used to grow cash crops or for urban and industrial development, the period from 1985 to 1995 marked the lowest increase in grain products. Grain output reached only 466.6 million tons, a 1.2 percent increase per year, despite the significant increase of unit yield due to improved agricultural technologies. In comparison, fruit output during the same period increased 3.3 times, and aquatic products increased 3.1 times (Press Office of the State Council 1996).

Farmland Abandonment

In the coastal provinces such as Guangdong, Zhejiang, and Jiangsu, increases in farmers' income are mostly from industrial and commercial activities. The large number of foreign-owned businesses, joint ventures, and village and township enterprises provide job opportunities to farmers, often with better pay. In those areas, the income from farming has become secondary. Based on the author's investigation, 80 to 90 percent of farmers' income in most counties of Zhejiang province is from nonfarming sources. In fact, farming is becoming a burden for farmers in the developed areas, because the return from farming is much smaller than that from other business activities. The main reason for this low return is that the land area is too small for individual farmers. For example, in Wenling County of Zhejiang province, on average each farmer has only about 0.1 hectare of farmland. It is impossible to make a decent profit from such a small piece of land. Many farmers abandon their farmland and subsequently either migrate to cities for higher-paying jobs or engage in small business activities. It is estimated that the number of migrant workers in China is currently at 150 million. Because farmers engage in nonagricultural activities, they have abandoned approximately 2,200 hectares of farmland in suburban Shanghai alone, about 0.5 percent of the total land area of Shanghai (Wu et al. 1998). Nationwide, the total amount of abandoned farmland is very large, although no national statistics are available.

Another major source of abandoned farmland is unused construction land. Since the implementation of the open-door policy, local governments want to industrialize and urbanize as soon as possible. Governments from the township level up to the central level set up economic development zones to attract domestic and foreign investments. These development zones are so numerous that almost every town in the coastal area has at least one. However, many of these zones do not have enough projects to occupy the land purchased from farmers for that purpose, and some land therefore remains idle. For example, in suburban Shanghai alone, such idled land accounts for more than 4,500 hectares, or about 1 percent of the total land area of Shanghai (Wu et al. 1998).

Arable Wasteland Reclamation

Although most of China's arable land has already been cultivated, there is still some left that can be converted to farmland, including wetlands and forest in Northeast China, grassland and desert in Western China, and coastal wetlands. The largest reclamation projects are centered in Northeast China, where farmland is being reclaimed from wetlands for growing corn and wheat. During the period of the eighth 5-year plan (1991–1995), about 700,000 hectares of forest, wetlands, grassland, and desert were reclaimed and converted into productive farmland annually (New China News Agency 1995b). In Heilongjiang province alone, 160,000 hectares of new cropland, most of which was reclaimed from wetlands, were cultivated in 1997 (Hu and Chen 1997). From 1978 to 1995, more than 1.33 million hectares of barrens had been converted to cropland, pasture, or orchards (Wang 1995a). China plans to continue to reclaim arable wasteland for agricultural use at a pace of 300,000 hectares per year for the next several decades (Press Office of the State Council 1996).

Deforestation and Reforestation

Many years of human economic activity and increasing population pressure have led to the cutting of much of China's forests, and the cleared land has been converted to farmland or become barren because of soil erosion following deforestation. As a result, only about 14 percent of China's land is currently covered by forest (CIA 1997). Because of the huge population of China, the per capita forest resources of China are only 14 percent of the world average (Wang et al. 1998).

Even with so little forest left, deforestation has continued. Reports of serious deforestation events appear frequently in official newspapers such as the *People's Daily* and on China Central TV. However, the nationwide statistics on deforestation are not publicly available. Based on the author's survey of those reports, there are three major reasons for deforestation. First, there is an increased commercial demand for lumber. Because of the intense economic development in China, the demand for wood has increased significantly, resulting in a substantial rise in lumber prices. This high demand encourages deforestation. For example, Northeast China is one of the major sources of commercial lumber in China. Currently, almost all the mature forest has been cut, and many mill workers have been laid off. Second, forest is being converted to orchards. In South China, where the climate is suitable for growing fruits such as orange and litchi, farmers have cut the forests and planted fruit trees with the hope of making more money. This type of land-use conversion was widespread in the late 1980s and early 1990s. The fruit trees have entered their peak production period in recent years. Because a large amount of forested land was converted to orchards, fruit supply in China has surpassed the market demand, causing some fruit markets to almost collapse in recent years. For example, the price for oranges dropped to less than 0.36 yuan (about $0.05) per pound in the harvest season of 1997, even in high-demand consumption centers such as Shanghai. Third, household demands for firewood have been high. In 1997, more than 50 million farmers still lived below the poverty line, and most of them lived in West China's deserts, semi-deserts, or mountainous areas (Chen 1998). These farmers are so poor that they cannot afford fossil fuel for their household use; their main fuel sources are crop wastes and firewood. Because they live in places where the ecosystems are extremely fragile, the destruction of forests for fuel has a serious impact on the environment.

Reforestation has been a major governmental initiative since the implementation of the open-door policy. The forest law requires every citizen to plant trees every year as a civic duty. Based on statistics, more than 2.2 billion person-days have been devoted to tree planting, with more than 11 billion trees planted (Wang et al. 1998). The Chinese government has also invested a large amount of money on many major reforestation projects, such as the famous "Three-North" (Northeast, North, and Northwest China) Protective Forest Project. The project covers 1.57 million square miles of land area, about 42.4 percent of China's total territory. From 1978 to 1995, the project had reforested 20 million hectares of land (Wang 1995a). At the Ninth People's Congress, Mr. Chen Jinghua, the Chairman of the National Planning Commission, reported that 4.71 million hectares of land were reforested in 1997 alone (Chen 1998). However, the effectiveness of these reforestation efforts remains unclear.

Desertification

Desertification is a serious environmental problem in China. Because of the geographic location and physical environment of China, a large portion (currently about 27 percent) of China's land is barrens (Wang et al. 1998). Of this, about 126,640 square miles were formed in the past 5,000 years, especially in the past 50 years, because of desertification (Qu 1995b). Currently, desertification is occurring at a rate of 950 square miles per year (Wang et al. 1998). There are two types of desertification in China: 1) desertification caused by climatic change, which is most serious in Northwest China; and 2) desertification caused by the destruction of vegetation cover and improper use of land (such as overgrazing and overfarming), mainly in Northwest China and South China. The latter is the major type of desertification. In Northwest China, the climate is so dry that vegetation cover is very hard to restore once destroyed. In South China, the climate is wet, but the soil layer is very thin on many mountain slopes, especially in the limestone mountain region. Once the forest is destroyed, soil erosion soon converts the land to bare rock. In addition, mining activities have also played a role in desertification. For example, activities at a mine in Gansu province desertified a 1.16 square mile area in recent years (Qu 1995a). Other than desertification, salinization of irrigated areas in North and Northwest China due to poor drainage systems has also degraded much farmland (Dai 1986).

Regional Patterns of Land Use and Land-Use Change

China's natural and socioeconomic conditions give rise to some unique features in its land-use patterns. Land use is extremely heterogeneous in many places, even within small areas, because of high population density, the widely scattered distribution of privately owned small-scale industries, and the lack of zoning. Labor-intensive farming also necessitates small field sizes (less than a quarter of an acre in many cases).

Because China is a large country with very diverse natural environments and socioeconomic conditions, its land-use patterns and changes vary significantly from region to region. Based on geographic location and natural environmental conditions, China can be divided into seven geographic regions, namely, East, South, North, Central, Northeast, Northwest, and Southwest China.

East China

The major land-use change in East China is the conversion of farmland to urban and industrial uses. This is one of the most developed and most densely populated regions in China. Population density exceeds 2,500 persons per square mile in rural areas. Available farmland per rural resident is less than 0.05 hectares and the typical field size is less than 1,000 square meters. This region has experienced one of the fastest economic growth rates over the past twenty years. Coupled with the high population density, the scattered distribution of large numbers of townships, villages, and privately owned small-scale industries has made the land use in this region extremely complex.

This area is also one of the major rice-growing regions in China. The productive farmland is harvested two to three times a year, sustaining yields as high as 15 tons per hectare. The rice produced in this region used to exceed the amount required to feed its residents. However, because of the large-scale depletion of farmland, it has become one of the largest food deficit regions in China.

South China

South China includes the Pearl River Delta, where China's economic boom originated. The major land-use change is the conversion of farmland to urban and industrial uses. Because of favorable climatic conditions, rice production used to greatly outweigh regional demand. However, because of the area's economic growth, large sections of farmland have been lost to industrial, urban, and transportation expansion. Many people have migrated into this area to seek better work opportunities. Thus, this region has the highest food deficit in China. The socioeconomic conditions and land-use patterns in this region are similar to those in East China.

North China

The major land-use changes in North China include the conversion of farmland to urban and industrial uses, the reclamation of salinized land for agricultural use, and the desertification of farmland and pastures. Because of its larger field sizes and a more homogeneous landscape, this region is easily distinguished from the previous two regions. The primary grain crop in this region is wheat. However, more land has been dedicated to cash crops in recent years. Advances in farming practices, such as applying thin plastic film to cover the land surface for increasing the soil temperature and reducing the loss of soil moisture, have had a positive effect on agricultural productivity. In addition, large areas of farmland have been converted into urban and industrial uses (Dai et al. 1995).

This region includes one of the largest plains in China, the North China Plain, which was formed by the Yellow, Huaihe, and Haihe Rivers. Because of the high groundwater table, the plain has large areas of salinized soil (Dai 1986). Most of the salinized land is idle. Some of it was cultivated, but the productivity was very low. In recent years, the Chinese government has invested heavily in this region to improve the irrigation and drainage systems. Scientific experiments have been conducted to improve soil quality. With the advances in agricultural technology and the improvement in infrastructure, large areas of salinized wasteland are now being cultivated with increased productivity. As a result, this region is now one of the largest commodity grain producers in China. Yet, a significant amount of land remains available for cultivation. The Chinese government believes that this region and the Three River Plain in Northeast China will become China's primary area for grain production in the next century (New China News Agency 1995a,b).

Central China

Central China is mostly rural and is one of the most important commodity rice producers in China. The major land-use changes in this region include the reclamation of wetland for agricultural uses and the conversion of farmland for urban and industrial uses. The largest land-use change in the region is happening in the Dongting Lake Basin. The lake used to be the largest freshwater lake in the country, but it has changed dramatically because of irrigation practices and siltation from upstream soil erosion caused by deforestation. The most visible impacts have been a 33 percent decrease in surface area and a 50 percent decrease in storage capacity over the past thirty years. More than 100 million tons of sand are deposited in the lake each year. It is expected that if the upstream soil erosion is not reduced, the lake will disappear completely. As the lake shrinks, it is less capable of handling summer floods, and harvest from the reclaimed land becomes unreliable. The frequency of significant flooding has increased substantially. One such event occurred in the summer of 1995, resulting in huge losses of property and human life.

Northeast China

Northeast China currently is the largest commercial grain producer in China. The major land-use change in this region is the conversion of wetlands to cropland and the deforestation of the mountain ranges. Land-use change has taken place mainly around the largest plain in the region, the Three River Plain, formed by the Heilongjiang (Amur), Songhuajiang, and Wusulijiang Rivers and consisting mostly of wetlands. A large-scale drainage project started in the mid-1960s and is still under way. This has resulted in the destruction of numerous wetlands ecosystems in the plain. Because local consumption is low due to low population density, the plain contributes significantly to China's food supply.

The main crops in the plain are wheat, corn, and rice. Because of climatic limitations, crops in this region are harvested usually once a year or twice every three years. The farms in this area are the largest and the most advanced in China. Most large farms are owned by the government, and large-scale machines are widely used. The field size in this area is comparable to that of the American Midwest.

Northwest China

Northwest China has experienced severe desertification caused by overfarming and overpopulation. The region also has the poorest economic conditions in China. Overpopulation and poor economic conditions have resulted in excessive exploitation of the landscape. For example, the local residents destroy trees, grasses, and roots for cooking fuel as fossil fuels are unavailable. Many parts of the region have very fragile ecosystems. The environment has deteriorated so drastically that it can no longer support the population, even at the minimum level. In recent years, the Chinese government has made an effort to restore the environmental balance by relocating residents from environmentally stressed areas to other areas, returning some farmland to grassland, and planting shelter forests.

Southwest China

Southwest China is the most biologically diverse region in China. Compared with the eastern regions, land-use change in Southwest China, in terms of farmland conversion, is not significant. Neither food production nor consumption in the region is significant to China's balance of food.

The major land-use change in this region is the conversion of forestland to farmland and barrens because of deforestation. For example, Xishuangbanna is a unique area in China because of its large expanse of rain forest. Although it constitutes only 0.22 percent of China's total land area, Xishuangbanna is home to 12 percent of all the plant species found in China (Kong 1995). However, the area has experienced tremendous change since the founding of the People's Republic in 1949. Although sparsely populated, the population has increased from about 300,000 to more than 800,000 people. Rain forest covered 60 percent of the land forty years ago, but it currently covers only 30 percent. The majority of the land has been converted to farmland for growing rice and corn. Some of the forested area is also used to grow rubber trees. The Chinese government realized the importance of preserving this natural treasure and established four natural reserves in 1957. In 1980, the reserves were expanded to cover more than 250,000 hectares of rain forest.

Within individual geographic regions, land use is also very diverse. In each region, there are one or several representative areas that either have experienced significant land-use change or are important to the environment and biodiversity. Table 2.1 summarizes the natural and socioeconomic conditions and the land-use patterns and changes for representative areas of each region.

TABLE 2.1 Characteristics of the Representative Areas in China's Geographic Regions

	Pearl Delta	Yangtze Delta	Suburban Beijing	Dongting Lake	Huang-Huai-Hai Plain	Three River Plain	North Shaanxi	Xishuang-banna
Region	South China	East China	North China	Central China	North China	Northeast China	Northwest China	Southwest China
Climatic zone	Tropic wet	Subtropic wet	Temperate moist	Subtropic wet	Warm temperate moist	Cool temperate moist	Temperate dry	Tropic wet
Land-use scale (Field size)	Small	Small	Small to medium	Small to medium	Small to medium	Large	Small to medium	Small, medium, large
Land-use pattern	Extremely hetero-geneous	Extremely hetero-geneous	Very hetero-geneous	Medium hetero-geneous	Medium hetero-geneous	Low hetero-geneous	Medium hetero-geneous	Medium hetero-geneous
Farmland productivity	High	High	Medium to high	Medium to high	Low to medium	Medium	Low	Medium to low
Major crops	Rice	Rice	Wheat, vegetables	Rice	Wheat	Wheat, corn, rice	Wheat, sorghum	Rice, corn
Major categories of land-use change	Farmland to industry/ urban	Farmland to industry/ urban	Farmland to urban/ industry, grain crop to cash crop	Lake to farmland, wetland to fish ponds	Salinized wasteland to farmland	Wetland to farmland	Farmland to desert, desert to farmland	Tropic rain forest to farmland/ rubber forest
Availability of arable wasteland	None to very little	None to very little	None to very little	Moderate	Abundant	Abundant	Moderate	Moderate
Current balance of food supply	Huge deficit	Large deficit	Deficit	Small surplus	Large surplus	Large surplus	Deficit	Deficit
Population per sq. mi.	> 2,500	> 2,500	1,500	1,000	500 ~700	250~500	< 250	< 250
Residents' education level	High	High	High	Medium	Low to medium	Medium to high	Low	Low
Residents' income	High	High	High	Medium	Medium	Medium to high	Low	Low to medium
Percent of farmers income from farming	Less than 25 percent	Less than 25 percent	25-50 percent	~75 percent	~75 percent	~75 percent	More than 90 percent	75~90 percent
Degree of industriali-zation/ urbanization	High	High	High	Low to medium	Low to medium	Medium to high	Very low	Low
Degree of socioeconomic development	High	High	High	Medium	Medium	Medium to high	Very low	Low

Implications of Land-Use Change

In the past two decades, the key land-use change in China has been the conversion of farmland to nonfarming uses. Of considerable importance is how such changes have affected and will continue to affect the agricultural landscape. During the period of the eighth 5-year plan (1991–1995), China experienced a net loss of approximately 700,000 hectares of farmland per year (New China News Agency 1995b). In recent years, the Chinese government has realized the importance of protecting the farmland for the sustainable development of the country. A strict policy of farmland protection has been implemented, and reclamation has been accelerated. Despite these changes, China has experienced a net loss of about 200,000 hectares of farmland annually in recent years (Wang et al. 1998).

China's most productive agricultural lands are located in the southeast and coastal areas, which coincidentally overlap the country's primary population centers. This has led to a dramatic loss of prime farmland. However, in the more sparsely populated north and northeast regions, farmland is being reclaimed for growing corn and wheat crops. The net result has been an increase in wheat production by 18 percent and corn by 34 percent, while rice production declined by almost 1 percent from 1985 to 1994 (New China News Agency 1995b).

The depletion of farmland becomes increasingly consequential when one takes into account the tremendous rise in population experienced throughout China. Despite the strict population controls enforced by the government, an increase of approximately 14 million people (a new Beijing!) is expected annually. At this rate, grain output will have to grow between 7 and 9 million tons a year to maintain current levels of food supply (People's Daily Editorial 1995; National Planning Commission 1995). Such demands will no doubt place tremendous pressure on the agricultural sector to generate more food from a reduced amount of farmland.

Other factors affecting agricultural production include weakening support from local government, aging irrigation systems, inadequate farm materials, and the small scale of farming operations (People's Daily Editorial 1995; Reuters 1995). In addition, many young farmers are migrating to cities and economically developed regions, which has left an aging population to work on farms. These factors have made agriculture the weakest link in China's sustainable economic development (Reuters 1995).

China has enjoyed good harvests since the mid-1990s. It produced more than enough grain for domestic consumption. However, the continued large-scale losses of prime farmland and the increasing population will eventually force China to rely on grain imports, according to several Western researchers. Lester Brown (1995a) of the Worldwatch Institute has claimed that China is likely to become the world's largest grain importer in the near future, putting a strain on world grain supplies and driving up food prices. Yet, the Chinese government insists that China will be able to feed its people well into the future by controlling population growth, improving agricultural infrastructure, reducing farmland depletion, reclaiming arable wasteland, and advancing agricultural science and technology (New China News Agency 1995a; Press Office of the State Council 1996). Nevertheless, many people around the world remain skeptical of China's ability to meet the food demands of its population (Brown 1995a,b).

Land-use change throughout China over the next decade will no doubt play a significant role in China's ability to sustain current levels of agricultural production. And if current predictions prevail, the potential fallout will become even more devastating as "environmentally fragile" land is reclaimed as farmland. In China, most of the potential arable land is forestland, grassland, or wetlands, all of which play an important role in sustaining ecological balance and biodiversity. The continued reclamation of such land for agricultural uses will have severe consequences to the ecosystem and China's living environment (Muldavin 1997). The deforestation in Southwest China has significantly accelerated the siltation of the lakes, reservoirs, and rivers in the Yangtze River Basin. Many experts worry that the Yangtze River will become the second Yellow River if the deforestation continues. Studies have also shown that the frequent flooding in recent years in China is caused partially by upstream deforestation and large-scale conversion of farmland to urban and industrial uses (Dai et al. 1998).

Although China's land-use changes since the implementation of the open-door policy are unparalleled in Chinese history, the implications of such changes have not yet been fully understood or manifested. Because the current pace of China's economic growth is expected to continue well into the twenty-first century, large-scale land-use changes are also expected to continue. The influence of such changes on China's, as well as the world's, socioeconomic development and environment will be profound. Further research on how land-use changes will occur in China and the consequences of such changes is clearly needed.

References

Brown, L. 1995a. *The State of the World 1996.* Washington, D.C.: Worldwatch Institute.
_____. 1995b. *Who Will Feed China? Wake-up Call from a Small Planet.* Washington, D.C.: Worldwatch Institute.
Chen, J. 1998. The report to the Ninth People's Congress on the execution results of 1997's national economy and socio-economic development plan and the draft plan of 1998 national economy and socio-economic development. *People's Daily,* March 7, p. 4.
CIA. 1997. *CIA World Fact Book.* Washington D.C.: Central Intelligence Agency.
Dai, C. 1986. The survey of low productive soil in the Huang-Huai-Hai Plain by using remote sensing. *Chinese Journal of Remote Sensing of Environment,* Vol. 1, No. 2.
_____. 1996. The analysis of China's potential for sustainable agricultural development by using satellite remote sensing. *China Aerospace,* No. 11, pp. 12–15.
Dai, C., L. Tang, and G. Chen. 1995. Monitoring of urban expansion and environmental change using satellite remote sensing data. *Chinese Journal of Remote Sensing of Environment,* Vol. 10, No. 1, pp. 1–8.
Dai, C., L. Tang, L. Di, and R. Brenkenridge. 1998. The characteristics of China's flood disasters and the mitigation strategy. Manuscript to be published.
Hu, X., and B. Chen. 1997. The eighth harvest year in Heilongjiang province. *Xinmin Evening News,* December 11, p. 2.
International Statistics Information Center, State Statistical Bureau of China. 1998. The major socio-economic parameters for which China is listed as one of the front runners in the world. *People's Daily,* March 3, p. 2.
Kong, X. 1995. Hopes in the rainforest. *People's Daily,* October 7, p. 8.
Li, J. 1998. China announced the number of cities: 666. *People's Daily,* February 28, p 1.
Muldavin, J. S. S. 1997. Environmental degradation in Heilongjiang: Policy reform and agrarian dynamics in China's new hybrid economy. *Annals of the Association of American Geographers,* Vol. 87, No. 4, pp. 579–613.
National Planning Commission. 1995. The ninth 5-year (1996–2000) and 15-year (1996–2010) socio-economic development plans. *People's Daily,* October 2, p. 1.

New China News Agency. 1995a. The answers of the agricultural minister to questions from reporters of New China News agency. *People's Daily*, October 4, p. 2.

_____. 1995b. The distribution of China's grain production has moved northward. *People's Daily*, October 24, p. 2.

People's Daily Editorial. 1995. Put agriculture at the top of the government's economic agenda. *People's Daily*, October 21, p. 1.

Press Office of the State Council, the People's Republic of China. 1996. *China's Food Issues*. Beijing, China.

Qu, Z. 1995a. Gansu achieved success in harnessing the desertification caused by mining. *People's Daily*, October 7, p. 3.

_____. 1995b. Chinese scientists investigate the desertification conditions in north China. *People's Daily*, October 24. p. 4.

Reuters. 1995. Agriculture is the weakest link in China's economy. October 22.

Wang, H., L. Wang, and X. Xu. 1998. Treasuring our common home. *People's Daily*, March 19, p. 4.

Wang, L. 1995a. The three-north protective forest project covered more than 500 counties. *People's Daily*, November 1, p. 8.

_____. 1995b. The degree, characteristics, and trends of China's urbanization. *People's Daily*, October 31, p. 4.

Wu, Z., X. Ji, and H. Gu. 1998. Why not reforest the abandoned land in suburban Shanghai? *Xinmin Evening News*, March 13, p. 6.

Yang, F. 1995. The rapid construction of tourist vacation zones. *People's Daily*, October 10, p. 8.

Zhang, W., M. Fu, and J. Shen. 1992. *Maps of China*. Beijing, China: China Map Press.

3

Diminishing Cropland and Agricultural Outlook

Clifton W. Pannell and Runsheng Yin

China's agricultural system and farm production are remarkable for their large size and scale and the breadth and variety of production. Chinese agriculture provides the food and fiber on which about 21 percent of the world's people depend. Moreover, despite ongoing structural shift in the economy and a gradual decline in the share of the labor force in agriculture, more than 330 million people work in China's agricultural sector. They account for almost half of the total labor force of the country (Ash 1992; Sicular 1993; National Bureau of Statistics 2002). The number of farmworkers peaked at roughly 350 million in the early 1990s and has decreased somewhat erratically since then in step with China's rapid economic growth. Farmwork acts as a sponge-like mechanism to absorb some of those who cannot find other employment.

One of the basic factors on which this agricultural system and China's huge population depend is cultivated cropland, the supply of which has been a topic of considerable debate in recent years. For example, in a provocative essay and a book, Lester R. Brown (1994, 1995) of the Worldwatch Institute claimed that China's demand for food grain is expanding rapidly due to economic and population growth, while its grain-producing capacity based on current technologies is nearing its peak production and may in fact soon begin to decline. This will cause China to require substantial future grain imports to feed its people. According to Brown, a number of factors account for this stagnating grain production. These include rapidly diminishing cropland, a slowdown in productivity gains, and environmental problems such as diminishing water supplies, soil impoverishment, and increasing soil erosion. Among all of these, the most serious is the diminishing supply of arable cropland.

Although Brown's assertions have been challenged and criticized in various quarters, and the reliability of his statistical sources may be called into question, his claims have served a useful purpose in highlighting the necessity for an accurate accounting of the quantity of China's farmland. This chapter presents some data and interpretations in order to establish a baseline for the quantity of cultivated farmland in China and to identify the trends in China's farmland supply and the regional aspects of farmland change. We discuss the various dynamics of farmland change and where and why these changes are occurring. Investigating this topic allows us not only to review China's supply of cultivated farmland but also to assay the validity and accuracy of Brown's thesis as we seek a better understanding of China's farmland dynamics. The investigation is done within the context of the larger issue of China's efforts to provide food and fiber for the world's largest and growing population.

We first review China's historic supply of farmland and the changes since the late 1980s using aggregated and disaggregated data from *China's Yearbook of Rural Statistics (1986–1993)* and recent editions of the *China Statistical Yearbook*. We then proceed to examine geographic variations in cropland changes at the provincial level and to identify and explain some of the causes for these changes, both structurally and regionally. The nationwide investigation is supplemented by a regional case study of western Shandong.

A Historical Note on China's Cultivated Land

Good farmland has always been important in China as the basis for the sustenance and support of the population and civilization. As far back as the Han Dynasty (207 B.C.–A.D. 220), the historical record tells of systematic efforts to measure and record the quantity of land available for farming and to categorize land into various types according to its use and quality. This was done in order to estimate the production base of the country on which the farming system depended as well as to assay the tax base on which the central imperium could draw for its support. However, the ability of various emperors to survey and record the supply of farmland waxed and waned over the centuries and dynasties, and some reigns were far more able to do this effectively and accurately than others (Ho 1959; Perkins 1969; Chao 1984).

The varying capacity and commitment of reigns and dynasties to survey farmland accurately means that the data on cultivated farmland available to the historian and scholar are fragmented and somewhat disjointed, which makes comparability of available cultivated farmland across the temporal horizon of China's long history problematic. This results from the varying size of the Chinese Empire during various periods of history, a lack of any standardized measuring units that survived over the centuries, and the varying methodologies that were employed. These factors had the result of skewing the measurements in varying degrees (Perkins 1969; Chao 1984). Finally, land ownership was directly tied to likely commodity production, which was in turn the basis for tax assessment; therefore, landowners as potential taxpayers had a strong incentive to underreport or not report at all the ownership of land in order to avoid tax assessment. Buck (1937) commented on such a pattern of underreporting in pre-communist China in his book *Land Utilization in China:* "The amount of cultivated land in China is even less definitely known than is the gross area. Official statistics are unreliable because land owners often evade taxes by misrepresenting their area, and newly broken lands may not be reported. Actual land surveys in several *hsien* (counties) have shown that the reported cultivated area differs from the actual by as much as one-third" (Buck 1937, pp. 162–163).

Underreporting persisted in times of socialist management, because it was in the interest of farmers and the leaders of collectivized farm units for the state to believe that they had less land than they actually did. It poses a major problem in attempting to compare the quantity of cultivated farmland in China over the centuries (Chao 1986, pp. 64–87).

Another complicating factor is that in many regions of China, farmland was allocated in such a way that most farm families had some good and some not-so-good land, and such land was often physically noncontiguous and fragmented. Therefore, knowing the precise quantity of land owned or controlled was problematic, and peasants had an interest in undercounting their land holdings and leases in the case of tax levies or commodity quotas. The principle of dividing land equally among all surviving sons led to increasing fragmentation as families sought to bequeath each succeeding generation and stem of the family some land of comparable quality.

Given the vast size and extent of the Chinese territory and the cost and technical difficulties involved, few nationwide systematic surveys of farmland were undertaken. Yet there was a clear need for this information, and local, provincial, and the imperial government all made good use of such land-use and ownership data as existed. Although the size of the empire varied over the centuries, the core area of the traditional provinces, south of the Great Wall and extending from today's Hebei province in the north, south to the South China Sea and west to Sichuan and Yunnan, has not fluctuated greatly. It was this area that contained the main agricultural lands or potential farmlands, and most of this land had been in the Chinese imperium throughout this long span of history. Only in the nineteenth century did territorial expansion (of Han Chinese into Manchuria in the northeast) occur on an extensive enough basis to incorporate new farmlands into China. The expansion and reclamation of new cultivated farmlands in frontier areas of the west in Inner Mongolia and Xinjiang took place mainly in the twentieth century.

Over the centuries, various schemes for allocating land were devised. Some schemes sought to provide equal shares of land to all who were able to farm, and the Song Dynasty (A.D. 960–1279), with its expanding commercial activities and intensified economic growth, placed an emphasis on accurate land surveys. In A.D. 1072, the great Song reformer, Wang An Shih, not only attempted a national survey but also sought to rearrange the land in square fields and to require an assessment of the land based on the quality of its soil, presumed to be an indicator of its productivity. Although Wang was unable to complete his reforms and surveys, efforts to improve the accuracy and completeness of surveying continued. Surveys during the Ming Dynasty (A.D. 1368–1644) improved the overall assessment and, according to Chao (1984), were among the most accurate ever recorded historically.

Historians have debated the validity of various conversion factors and methods used in these historical surveys and disputed the rationale for the surveys. Although it is not known whether they were simply cadastral surveys for property delineation or they were in fact done for taxing purposes, one central fact emerges: The supply of China's cultivated farmland increased substantially over the years and roughly doubled from Han times, at approximately the time of the beginning of the Christian era, to Republican China in 1930 (Chao 1984, p. 87). We have elected to use the baseline data provided by Chao (Table 3.1). His data are drawn from surveys originally compiled in the studies of Buck (1937), who used a team of agricultural scientists from Nanjing University to survey more than 16,000 farms in twenty-two Chinese provinces. This was the first modern, large-scale, systematic attempt to inventory and analyze the farming systems of China. Chao adjusted Buck's results for what he considered underreported quantities. These data have been standardized into quantities that can be measured and compared against the quantities in common use today, the *mou* and the hectare. The hectare is a metric measure that equals approximately 2.47 English acres; it has the added advantage of being equal to 15 Chinese *mou*, which is a traditional, although historically nonstandardized, unit of areal measurement.

The data in Table 3.1, though only rough approximations of reality during the various time periods noted, indicate that cultivated farmland from Han to Song times, a period of a millennium, did not rise. During the Song Dynasty, with its expanded commercial and economic growth, the farmland began to expand and roughly doubled during the next 900 years. Such growth seems logical, for during the Ming (A.D. 1368–1644) and Qing (A.D. 1644–1911) Dynasties, China's population grew substantially despite prolonged episodes of serious political and social instability. The expansion of farmland, along with gains in farm productivity due to improved cultivation and husbandry techniques, no doubt accounts for the increase in food grain essential to support a fast-growing population. The twin forces of more available farmland and improvements in productivity continued to drive increases in output of food grain until well into the twentieth century. However, in the socialist period, the amount of good-quality arable land expanded little, if at all, while the population more than doubled.

A final note of historical significance drawn from both Ming and Qing land surveys as well as from Buck's surveys in the 1930s is that the amount of officially reported farmland was roughly 20 percent less than what was surveyed and recorded in the official registers of cultivated farmland. The figures drawn from the sixteenth, nineteenth, and twentieth centuries are, again, approximations, but they are based on what were increasingly reliable and accurate surveys that provide an interesting and

TABLE 3.1   China's Cultivated Land, A.D. 2–1930

Dynasty and year	Original data (million mou)	Adjusted Acreage (million mou)	(million hectares)
Western Han, 2	827	571	38
Eastern Han, 105	732	535	35.1
Eastern Han, 146	694	507	33.8
Northern Song, 976	295	255	17
Northern Song, 1072	462	666	44.4
Ming, 1393	850	522	34.8
Ming, 1581	701	793	52.9
Qing, 1662	549	570	38
Qing, 1784	761	886	59.1
Qing, 1812	789	943	62.9
Qing, 1887	925	1,154	77
Republic, 22 provinces, 1930	1,143	1,143	76.2

SOURCE: Chao (1984, p. 87), who used original data from Buck (1937, p. 164) and adjusted the numbers for underreporting. The data do not cover Northeast China.

seemingly significant trend of systematic underreporting of cultivated farmland (Chao 1984, p. 86). The underreporting is certainly no surprise, as landowners historically are known to have underreported or failed to report their farmland in order to evade or minimize their taxes. The interesting coincidence is the approximate 20 percent discrepancy between reported cultivated farmland and surveyed cultivated land that apparently recurs over the centuries. This numerical discrepancy at least provides some historical guidance for us to make sense of and provide an accurate estimate of the cultivated farmland available in contemporary China.

Land Supply and Utilization

China displays a great environmental diversity, which has had a far-reaching impact on its land productivity. Moisture, temperature, and soil quality are the three most important physical variables. China's most productive agricultural lands have traditionally been located in the humid eastern flank of the country, especially the southeastern quadrant and the Yangtze River Basin with its abundant precipitation, extensive alluvial lowlands, and mild winter temperatures.

According to Chao (1970), 110.6 million hectares, or 11.45 percent, of China's surface area was thought to be cultivated in 1963. That was a substantial increase from the estimated 94 million hectares of cultivated land in the 1930s that Buck (1937, p. 165) argued was probably the most accurate compilation available to him. Buck's figures were compiled from various sources, however, and did not include the northeastern provinces, then controlled by the Japanese, nor Xinjiang nor the western part of Tibet, so they are not comparable to the figures used after 1950. The increase was thought to have resulted from field consolidation, reduction of fragmented plots, expansion of marginal and slope land not previously cultivated, and removal of grave sites. Buck (1937, p.178) estimated that grave sites amounted to 1.1 percent of China's total land suited for cultivation, enough to support more than 400,000 farm families. Land reclamation in Northeast China and in arid areas was also very important (Kuo 1972). By contrast, the forty-eight conterminous states in the United States have 156 million hectares of land suitable for cultivation, although in 1964, only 116 million hectares of the total were cultivated. Thus, China and the United States, with approximately the same national areas, have different land-use resources and utilization practices.

Wu Chuan-chun, a Chinese geographer working at the Institute of Geography in Beijing, estimated (1981) that China's 1978 total cultivated land supply was 100 million hectares, or 10.4 percent of the total land stock. Based on the various estimates over the past forty-five years, the supply of cultivated land in China has decreased by more than 10 percent. Zhao's (1990) figure indicated a total of 99.3 million hectares under cultivation. This appears to be outdated. The State Statistical Bureau (1992) reported a decrease to 95.7 million hectares by 1991. Much of the decline is accounted for by the recent reforestation of poor and marginal farmland (Yin 1994). Although there may have been a modest increase after the end of World War II, there appears to have been a decrease in recent years despite an active and vigorous land reclamation program. Land reclamation has been the greatest in the northeast (especially Heilongjiang province) and in arid lands of western China (Yin 1994, 1995). This land reclamation, however, has been offset by the loss of farmland to China's growing cities, to expanding transport lines, to water conservancy and afforestation projects, and to burgeoning industrial areas. Wu (1981) claimed that approximately 20 million hectares, a tremendous amount, have been lost to such conversions. Any major increase in agricultural output, it appears, must come from increased yields, for the net amount of cultivated land is not likely to increase much in the years ahead and may very well decrease further.

In 1997, the U.S. Government is reported to have commissioned a major study of the arable/cultivated land supply in China, given that there has been so much uncertainty and indeed controversy over the amount of arable land that is available in China for intensive farming. Methodology used in this study reportedly was based on remote sensing, coupled with sophisticated mathematical and statistical analysis. Although the study has not officially been released into the public domain at the time of writing, unofficial reports of it indicate that its findings conclude that China has approximately 40 percent more arable land than is usually reported in Chinese government statistical sources. Many Chinese scholars and government officials with whom we have discussed the matter admit that there is considerably more arable land than is conventionally reported, but they dispute the claim that it is in fact as much as 40 percent greater.

Other land uses provide an indication of the amount and potential development of China's land stock. For example, in 1963, approximately 8 percent of China's land area was estimated to be forestland. Wu (1981) indicated that 12.7 percent of China's land area in 1978 was composed of forestland. Zhao (1990) reported that 115.3 million hectares (12 percent of the total) was forests. According to Chao (1972), 30 million hectares of land is suitable for reforestation, although there is no verification of that figure. Some of this land may be former cropland degraded by severe erosion, as found, for example, in the grassland areas of Shaanxi, Gansu and Ningxia. Grasslands accounted for approximately 37 percent of the total, while deserts, high mountains, and wasteland totaled approximately 18 percent. China's land is vast, but it has not always been managed wisely and well. Today much of it is degrading (Smil 1984, 1993).

One of the most pressing issues facing the Chinese government is how to use and manage China's large but qualitatively variable land resources to support China's enormous population. Whether China has a looming crisis of agricultural and forest land is open for debate, but it is clear that there have been serious reductions of prime agricultural and forest land (Pannell 1982; Smil 1984, 1993; Yin 1994). At the same time, if there has been a continuing, systemic underreporting of cultivated land, the situation may not be as serious as some have claimed, and it is not approaching any crisis of land shortage at this time (Brown 1995).

We base this conclusion on our attempt to estimate roughly the real amount of cultivated land in China today. The official number for cultivated land in 1995 is 94.97 million hectares, but if one considers the U.S. government-sponsored 1997 study finding that the actual amount may be 40 percent greater, one could add approximately 38 million hectares to the officially reported total. This is a remarkable increase above both the Chinese government estimates and indeed any previous estimates. However, if we take Buck's 1937 estimate of 94 million hectares and add to it the roughly 20 million hectares of farmland in the provinces he excluded from his calculations, our total is 114 million hectares. Another rough estimate would be to adjust the official figure using the 20 percent discrepancy rate that Chao (1984) used in his analysis of the cultivated farmland during the Ming and Qing Dynasties. That adjusted figure is also 114 million hectares. In recent years, China's National Bureau of Statistics (formerly the State Statistical Bureau) has adopted the cropland data from the land-use survey in its yearbooks. The cultivated land area in 2002, as published in the 2002 yearbook, is 130 million hectares, which is 36.9 percent higher than the published 1995 number (National Bureau of Statistics 2002). Incidentally, this discrepancy is very close to the 40 percent difference estimated by the 1997 U.S. study.

Recent Changes in China's Cropland Supply

The losses of cropland have continued during the recent period of economic reform and growth. The data drawn from the Chinese official sources paint a somewhat different picture of cropland change than Lester Brown did. Brown (p. 13, 1994) claims losses of 1 million hectares per year during 1990–1993, although it is not clear whether he is referring to gross or net losses. Our data indicate a total net loss of 552,400 hectares for the same period. Although the losses revealed by the official data are serious, they are not nearly as dramatic as those described by Brown. From 1986 to 1995, China's cropland decreased from 96.23 million hectares to 94.97 million hectares (National Bureau of Statistics 1997).

Cropland changes showed an interesting and intriguing pattern that relates to regional economic changes (Tables 3.2, 3.3). Table 3.2 reports the percentage changes in cropland for each province for the period 1987–1995. The numbers reflect the changes that took place a decade after the agricultural reforms began. Following substantial gains in agricultural production and farm incomes in the late 1970s and early 1980s, grain production stagnated due to diminishing returns. In the meantime, urban reforms unfolded after the early 1980s. These events have had a far-reaching effect on the use and conversion of farmland. For example, the greatest losses in cropland occurred in areas that experienced rapid economic growth, mainly along the coast.

Cropland conversion to nonagricultural uses (for example, roads, highways, housing, and factories) took place mainly in and around middle-sized and small cities and towns but also in rural areas (particularly the so-called peri-urban zones). Inexpensive land, low construction costs, and few, if any, environmental regulations make peri-urban locations very attractive. Table 3.3 lists the amount of cultivated land losses due to three types of construction (state construction projects, rural collective projects, and rural housing).

Regions with significant net gains in cropland are typically located in the interior provinces of the country such as Inner Mongolia, Qinghai, Yunnan, and Xinjiang (Table 3.2), usually in places with water available for irrigation. These regions have been less involved in the global economy and less affected by the economic reforms and remain more committed to national policies that have stressed expansion of the farm economy as the fundamental bedrock of China's economy and grain production as the key link in that farm economy. The interior provinces are also sparsely populated with much land available for agricultural development. Both locational and environmental factors control the land conversion processes at the regional level and account for their regional variations.

Some regions do not appear to fit the above general regional pattern, but these may be accounted for by special circumstances. One such example is Shaanxi province, which saw a 5.4 percent decrease in cropland. Shaanxi is an interior province but has chronically suffered from the most serious soil erosion in China. Another example, Hebei, is a coastal province but had a cropland decrease of only 1.1 percent. Much of the cropland decline in this province may have been included in the quantities for Beijing and Tianjin, which are spatially and functionally linked to Hebei. Moreover, much of Hebei's coastal region is not developed because of its low-lying deltaic environment.

TABLE 3.2 Cropland Gains and Losses by Province, 1986–1995 (% change)

Shanghai	-12.9	Jiangsu	-3.1	Jilin	- 0.5
Guangdong	-10.4	Anhui	-2.8	Tibet	- 0.2
Zhejiang	-7.7	Henan	-2.8	Gansu	0.1
Shaanxi	-5.4	Hunan	-2.4	Heilongjiang	1.4
Hubei	- 5.3	Sichuan	-2.4	Guangxi	1.9
Beijing	-4.6	Shanxi	-2.3	Ningxia	2.0
Shandong	-3.9	Jiangxi	-2.3	Xinjiang	2.6
Liaoning	-3.9	Guizhou	-1.2	Yunnan	3.4
Fujian	-3.7	Hebei	-1.1	Qinghai	4.7
Tianjin	-3.5	Hainan	-0.6	Inner Mongolia	12.2

SOURCES: *China's Yearbook of Rural Statistics* 1986–1993; *China Statistical Yearbook* 1996.

TABLE 3.3 Decreases in Cultivated Land by Cause, 1978–1995 (1,000 ha)

	Total Decrease	State Construction	Village Collective Construction	Rural Housing Construction
1978	800.9	144.5	–	–
1985	1597.9	134.3	92.3	97.0
1989	517.5	70.1	34.6	27.4
1992	732.3	161.0	64.1	23.9
1995	621.1	111.9	84.9	31.6

NOTE: Other causes listed include afforestation and pasturing.
SOURCE: *China Statistical Yearbook* 1996.

A Case Study of Western Shandong

To deepen our understanding of the dynamic changes in China's farm sector, we conducted a case study. We chose four prefectures (Linyi, Liaocheng, Jining, and Heze) in western Shandong province as our study site. This region is located in the North China Plain and is one of the major grain producers in China. The case study sheds light on some future trends in agriculture. Table 3.4 summarizes the basic statistics of the study area, collected from *Shandong Statistics Yearbook* (1978–1992). While the

TABLE 3.4 Agriculture in Western Shandong

	Cultivated Land (1,000 ha)	Area Sown (1,000 ha)	Total Population (1,000)	Farm Labor (1,000)	Grain Output (1,000 tons)	Fruit Production (1,000 tons)	Machinery (1,000 kwh)	Fertilizer Use (1,000 tons)
1978	281.64	405.28	24,952.30	8,798.60	7,320.95	326.27	3,224.90	286.44
1979	280.08	403.01	25,474.00	9,334.20	7,567.26	408.19	3,699.80	326.99
1980	279.28	403.10	25,559.60	9,659.20	7,196.62	347.97	3,995.90	395.39
1981	278.49	396.90	25,967.20	9,925.90	7,733.98	420.73	4,407.70	460.04
1982	277.72	396.37	26,338.80	10,001.30	8,854.17	332.27	4,877.90	539.30
1983	277.01	407.07	26,625.50	10,056.50	10,274.58	469.58	5,837.30	607.47
1984	275.92	421.59	26,965.10	10,174.20	11,069.70	355.73	6,277.40	636.09
1985	272.16	431.47	27,176.90	10,119.00	11,429.13	444.65	7,092.30	682.33
1986	269.51	435.87	27,477.20	10,143.50	11,949.82	412.88	7,812.80	661.86
1987	268.38	434.15	27,874.60	10,235.00	12,327.04	517.97	8,459.30	648.16
1988	267.74	435.66	28,484.30	10,449.70	11,559.97	558.13	9,383.70	723.18
1989	267.12	432.61	29,488.10	10,650.50	12,478.47	554.16	9,733.40	790.01
1990	256.58	422.73	30,662.00	10,463.10	12,584.20	641.58	10,034.70	856.50
1991	256.02	418.51	31,259.60	11,167.70	14,016.05	727.01	10,359.20	926.10
1992	255.23	416.54	31,492.00	11,212.50	14,168.58	884.78	10,623.10	991.57

NOTE: The study area includes Jining, Liaocheng, Linyi, and Heze prefectures.
SOURCE: *Shandong Statistical Yearbooks, 1986–1992.*

total population of the region grew at an annual rate of 1.68 percent during 1978–1992, its cultivated land decreased by 0.7 percent a year. Per capita cultivated land decreased from 0.113 hectares in 1978 to 0.081 hectares in 1992. However, the region's total crop-sown area remained above 400,000 hectares due to the extensive practice of multiple cropping. The multiple cropping index is the ratio of total sown area over total cultivated area, and it increased from 143.9 percent to 163.2 percent.

Because of the continuing decrease in per capita cultivated area in western Shandong, labor density per hectare of sown land has reached 4.39 persons. This has become a critical constraint to increasing farm productivity. During 1978–1984, grain production grew by 7.13 percent each year, but the growth slowed significantly afterward. After 1985, the growth rate was only 3.13 percent. A closer look suggests that this moderate growth was largely driven by intensive use of farm machinery, fertilizer, and improved seeds (Yin 1995). Per hectare yield more than doubled, from 5,727 pounds in 1978 to 12,227 pounds in 1992, a remarkable achievement. But it will be harder to increase productivity further without breakthroughs in seed breeding.

Another trend is that more profitable cash crops have increasingly attracted farmers' attention. For example, fruit output per capita increased from 29 pounds in 1978 to 62 pounds in 1992. Much of this increase was achieved through intercropping. While generating more revenue for farmers, this change may adversely affect grain production.

The main conclusion from this Shandong case study is that in densely populated regions of eastern China, available cropland is likely to shrink in the future, and the situation will only be exacerbated by growing population in these areas. Whatever gains are achieved in agricultural output will come not from expansion of cropland but from increased productivity. Boosting productivity, however, will become more difficult as production approaches the point of diminishing returns under current agricultural technologies.

Conclusion

Cropland loss in China is clearly a serious problem. Rapid population and economic growth, coupled with expansion of infrastructure, have placed a great pressure on the available land resources because people are concentrated in the midst of the most productive agricultural land. But the problem has not become a crisis, as projected by Lester Brown, and our data for cropland losses, admittedly drawn from questionable government statistics, indicate considerably more modest reductions in cropland than Brown projects.

The greatest losses of cultivated farmland have occurred in areas where there is intense economic development, such as the coastal areas of Zhejiang and Guangdong, whereas some sizable farmland expansions have occurred in interior border provinces such as Inner Mongolia and Xinjiang. Since land productivity differentials between the regions of shrinking farmland and the regions of expanding cropland are substantial, with losses occurring in the much more productive rice land in the coastal provinces, the productivity-weighted rate of land decline is definitely greater than the overall rate of annual decline.

Using reclaimed farmland to make up for the farmland lost to development will only solve part of the problem. As China strives to create more employment opportunities for its massive population, preserving highly productive farmland will be one of China's great challenges.

Various government estimates of farmland decrease are at odds with one another. Our national data source suggests that the rate of loss is smaller than that reported by the State Planning Commission (1994)—0.3 percent annually (or 360,000 hectares annually in the decade from the mid-1980s to the mid-1990s), while our regional analysis indicates a rate of decrease of less than 0.2 percent per annum. The discrepancy among the government statistics poses the possibility that China's farmland may be decreasing at a faster pace due to rapid expansion of industrial activity, housing, transportation infrastructure, and other urban and rural development activities.

In addition, our regional case study reveals that since the impact of one-time institutional changes was exhausted after 1984, the growth of grain output has been largely driven by increased uses of machinery and fertilizer as well as other technological inputs such as improved seeds. The shrinking or static cultivated land base and the large labor force may have contributed little to increasing productivity. These findings have profound implications for the future.

Obtaining accurate estimates of cultivated land in China is not an easy matter. Even such a putatively reputable source as the *China Statistical Yearbook, 1997* (p. 56), for example, states plainly and clearly that its published data on cultivated land are not accurate, although more recent sources agree on 113 million hectares. Thus, the reader must understand that there are serious problems in the use of the official Chinese statistical data even on a topic as mundane as cultivated farmland. A positive aspect is that there appears to be more cropland than is officially reported. This was true historically, and it appears to be still the case, even in the face of modern, socialist accounting procedures. Although the U.S. government has indicated that there may be as much as 40 percent more cultivated land than the official Chinese estimate, our analysis concludes that the cultivated area is roughly 20 percent greater than the official estimate. In either case, China has a bit more flexibility in its agricultural production potential, and the pressure on existing farmland resources may not be as great as is frequently lamented. At the same time, given that we do not know how great the discrepancy between the actual amount of cultivated land and the officially reported quantity, caution is urged on all who seek to analyze and appraise the Chinese farmland and commodity production situation.

References

Ash, R. 1992. The Agricultural Sector in China: Performance and Policy Dilemmas During the 1990s. *China Quarterly* 131: 545–576.

Brown, L. R. 1994. Who Will Feed China? *Worldwatch* (September-October): 10–19.

_____. 1995. *Who Will Feed China?* New York: W. W. Norton.

Buck, J. L. 1937. *Land Utilization in China.* Nanking: University of Nanking.

Chao, K. 1970. *Agricultural Production in Communist China, 1949–1965.* Madison: University of Wisconsin Press.

_____. 1984. *Man and Land in Chinese History: An Economic Analysis.* Stanford: Stanford University Press.

China's Yearbook of Rural Statistics. 1986–1993. Beijing: Statistical Press.

Ho, P.-T. 1959. *Studies on the Population of China, 1368–1953.* Cambridge: Harvard University Press.

Kuo, L.T.C. 1972. *The Technical Transformation of Chinese Agriculture.* New York: Praeger.

National Bureau of Statistics (formerly State Statistical Bureau). 1992, 1996, 1997, 2002. *China Statistical Yearbook.* Beijing: China Statistics Press.

Pannell, C. W. 1982. Less Land for Chinese Farmers. *Geographical Magazine* 44 (6): 324–329.

Pannell, C. W., and L.J.C. Ma. 1983. *China: The Geography of Development and Modernization.* London: Edward Arnold.

Perkins, D. 1969. *Agricultural Development in China, 1368–1968.* Chicago: Aldine Press.

Shandong Statistical Yearbook. 1985–1992. Jinan: People's Press.

Sicular, T. 1993. Ten Years of Reform: Progress and Setbacks in Agricultural Planning and Pricing, in *Economic Trends in Chinese Agriculture,* ed. Y. Y. Kueh and R. Ash. Oxford: Oxford University Press.

Smil, V. 1984. *The Bad Earth: Environmental Degradation in China.* Armonk, N.Y.: M. E. Sharpe.

_____. 1993. *China's Environmental Crisis: An Inquiry into the Limits of National Development.* Armonk, N.Y.: M. E. Sharpe.

Wu, C.-C. 1981. The Transformation of Agricultural Landscape in China, in *The Environment: Chinese and American Views,* ed. L. J. C. Ma and A. Noble, pp. 35–43. New York: Metheun.

Yin, R. 1994. China's Rural Forestry Since 1949. *Journal of World Forest Resource Management* 7: 73–100.

_____. 1995. *Forestry and the Environment in China.* Paper presented at the 22nd PAFTAD Conference, Environment and Development in the Pacific, Ottawa, Canada, September.

Zhao, Q. 1990. Land Resources in China, in *Recent Development of Geographical Science in China,* ed. Geographical Society of China, pp. 41–54. Beijing: Science Press.

4

Agricultural Growth and Food Supply

Jianfa Shen

China may soon emerge as an importer of massive quantities of grain—quantities so large that they could trigger unprecedented rises in world food prices. If it does, everyone will feel the effect, whether at supermarket checkout counters or in village markets.

—Lester Brown

Who will feed China? Can China feed itself? These are important questions facing the world and China in the future. There has been a hot debate on these issues, which was started by Lester R. Brown of the Worldwatch Institute in a series of articles and talks after 1994 (Brown 1995; Liang 1996). Citing the industrialization experiences of Japan, South Korea, and Taiwan, which changed from being self-sufficient in grain to importing up to 66–76 percent of their grain today, Brown argued that China's food supply will decrease along with its rapid industrialization, and China will import a huge amount of grain from the world market. Brown was not trying to discourage China from moving toward industrialization or blame China for the problems that are likely to arise but was alerting politicians everywhere that the world is on a demographic and economic path that is environmentally unsustainable (Brown 1995:19).

Whether China's land can support its population has generated much interesting research. Several issues need to be examined to correctly assess the consequences of China's changing food production and supply.

First, how many people will China have in the future? China's population was 1.27 billion in 2000, and some have estimated that it will peak at around 2030–2040 with 1.6 billion people (Brown 1995; Shen 1994). China's grain production will need to expand by at least 25–33 percent, even if the current level of food consumption per capita is maintained. Considering the rising level of food consumption per capita and the possible decrease in arable land, greater efforts are needed to ensure that the demand for food is met.

Second, how rapidly will China's food consumption per capita increase? In 1990, the average American food grain consumption per capita was 1,762 pounds per year, whereas the average Chinese food grain consumption per capita was 661 pounds (Brown 1995). Within China, the average rural food grain consumption per capita was 533 pounds, while the average urban food grain consumption per capita was only 392 pounds. Rapid urbanization in China may ease the pressure on food demand. However, overall demand for grain may rise due to increased meat consumption as incomes rise with industrialization.

Third, will food production decrease due to a decrease in arable land and insufficient capital input? Or will further advancements in agricultural technology continue to increase grain production? Views differ greatly on this aspect.

Fourth, if China increases its grain imports, will that stimulate world grain production to meet the demand, or will there be shortages that result in higher world grain prices, resulting in food problems for poor countries? Can the world's land resources be better utilized to increase grain production? Up to now, most studies have focused on China's food demand and supply. Little work has been done to examine the capacity of *world* food production, although some scholars have suggested that the current world food problem lies mainly in distribution rather than in production (Bradley and Carter 1989). For the most part, poverty and population growth are considered to be the causes of food problems in the world (Crigg 1985).

This chapter does not attempt to address all the issues mentioned above, but it will examine the fundamental issues related to China's population growth, land resources, agricultural development, and grain production. It assesses China's food demand and supply in the future.

Population and Land Resources

In 2002, the People's Republic of China had 1.29 billion people, accounting for about 21 percent of the population in the world (6.24 billion). After the communist takeover in 1949, China's population grew rapidly, from 542 million in 1949 to 600 million in 1954, and by 1974 China had 900 million people. Population growth after the mid-1970s slowed down slightly as a result of the family planning campaign, but it still added 100 million people every six to seven years. The government's goal of keeping China's total population at 1.2 billion by the end of the century was exceeded in February 1995. Improved health care greatly reduced the mortality rate, while the fertility rate was not significantly decreased until the early 1970s when large-scale family planning programs were introduced. Its agricultural and industrial output, for the most part, has been able to sustain an increasing population. Only in the early 1960s did China face a real food crisis that caused widespread famine and negative population growth (−0.46 percent in 1960) (Shen 1994; Jowett 1989).

In 1994, China's population density was 324 persons per square mile, while the world average was only 106 persons per square mile (SSB 1996). Moreover, China's population is heavily concentrated in the eastern part of the country, where population density exceeds 1,036 persons per square mile, compared with only 228 persons per square mile in the west (Shen 1996). Heilig (1997) reports that nearly 115 million people in China live in areas with a population density of 6,283 persons per square mile; 50 percent of the total population lives on about 8.2 percent of the land with a population density of 1,917 persons per square mile; over 90 percent of the population lives on about 30 percent of the land with a population density of 917 persons per square mile.

According to the official estimate, less than 10 percent of China's land is arable (SSB 1996). About 3.68 percent of undeveloped land may also be cultivated. In comparison, 27.38 percent of the land is forest, and 41.67 percent is prairie. Desert and other land types account for 9.81 percent of China's total land. It has been recognized that the official statistics on arable land are underreported (Heilig 1997). SSB (1996) also points out that the figures need to be adjusted according to recent land surveys, which revealed a total arable land of 133 million hectares (*People's Daily* 1996). Wu and Guo (1994) report a similar estimate of arable land in their book on China's land use. They also show that urban and industrial land amounted to 1.9 million hectares; rural settlements accounted for 23.5 million hectares, and infrastructure such as highways accounted for 7.1 million hectares. It is the conversion of arable land to these nonagricultural uses that is the greatest threat to China's grain production capacity.

Although the recent estimate of China's arable land is much greater than the previous official figure of 95 million hectares, arable land per capita (about 0.1 hectare) is still much lower than the world average (about 0.4 hectare). China's limited land resources make it necessary to control population growth and to substantially raise land productivity in order to maintain and improve the living standard of the Chinese people.

Agricultural Development

Being the most populous country in the world, China has always paid substantial attention to agriculture and grain production. It made significant advances in agricultural production in the pre-reform period. Not only was the total grain output significantly increased to keep pace with the rapid population growth, but grain output per capita was also slightly raised from 627 pounds in 1952 to 698 pounds in 1978. This was achieved mainly by raising the grain yield per hectare through increasing input and improving agricultural technology (Table 4.1). From 1952 to 1978, China's total grain output increased from 163.92 million tons to 304.77 million tons, and grain yield per hectare increased from 2,911 pounds to 5,590 pounds, despite a decrease in grain-sown area (Table 4.2).

The economic reform launched in 1978 introduced a "household responsibility system," which allowed rural households to manage their own production to achieve

TABLE 4.1   Agricultural Input in the Pre-reform and Post-reform Periods

Year	Irrigated area (million ha.)	Machinery power (million kw)	Use of fertilizer (million ton)	Use of electricity (million kwh)
1952	19.96	0.18	0.08	50
1957	27.34	1.21	0.37	140
1962	30.55	7.57	0.63	1,610
1965	33.06	10.99	1.94	3,710
1978	44.97	117.50	8.84	25,310
1985	44.04	209.13	17.76	50,890
1990	47.40	287.08	25.90	84,450
1996	50.38	388.16	38.28	165,550

SOURCE: SSB 1991, 1997.

TABLE 4.2   Grain Production in the Pre-reform Period

Year	Total Grain Output (million ton)	Grain-Sown Area (million ha.)	Grain Yield per Hectare (kg)	Population (million)	Grain Output per Capita (kg)
1952	163.92	123.98	1,320	574.82	285
1957	195.05	133.63	1,470	646.53	302
1962	160.00	119.63	1,320	672.95	238
1965	194.53	119.63	1,635	725.38	268
1978	304.77	120.59	2,535	962.59	317

SOURCE: SSB 1991.

maximum output (Ash 1992; Hussain 1994). After meeting the state's contract quota and their own needs, rural households may sell their extra output at a higher price either to the government or at the local free market. Thus, rural households are operating, at least marginally, on a market basis. Many rural workers were released from the agricultural sector and transferred to nonagricultural sectors in township industries and urban areas (Taylor and Banister 1991; Chan 1994; Shen 1995). As a result, significant increases in agricultural output have been achieved since 1978. Grain production and many other agricultural products have significantly increased (Table 4.3).

Figures 4.1, 4.2, and 4.3 present the growth of total grain output, grain yield per hectare, and grain output per capita in China over the period 1978–1996. The total grain output increased from 304.77 million tons in 1978 to 490 million tons in 1996. The most rapid growth occurred over the period 1981–1984, when annual grain output jumped from 325 million tons to 407 million tons in just three years. The 1984 record was not surpassed until the early 1990s. In the late 1980s, China's total grain output began to overtake that of the United States, making China the largest grain producer in the world. Again, the main cause of the increase in grain output in this period was the increase in grain yield per hectare. Over the period 1978–1996, the

TABLE 4.3   Growth in Agricultural Products, 1978–1996

Year	Vegetable Oil (million tons)	Cotton (million tons)	Fruit (million tons)	Meat[a] (million tons)	Aquatic Products (million tons)
1978	5.22	2.17	6.57	8.56	4.66
1980	7.69	2.71	6.79	12.05	4.50
1985	15.78	4.15	11.64	17.61	7.05
1990	16.13	4.51	18.74	25.14	12.37
1995	22.50	4.77	42.14	42.65	25.17
1996	22.07	4.20	46.52	47.72	30.85
1996/1978	4.23	1.94	7.08	5.57	6.62

[a]Includes only pork, beef and mutton.
SOURCE: SSB 1996, 1997.

average grain yield increased from 5,584 pounds per hectare to 9,606 pounds per hectare, and grain production per capita increased from 698 pounds to 882 pounds.

The fast and almost across-the-board agricultural growth has been made possible by the introduction of market mechanisms in production and distribution. The government significantly increased its purchase prices for grains. For example, the state's contracted purchase price for rice increased from 0.16 yuan (about 2 cents in U.S. dollars) per pound in 1985 to 0.40 yuan in 1994; its noncontract purchase price for rice increased from 0.16 yuan per pound to 0.52 yuan per pound over the same period (Liang 1996). Major increases in the rice purchase price took place in 1989 and 1994. The retail price of milled rice was also increased from 0.38 yuan per pound in 1991 to 0.95 yuan per pound in 1994. However, the impact of increased grain purchase prices on production has been offset somewhat by the rising prices of various agricultural inputs. The grain purchase price index for 1996 was 748.3 in terms of 1978, while the price index for agricultural inputs stood at 550.1. Thus, between 1978 and 1996, farmers still gained from the grain price increases. However, the 1995–1996 price index for agricultural inputs (108.4) was greater than the grain purchase price index (104.2), meaning that farmers' price gains were offset by the greater costs of agricultural inputs. Table 4.1 also shows that irrigated area did not increase significantly before 1985, but went up from 44.04 million hectares in 1985 to 50.38 million hectares by 1996. Significant increases were also registered in the total power consumed by agricultural machinery, fertilizer usage, and electricity consumption in rural areas.

Grain Production in Transition

This section analyzes the changing grain output in thirty provincial units in mainland China over the period 1990–1995. Main trends of regional grain production and underlying forces are identified to provide a clear picture of the spatial dynamics of grain production at the provincial level. The data for the years 1990 and 1995 have been collected from the official Chinese sources (SSB 1991, 1996). The underreported

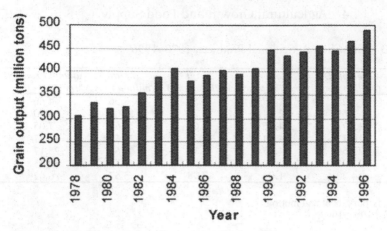

FIGURE 4.1   Grain Output in China, 1978–1996

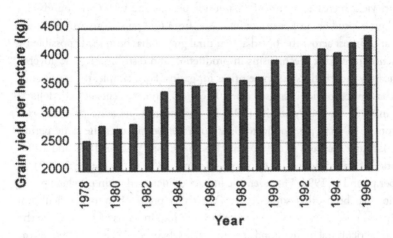

FIGURE 4.2   Grain Yield Per Hectare in China, 1978–1996

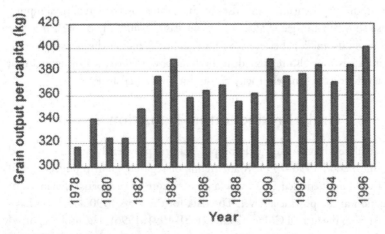

FIGURE 4.3   Grain Output Per Capita in China, 1978–1996

official arable land estimate has been revised according to the correction factor cited in Heilig (1997). The underreporting was most serious in Guizhou, Yunnan, and Ningxia provinces, where less then 50 percent of the actual arable land was reported. These regions are not major grain producers in China. Their grain outputs in 1995 ranked eighteenth, sixteenth, and twenty-seventh respectively among China's thirty provincial regions.

Grain output in China reached 446.24 million tons in 1990. Declining grain outputs in 1991 and 1994 raised concern over China's grain production capacity and its long-term implications for food supply and security. Various factors, including decreases in arable land, smaller proportions of cropland devoted to grain production, declining capital investment in agriculture, and the transfer of rural labor to nonagricultural sectors, affected grain production. These factors are related to the processes of industrialization, urbanization, and economic development. The varying pace of regional development, together with distinctively different regional resources for agricultural development, are now shaping grain output, which has important implications for internal redistribution of grain and the destination of imported grain. It has been reported that the traditional direction of south-to-north grain transfer in China has been reversed in the 1990s.

Four factors determine grain output: the amount of arable land, the cropping index (the ratio of total cropped area to the arable land, reflecting land use intensity), the proportion of cropland devoted to grain production, and the yield per hectare of cropland. Yield is affected by such factors as soil properties, labor input, and fertilizer used.

The ten provinces with 20 million tons or more of grain output were the same in 1990 and in 1995. They are Sichuan, Shandong, Henan, Jiangsu, Hebei, Hunan, Anhui, Heilongjiang, Hubei, and Jilin. Jiangsu is the only coastal province. Hubei and Jilin experienced a slight decrease in annual grain output over 1990–1995, while the grain output in the other eight regions increased in the same period. Most interestingly, Hebei increased its rank from ninth in 1990 to sixth in 1995, as its grain output increased from 23 to 27 million tons. On the other hand, Hubei dropped from sixth place in 1990 to ninth in 1995 with a slight decrease in its annual grain output. Except for Jiangsu, grain yield per hectare in these major grain-producing provinces increased during this period.

The bottom nine provinces and municipalities, all with annual grain output of less than 7.2 million tons, also remained unchanged over the period 1990–1995. They include Xinjiang, Gansu, Beijing, Shanghai, Tianjin, Ningxia, Hainan, Qinghai, and Xizang. Three of them, Gansu, Beijing, and Shanghai, experienced a decrease in grain output, while Tianjin and Hainan experienced a decrease in grain yield.

Significant changes took place in grain production conditions and outputs in China over the period 1990–1995. The total arable land decreased from 137.8 million hectares to 136.8 million hectares. The cropping index increased slightly from 1.55 to 1.58. The proportion of cropland devoted to grain production dropped from 76 percent to 73 percent. However, the grain yield per hectare of cropland increased from 6,014 pounds to 6,477 pounds, and the total grain output increased from 446 million tons to 467 million tons. It is possible to ascertain numerically the degree to which each of the four factors mentioned above have contributed to this increase in grain output. The method involves estimating the impact of each factor's change over the period 1990–1995, while holding the other factors constant at either the 1990 or the 1995 values. The estimation takes place recursively; those factors whose impacts have been estimated will use the 1990 value, while the other factors will use the 1995 value. The regional and national impacts of these factors can be estimated by examining their change in various provinces. The results in Table 4.4 present a clear picture of how the national and regional grain outputs have been affected by each of the four factors.

China's grain output increased by 33.77 million tons due to increased grain output per hectare and by 7.836 million tons due to an increased cropping index over the period 1990–1995. But output was decreased by 17.957 million tons due to a reduced proportion of cropland sown to grain production and by 3.274 million tons due to the loss of arable land. Thus, most of the gains in grain output came from the increase in land productivity, and the losses were largely due to the conversion to higher-value nongrain crops rather than from the decrease in arable land. However, different regions have experienced different changes in grain production. Grain yield and cropping index generally had positive impacts on grain output. Only five provinces (Shanxi, Liaoning, Jilin, Shaanxi, and Gansu) experienced a decrease in grain yield per hectare over 1990–1995. Nine provinces had a smaller cropping index in 1995 than in 1990, while twenty-one provinces experienced higher cropping indices. Reductions in the proportion of land sown to grain production and decreases in arable land negatively impacted grain output in most provinces. Only four provinces—Hebei, Jilin, Heilongjiang, and Shandong—increased grain output by increasing the proportion of cropland devoted to grain production. Nei Mongol, Jilin, Heilongjiang, Guangxi, Yunnan, Gansu, Qinghai, Ningxia, and Xinjiang managed to increase their arable land.

Eleven provinces had a smaller grain output in 1995 than in 1990. The main contributing factor in Shanghai, Zhejiang, and Guangdong was a decrease in the amount of arable land (Figure 4.4). Rapid industrialization and urbanization in these regions are taking substantial amounts of arable land out of production.

The culprit in Jiangxi and Hubei was a decline in the proportion of cropland sown to grain production, reflecting the impact of nongrain crops on grain production (Figure 4.5). In Jilin, Shaanxi, and Gansu, the main reason for the decrease in grain output was the fall in grain yield per hectare (Figure 4.6). Three or four factors were responsible for the decrease in grain output in Beijing (though grain yield per hectare increased), Shanxi, and Liaoning (despite a favorable cropping index).

For sixteen out of nineteen provinces whose grain outputs increased over 1990–1995, the main contributing factor was the rising yield. In Fujian and Guizhou, an increased cropping index was also an important factor (Figure 4.7). An increased cropping index was the main contributing factor for the increase in out-

TABLE 4.4 Change in Regional Grain Output by Factor, 1990–1995

Region	Grain Yield (per ha.)	Grain Land Ratio	Cropping Index	Arable Land	Total Grain Output Change
Beijing	227	-109	-81	-85	-48
Tianjin	246	-58	22	-23	186
Hebei	4,615	178	-38	-135	4,621
Shanxi	-110	-117	-167	-124	-519
Nei Mongol	149	-61	-294	1,030	824
Liaoning	-278	-454	354	-334	-712
Jilin	-837	195	29	72	-541
Heilongjiang	2,146	10	-190	430	2,396
Shanghai	129	-83	-92	-246	-291
Jiangsu	3,641	-1,716	-593	-777	555
Zhejiang	641	-523	-698	-972	-1,552
Anhui	2,783	-1,668	539	-419	1,235
Fujian	671	-553	517	-231	403
Jiangxi	344	-1,406	845	-290	-508
Shandong	8,999	56	627	-767	8,915
Henan	3,423	-2,481	1,294	-607	1,628
Hubei	1,903	-2,191	1,022	-846	-112
Hunan	1,638	-864	129	-500	402
Guangdong	866	-1,259	329	-1,558	-1,621
Guangxi	1,366	-1,517	1,507	96	1,451
Hainan	300	-78	124	-23	322
Sichuan	522	-783	1,984	-741	982
Guizhou	1,368	-347	1,313	-55	2,279
Yunnan	1,256	-1,038	1,005	94	1,317
Xizang	155	-25	15	0	145
Shaanxi	-726	-47	-376	-423	-1,573
Gansu	-594	-181	298	12	-465
Qinghai	48	-96	26	24	2
Ningxia	30	-43	116	27	131
Xinjiang	1,369	-1,004	68	90	523
TOTAL	33,770	-17,957	7,836	-3,274	20,375

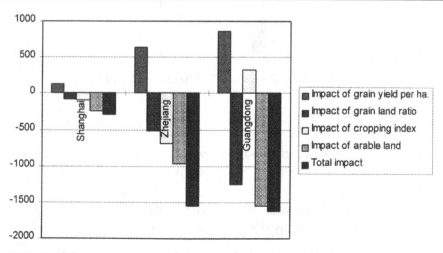

FIGURE 4.4 Grain Output Change in Shanghai, Zhejiang, and Guangdong, 1990–1995 (1,000 tons)

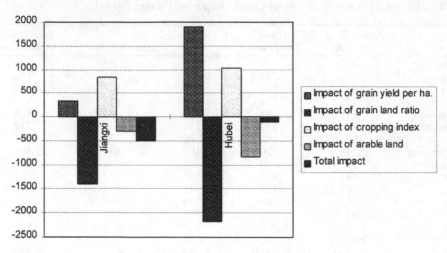

FIGURE 4.5   Grain Output Change in Jiangxi and Hubei, 1990–1995 (1,000 tons)

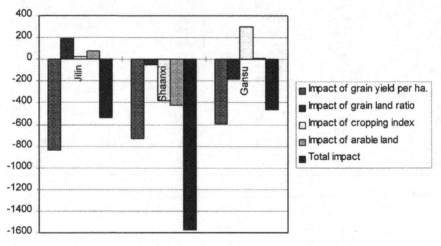

FIGURE 4.6   Grain Output Change in Jilin, Shaanxi, and Gansu,
1990–1995 (1,000 tons)

put in Sichuan and Ningxia, while increased arable land was the main factor responsible for the grain output increase in Nei Mongol (Figure 4.7).

Assessing Future Food Supply and Demand

China is the largest grain producer in the world, and its grain output per capita is above the world average. Nevertheless, with the rapid industrialization, urbaniza-

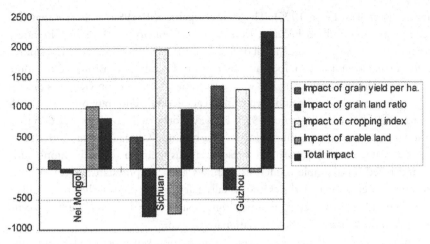

FIGURE 4.7   Grain Output Change in Nei Mongol, Sichuan, and Guizhou, 1990–1995 (1,000 tons)

tion, and economic growth in China in the 1990s, there has been a growing concern over its future food supply and demand, particularly as its population is expected to continue to grow until the 2030s–2040s. Among various factors that may affect its future grain supply and demand, population growth, declining arable land, falling cropland for grain production, rising demand for meat and feed grain are important and need to be carefully assessed. Different assumptions about the amount of China's arable land, grain yield per hectare, and grain consumption per capita will produce significantly different estimates of China's grain supply and demand (Huang, Rozelle, and Rosegrant 1997).

Based on the experience of Japan after 1960 and on the changes in cropland devoted to grain production and in grain output in China in 1990–1994, Brown (1995) projected that China's grain production will fall by at least 20 percent by the year 2030 and that it will need to import 207 to 369 million tons of grain annually, assuming constant or increasing per capita consumption of food. Huang, Rozelle and Rosegrant (1997) made a more modest projection of China's future grain imports using a supply and demand projection model. According to their baseline scenario, China may need to import 24, 27, and 25 million tons of grain in 2000, 2010, and 2020 respectively. Other scenarios produced quite different estimates. For example, the estimate for 2020 ranged from exporting 20 million tons of grain if low income growth is assumed to importing 78 million tons of grain under the assumption of high income growth.

Brown's estimates of China's future grain output are probably too low. Brown (1995) asserts that China's arable land decreased from 90.80 million to 87.40 million hectares between 1990 and 1994, a decrease of 1 percent or 0.85 million hectares annually (see, also, Liang 1996). The official figure (SSB 1996) indicates that the decrease was 0.76 million hectares, from 95.67 million to 94.91 million hectares, over the same period, with an annual decrease of 0.19 million hectares. Furthermore, the loss of arable land was compensated by reclaimed arable land. Brown's estimation of China's ability to raise grain yield was conservative. China's grain yield was probably overestimated due to the underreporting of arable land. The grain output in 1994 was also the lowest over the period 1993–1996. Unfortunately, Brown's assertion that the long-term decrease of grain production has begun in China is based on the grain output in 1994. In this section, we present various scenarios of grain supply and demand based on a simple framework. China's grain output statistics cover more crops than international statistics and usually refer to raw grain rather than milled grain. Thus, China's official grain output figures, used in this section, are 25 percent higher than those based on the international definition.

Population growth demands increased grain supply even if per capita grain consumption does not increase. A series of urban-rural population and urbanization projections have been made for China, assuming different fertility rates in urban and rural areas (Shen and Spence 1995, 1996). Figure 4.8 presents the main projections based on the 1987 fertility rates in urban and rural areas. Projections of urban and rural populations depend on how these populations are defined. Several definitions for "urban" and "rural" exist in official statistics and in the literature. In these projections, a slightly modified definition based on actual urban nonagricultural employees was used, which is similar to what was used in China's 1990 census.

Figure 4.8 shows that China's population will increase from about 1,229 million in 1995 to 1,604 million by 2040. About 400 million more people will be added to the current population, though the two alternative projections give slightly slower population growth over the same period (Shen and Spence 1995, 1996). Due to industrialization and urbanization, rural population will decrease from 779 million to 481 million, while urban population will expand rapidly from 450 million to 1,123 million over 1995–2040. The percentage of urban population will rise from 37 percent in 1995 to 70 percent by 2040. Our analysis uses the total population as the basis to estimate grain demand in the future.

The direct grain consumption per capita in 1993 was estimated at 518 pounds, and the feed grain per capita for the production of other foods was estimated at 145 pounds (Liang 1996). This gives a total grain consumption per capita of 663 pounds. The actual grain per capita available was 835 pounds, part of which was used as seeds. Thus 26 percent more grain is needed than the actual amount of grain consumed. This parameter is used in estimating grain demand.

Grain output is estimated using assumed arable land, the cropping index, the proportion of cropland devoted to grain production, and grain yield per hectare. The projected net import is the difference between grain demand and grain output. Four scenarios have been devised to reflect the range of China's future grain imports from the world market (Table 4.5).

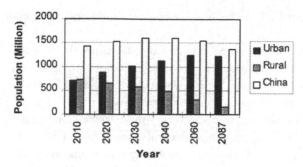

FIGURE 4.8   Population Projection for China, 2010–2087

TABLE 4.5   Grain Import Projections (million tons), Scenarios A–D

Year	Population (million)	A	B	C	D
2010	1,427.61	74.46	31.38	30.05	93.74
2020	1,530.22	113.35	55.27	34.17	168.13
2030	1,594.61	137.75	76.73	24.38	184.44
2040	1,603.60	141.16	89.80	-10.56	178.38
2060	1,540.42	117.21	124.80	-126.16	129.40
2087	1,369.55	52.45	227.60	-349.02	16.75

Scenario A assumes that grain consumption per capita and production parameters remain at the 1995 level, with no decrease in arable land or increase in grain yield. Any increase in grain demand is the result of population change. Thus, China's grain output will remain at the 1995 level of 466.62 million tons per year. Grain demand will increase to 497.26 million tons in 2020 and 607.78 million tons in 2040. In 1995, the grain demand and production were more or less balanced under the grain consumption level of 663 pounds per capita or 835 pounds available grain per capita. Indeed, China's net imports of grain were 19.85 and 9.59 million tons in 1995 and 1996 respectively. The net import means that the grain consumption per capita in these two years might have been higher than the assumed 1993 level or that grain stocks were increased. For long term simulations, grain stocks may be assumed to remain constant for simplification.

According to Scenario A, China's net import of grain will be 113.35 million tons in 2020 and will reach the peak of 141.16 million tons in 2040 due to population growth. The 2040 figure is equivalent to 70 percent of the current total grain export in the world. This scenario clearly shows the consequences of further population growth in China. However, the grain production parameters and per capita grain consumption are likely to change in the future—considerations incorporated in other scenarios.

The grain demand assumption for Scenario B remains the same as for Scenario A, but the four production parameters—arable land, the cropping index, the proportion of cropland devoted to grain production, and grain yield per hectare—will change at the respective rates found to occur in the period 1990–1995. Arable land will decrease by 0.20 million hectares per year, with 127.66 million hectares of arable land by the year 2040. The cropping index will increase from 1.58 in 1995 to 1.83 in 2040. The proportion of cropland devoted to grain production will decrease from 73 percent in 1995 to 46 percent in 2040. Grain yield per hectare will increase from 2.94 tons per hectare in 1995 to 4.83 tons per hectare in 2040. Under these assumptions, China's grain output will increase from 466.62 million tons in 1995 to a peak of 527.64 million tons in 2030 and then decrease to 517.98 million tons in 2040. China's net import of grain will be 55.27 and 89.80 million tons in 2020 and 2040 respectively. The projected imports are smaller than those in Scenario A.

However, it may not be realistic to assume constant rates of change in grain production parameters and a constant level of grain consumption. Thus, in Scenario C, the proportion of cropland devoted to grain production is set at 70 percent (the actual 2000 number was 69.39 percent, which was an unusual near 3 percentage point drop from the 1999 level) and remains unchanged afterward. Other production parameters are assumed to change at the same constant rates as in Scenario B. A study conducted by the Chinese Academy of Agricultural Sciences in the mid-1990s estimated that direct grain consumption per capita in China would drop to 491 pounds (223 kg) in 2000 (the actual figure is 435 pounds for the whole country, 184 pounds for urban residents and 550 pounds for rural residents) and 471 pounds (214 kg) in 2010 and that feed grain per capita would increase to 184 pounds and 249 pounds respectively (Liang 1996). These projections are used in Scenario C as the basis for estimating China's grain demand. The constant increase rate of 4.4 pounds per year of the total grain consumption per capita in the period 2000–2010 is also applied to the rest of the projection period. Thus, grain consumption per capita is projected to be 764 pounds in 2020 and 852 pounds in 2040. Based on Scenario C, China's grain demand will increase to 436.97 and 487.37 million tons in 2020 and 2040 respectively, and grain output will be 634.49 and 792.11 million tons in these two years. China's grain imports will therefore be much smaller. China may even become a major grain exporter by 2040. Grain imports will peak at 34.172 million tons in 2030. This indicates that if a high proportion of cropland (70 percent) can be devoted to grain production in the future, China will need to import only a modest amount of grain. As mentioned earlier, the major negative force in grain production over the period 1990–1995 was the decreasing proportion of cropland devoted to grain production rather than a decrease in arable land.

Scenario D reflects a situation of decreasing arable land. It assumes that arable land will decrease by 0.4 million hectares per year and that the proportion of cropland devoted to grain production decreases to 58 percent in 2020 before it stabilizes, while maintaining the other assumptions used in Scenario C. The projected grain imports are 168.13 and 178.38 million tons in 2020 and 2040 respectively.

Conclusion

This chapter has examined agricultural change in China, focusing on food demand and supply in the post-reform period. China has made significant improvements in agricultural technology and agricultural production since the reform. Yet China must make great strides in its agricultural production due to growing population, limited land resources, and diminishing arable land. The question of how much grain China will have to import from the world market depends very much on what kind of agricultural strategy is adopted in the future.

Four simulation scenarios show that many options exist for China, from importing a small amount of grain to importing vast amounts from the world market. Further increases in grain yield per hectare are essential to feed its still expanding population. Loss of arable land and the transfer of arable land from grain production to other agricultural uses should also be controlled to avoid importing large amounts of grain from the world market.

As the upper limit, China's annual loss of arable land should be kept below 0.4 million hectares, and more than 60 percent of cropland should be devoted to grain production in the long term. Only a modest increase in per capita grain consumption should be allowed. To this end, grain prices must be raised to control grain consumption and to encourage grain production.

China's future food problem may not result from the need to support its population at the subsistence level but from the increasing food consumption accompanying economic growth. In this case, China will not face a severe food security problem, but inadequate grain supply will adversely affect economic development and may even prevent China from becoming a modernized country. China may need to adjust its grain prices continuously to ensure profitable grain production. In an open world market, rising grain prices in China may push up world grain prices. The grain-producing countries will not suffer but will benefit from rising world grain prices, and consequently world grain output may be stimulated. Poor, food-importing countries may find it difficult to obtain a sufficient amount of grain. The international community may need to make arrangements to relieve food problems in those countries. For example, food-exporting countries could establish international aid funds with some of their gains from higher grain prices. The mechanism to achieve this may be complicated, but the world as a whole will be better off with rising grain prices and expanding grain output. Once again, the world food problem will mainly be a problem of grain distribution rather than grain production.

Note

The research in this chapter was supported by a Direct Research Grant of the Chinese University of Hong Kong, Project code 2020400.

References

Ash, R. F. 1992. The Agricultural Sector in China: Performances and Policy Dilemmas During the 1990s. *China Quarterly* 131: 545–576.

Bradley, P. N., and S. E. Carter. 1989. Food Production and Distribution—and Hunger. In *A World in Crisis: Geographical Perspectives*, ed. R. J. Johnston and P. J. Taylor, pp. 101–124. Oxford: Basil Blackwell.

Brown, L. 1995. *Who Will Feed China? Wake-up Call for a Small Planet.* London: Earthscan Publications.

Chan, K. W. 1994. Urbanization and Rural-Urban Migration in China Since 1982: A New Base Line. *Modern China* 20: 243–281.

Crigg, D. 1985. *The World Food Problem.* Oxford: Basil Blackwell.

DPES. 1996. *China Population Statistics Yearbook 1996,* Department of Population and Employment Statistics, State Statistical Bureau. Beijing: China Statistics Press.

Heilig, G. K. 1996. World Population Prospects: Analyzing the 1996 UN Population Projections. Working Paper WP–96–146. International Institute for Applied Systems Analysis, Laxenburg, Austria.

———. 1997. Anthropogenic Factors in Land-Use Change in China. *Population and Development Review* 23 (1): 139–168.

Huang, Jikun, Scott Rozelle, and Mark W. Rosegrant. 1997. China's Food Economy to the Twenty-First Century: Supply, Demand, Trade. *Food, Agriculture, and the Environment.* Discussion Paper 19. Washington, D.C.: International Food Policy Research Institute.

Hussain, A. 1994. The Chinese Economic Reforms: An Assessment. In *China: the Next Decade*, ed. D. Dwyer, pp. 11–30. London: Longman Group.

Jowett, A. J. 1989. China, the Demographic Disaster of 1958–1961. In *Population and Disaster,* ed. J. I. Clarke, P. Curson, S. L. Kayastha, and P. Nag, pp. 24–33. Oxford: Basil Blackwell.

Liang, Y., ed. 1996. *Can China Feed Herself?* Beijing: Economic Science Press (in Chinese).

People's Daily. 1996. What Do We Know About the Arable Land in This Country? June 24, 1996.

Shen, J. 1994. Analysis and Projection of Multiregional Population Dynamics in China, 1950–2087. Unpublished Ph.D. Dissertation. University of London.

———. 1995. Rural Development and Rural to Urban Migration in China, 1978–1990. *Geoforum* 26: 395–409.

———. 1996. Internal Migration and Regional Population Dynamics in China. *Progress in Planning* 45 (part 3): 123–188.

Shen, J., and N. A. Spence. 1995. Trends in Labour Supply and the Future Employment in China. *Environment and Planning C, Government and Policy* 13: 361–377.

———. 1996. Modelling Urban-Rural Population Growth in China. *Environment and Planning A* 28: 1417–1444.

SSB. 1991. *China Statistics Yearbook 1991*. State Statistical Bureau. Beijing: China Statistics Press.

_____. 1996. *China Statistics Yearbook 1996*. State Statistical Bureau. Beijing: China Statistics Press.

_____. 1997. *A Statistical Survey of China 1997*. State Statistical Bureau. Beijing: China Statistics Press.

Taylor, J. R., and J. Banister. 1991. Surplus Rural Labour in the People's Republic Of China. In *The Uneven Landscape: Geographical Studies in Post-Reform China,* ed. G. Veeck, pp. 87–120. Geoscience and Man Series, Department of Geography, Louisiana State University, Baton Rouge.

Wu, C., and H. Guo, eds. 1994. *Land Use of China*. Beijing: Science Press.

5

Agricultural Surplus
Labor Transfer

Sun Sheng Han

Unemployment is a serious problem China faces in its current economic transition. Millions of urban workers have been laid off or furloughed, and some estimates put the number of unemployed people in rural areas at 175 million, or 34.8 percent of the rural labor force. The potential threat of unemployment to social stability has been a primary concern of the Chinese government in determining the pace of economic reform. Whether the unemployed can be accommodated properly will in part determine the success or failure of the plans to reform China's inefficient state enterprises (*People's Daily*, January 13, 1998).

This chapter examines the concept, characteristics, and impacts of agricultural surplus labor transfer in China. While agricultural surplus labor is a product of rural reform, its surge has generated an active force toward the formulation of innovative development policies. As such, the transfer of agricultural surplus labor has shifted its emphasis from passive containment to active engagement in promoting regional economic growth. The first section of this chapter explores the definitions of agricultural surplus labor and its brief history in China. The second examines the main forces and approaches that influenced the formation and transfer of agricultural surplus labor in the 1980s. The chapter then evaluates the status of agricultural surplus labor transfer in the 1990s, followed by a discussion of the forces and approaches affecting the transfer.

What Is Agricultural Surplus Labor?

Different terms have been used in China to refer to agricultural surplus labor, such as agricultural surplus labor, rural surplus labor, rural surplus population, and the rural unemployed. Chinese concepts of rural area, agriculture, and agricultural population deserve some elaboration in order to understand the above terms.

Rural area is a geographic concept and refers to any territory that is designated as a county or anything below a county in the political hierarchy. Its opposite is urban area. Agriculture is an economic concept that refers to an economic sector, parallel to manufacturing, transportation, or commerce. Agricultural population is neither the population working in the agricultural sector nor the population residing in the rural area. Rather, it is a registration term that stands in contrast to nonagricultural population. Registered agricultural population can be found in both rural areas and suburbs of cities. Most of the agricultural population work as farmers in the agricultural sector. Others may work in nonagricultural establishments such as village industries, without converting their registration status (Figure 5.1).

Agricultural surplus labor is a part of the agricultural population, that is, the labor in excess of the employment provided by agriculture. Those people who work in rural nonagricultural sectors are part of the agricultural surplus labor. Rural surplus labor is the labor in excess of employment provided by farm and nonfarm establishments, or the sum of the hidden unemployed in agricultural and nonagricultural sectors. An alternative term for it is rural unemployed. Rural surplus population refers to agricultural workers who have migrated from their home villages (Yang 1994). It includes part of rural surplus laborers and their dependents. They were known as *mang liu* (blind migrants) in the 1950s and the "floating population" in the 1980s.

There is no systematic record on the number of various surplus laborers. Estimates vary widely due to the lack of accurate data and different methods used. The State Statistical Bureau used a sample survey and an index for the number of laborers that each unit of arable land could support to calculate the magnitude of agricultural surplus labor. It concluded that there were about 60 to 80 million agricultural surplus laborers in 1988. The Chinese Academy of Social Sciences conducted a survey on agricultural surplus labor in 222 villages over eleven provincial administrative areas, which yielded a result of 100 million agricultural surplus laborers for 1988. Other researchers, using the same survey data but incorporating the average productivity in the agricultural sector, found that about 80 to 150 million agricultural surplus laborers existed in 1988 (Nie and Bao 1996). In 2000, there were an estimated 200 million agricultural surplus laborers. This estimate might be conservative, as only the current level of agricultural productivity was taken into account. Indeed, some sources indicate that the number of agricultural surplus laborers exceeded 200 million in 1997 (*Liao Wang*, August 5, 1997).[1]

Over the past fifty years, China's agricultural surplus labor went through three stages. During the 1950s, flooding and other natural disasters uprooted tens of thousands of peasants and pushed them to cities. They were also pulled by planned industrial development in selected cities. Cities in western China, such as Baotou and Lanzhou, as well as coastal cities, such as Shanghai and Guangzhou, became

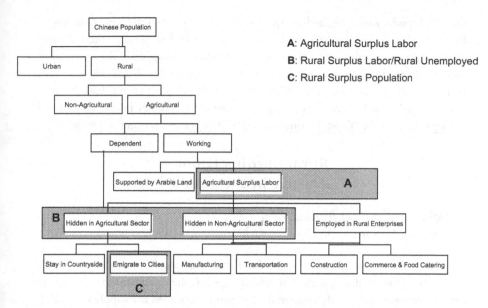

FIGURE 5.1   Categories of Chinese Rural Population and Surplus Labor

major receiving areas of free migrants. In Guangzhou alone, for example, there were about 350,000 to 400,000 peasant migrants between 1949 and 1955 (*Nan Fang Daily*, December 30, 1955). Free rural-urban migration caused shortages in grain supplies in the receiving cities and thus posed a threat to the newly established planned system (Han 1994).

Agricultural surplus labor became invisible in the 1960s and 1970s. In the mid-1950s, the State Council issued a number of instructions in order to stem blind migration. Centralized hiring procedures for employment were introduced in cities, which prevented rural migrants from finding jobs. Village registration of individuals reduced the mobility of peasants. In addition, checkpoints were set up in major railway stations to curb free rural-urban migration. In 1958, the household registration system was officially introduced by the Public Security Bureau, which made it easy to detect and control free migration. The use of rationing coupons in the distribution of foodstuffs and consumer goods eventually brought free migration to a halt. Thus, planned transfer from the agricultural to the nonagricultural population was the only channel to relieve agricultural surplus labor. However, planned transfer was small in volume and sometimes in the opposite direction, for example, the rustification of urban youth. Hidden unemployment, a concept that many Chinese researchers have used to describe the employment situation in China's countryside, was at its peak in these two decades.

Economic reform in the 1980s made rural surplus labor not only visible again but also sizable. After 1978, the household responsibility system was introduced in the countryside, which made clear the sharp contrast between the large rural population and the small amount of arable land. The gradual removal of the rationing coupons, the dismantling of rural communes, and the opening up of the urban labor market all contributed to freeing the agricultural surplus labor. Rural migrants with their belongings, sitting in front of railway stations or along streets, have become common sights in Chinese cities. Regional migration between inland and coastal areas gave rise to the "floods of rural migrants" *(min gong chao)* in the 1990s.

Rural Surplus Labor
Transfer in the 1980s

Policy to Contain Agricultural
Surplus Labor in the Countryside

In 1978, rural reform policies were introduced with the implementation of the household responsibility system. The reform repealed the egalitarian distribution in rural communes and greatly improved rural productivity. As the supply of agricultural products increased, rationing coupons for foodstuffs and other agricultural necessities decreased in value. For potential migrants, this meant that they could support themselves without necessarily acquiring rationing coupons from the government as long as they had cash in hand. In addition, the responsibility system weakened the role of communes in administering rural affairs. As a result, a pool of agricultural surplus labor was ready to leave the agricultural sector.

In cities, however, urban reform did not start until 1984. There were few employment opportunities for potential migrants, as the job market was still dictated by centralized planning, and private-sector jobs were virtually nonexistent. Even worse, cities had problems accommodating their own residents. Employment priorities were given to the urban youth returning from the countryside and to new high school graduates. Early migrants could work only in informal sector activities such as shoe repairing, garbage collecting, and recycling. Even in the late 1980s, the majority of peasant workers in cities were still engaged in dirty and dangerous work. In Beijing, for example, 94 percent of the peasant workers were working in construction, garbage collection, street and public toilet maintenance, and so forth (Li and Hu 1991: 16).

The time lag between rural reform and urban reform generated a situation in which agricultural surplus labor was being pushed from the countryside but being blocked in cities. Rural reform helped to generate agricultural surplus labor, but the largely untouched planned economic system made the potential receiving sectors, that is, state and collectively owned nonagricultural establishments in cities, visible but inaccessible.

This situation was further reinforced by the government's perception of the ability of cities to support rapid urban population growth. Over the years, a development strategy focused on heavy industry had neglected the provision of basic urban infrastructure. Chinese cities were facing significant problems, including a shortage of housing, deteriorating facilities for power generation and water supply, and poor road and sewer systems. Naturally, a further increase in urban population would exacerbate the burden on the already poor urban infrastructure. It was thus necessary to prevent rural surplus labor from migrating to cities and to contain them in the countryside.

The containment of rural surplus labor was embodied in two government policies. The first was the Urban Policy of 1980, which aimed at containing the growth of large cities, rationally developing medium-sized cities, and encouraging the growth of small cities (Han and Wong 1994: 544). Proponents of this policy argued that containing the growth of large cities would help to curb the infrastructure problems and reduce the socioeconomic disparities between China's countryside and the cities. The development of medium-sized cities would be feasible for utilizing scale economies with little additional infrastructure costs, while active development of small cities would help to integrate cities with the countryside, which was an ultimate goal of socialism. While the effectiveness of this policy in directing population settlement was debatable, it showed the desire of the government to contain agricultural surplus labor at the bottom of the urban hierarchy.

The second policy focusing on rural surplus labor transfer was known as *li tu bu li xiang* (Leave the farm but not the village). Unlike the Urban Policy of 1980, *li tu bu li xiang* did not emerge as a new policy statement. Rather, it was deeply rooted in the communist ideology of rural-urban dichotomy and had appeared in the speeches of key officials since the 1950s. Mao Zedong, for example, pointed out that the problem of blind migration should be resolved by maintaining free migrants in the countryside (Han and Wong 1994: 545). For him, surplus labor could survive easily by gathering foodstuffs from farms, but not in cities where grain supplies were clearly restricted. In the 1980s, *li tu bu li xiang* was a slogan that was frequently incorporated in instructions issued by the State Council and the Central Committee of the Chinese Communist Party.

Approaches to Agricultural Surplus Labor Transfer

The development of village and township enterprises was a direct result of the containment policy aimed at curbing rural-urban migration in the 1980s. Major industries of the latter include manufacturing, construction, transportation, and commerce and food catering. Table 5.1 shows the changes in output values of the four main industries of rural enterprises. State statistics show that manufacturing was the main industry that led the development of nonagricultural establishments during the late 1970s. Although manufacturing continued to be a leading sector in output value in 1985 and 1990, commercial activities caught up in the mid-1980s. The number of rural enterprises increased from 1.52 million in 1978 to 18.68 million in 1989. The total employment in rural enterprises tripled during the same period, from 28.26 million to 93.67 million (State Statistical Bureau 1996a, pp. 327, 329).

TABLE 5.1   Changes in Output Value in Rural Nonagricultural Sectors (billion yuan)

	1978	1980	1985	1990
Manufacturing	39.65	54.4	175	672
Construction	13.47	18.00	51.05	97.85
Transportation	3.45	4.71	19.04	57.96
Commerce and				
Food Catering	7.48	9.85	269.6	679.3

SOURCE: State Statistical Bureau 1996a, p. 35.

Other approaches to agricultural surplus labor transfer in the 1980s included peasant settlement in cities and towns, seasonal migration to other rural areas, and migration overseas. According to a survey of eleven provinces in 1986, a total of 15 percent of rural labor migrated, almost half of which went to other rural areas. About 17 percent migrated to towns and townships, while about 30 percent migrated to small cities. Less than 4 percent moved to large cities. The rest, about 0.6 percent, went overseas (State Statistical Bureau 1989).

The desire to contain agricultural surplus labor in the countryside, and various proposals for doing so, gave rise to the question of what was the optimum approach to agricultural surplus labor transfer. Three arguments were developed (Yang 1993). The first was to "transfer without relocation" *(jiu di zhuan yi)*. Agricultural surplus labor would be absorbed within the same geographical unit (that is, townships or counties) through adjustments in the agricultural structure, including the expansion of cropping varieties, the processing of agricultural products, and the development of rural enterprises. It was argued that transfer without relocation fit China's reality because 1) Chinese cities had little demand for rural labor, as their population increase was enough to meet all urban labor demand; 2) the countryside had the potential to generate sufficient employment opportunities; 3) the development of rural enterprises could be an important source of government income and required little in the way of state investment; 4) transfer without relocation would benefit the agricultural sector by satisfying seasonal demand for farmworkers; and 5) transfer without relocation would not overburden the already poor infrastructure in cities.

Others argued for "transfer with relocation" *(yi di zhuan yi)*. In contrast to transfer without relocation, proponents of transfer with relocation called for a full utilization of cities in absorbing agricultural surplus labor. The rationale for this position was based mainly on criticism of the first argument. It was believed that transfer without relocation overlooked the agglomeration economy and thus caused inefficiencies by dispersing industries over a large geographical area in the form of rural enterprises. The latter would further compete with urban industries for raw materials and thus undermine the leading role of the modern industrial sector. The scattered location of rural enterprises made it difficult to control pollution. The development of rural enterprises would undermine agricultural investment. Finally, many urban employment opportunities remained unfilled by city dwellers, because they had higher expectations and were more selective than agricultural surplus labor.

The third argument was for a "multiple absorption" *(duo yuan fu he zhuan yi)*. Proponents of this argument believed that transfers with and without relocation both had pros and cons. A combination of these two approaches would be more suitable for the various conditions in China. By multiple absorption, agricultural surplus labor would be absorbed by moving to towns, migrating to cities, resettling in the western provinces, and emigrating to overseas labor markets.

Impact on Development

Agricultural surplus labor transfer in the 1980s generated positive as well as negative impacts on China's development. On the positive side, the pushing and blocking forces on agricultural surplus labor generated pressure on local administrations to work out innovative solutions for the transfer. The result was rural industrialization, which was an unintentional product in the course of searching for solutions to transfer rural surplus labor, rather than a predetermined goal in China's development. As Deng Xiaoping pointed out, "the greatest outcome of rural reform that we had not foreseen is the development of rural enterprises. This is not an achievement of our central government. The only contribution by our central government to this, if any, is the reform policy" (Deng 1987, p. 429).

Although rural industrialization was a by-product of agricultural surplus labor transfer in the 1980s, its implications were far greater than just absorbing surplus labor. For the Chinese government, rural industrialization represented a new dimension in the Chinese model of development and for Chinese-style socialism. It continues the much-publicized rural focus in China's development policies and contributes to the Chinese model of urbanization, that is, rapid industrialization without a massive transfer of population to cities.

Various models were identified in promoting rural industrial development. According to the origins of the models, they were called the Sunan Model, the Wenzhou Model, the Fuyang Model, the Zhujiang Sanjiaozhou Model, and so on (see Table 5.2). In these models, rural enterprises were started in different ways. In Sunan, for example, rural industries started with manufacturing that served the agricultural sector. Consequently, the development of the agricultural sector and other nonagricultural activities were stimulated. By 1993, the Sunan area was well industrialized, with 90 percent of its total output value generated by rural manufacturing. Unlike Sunan, Wenzhou started with the development of specialized markets in its industrialization process. The specialized markets further spurred the growth of manufacturing that supplied the various markets.

TABLE 5.2   Models of Rural Industrial Development

	Place of Origin	Approach
Sunan Model	Southern Jiangsu Province	• Started with manufacturing that served the agricultural sector • Built up the strength of collectives, which were the main investors
Wenzhou Model	Wenzhou City, Zhejiang Province	• Started with specialized markets, such as markets for clothes, buttons, etc. • Promoted the development of household-based manufacturing that supplied the specialized markets
Fuyang Model	Fuyang City, Anhui Province	• Started with agricultural processing industries • Organized on the bases of households and corporations formed by two or more households
Zhujiang Sanjiaozhou Model	Pearl River Delta, Guangdong Province	• Started with export-oriented manufacturing • Utilized the linkages with and investment of overseas Chinese
Quanzhou Model	Quanzhou City, Fijian Province	• Started with import substitution-based manufacturing • Utilized the linkages with and investment of overseas Chinese
Pingding Model	Pingding County, Shanxi Province	• Started with mining and processing • Utilized the advantage of rich resource endowment
Guanghan Model	Huanghan City, Sichuan Province	• Started with agricultural productivity. This is the county where rural communes were first dismantled and the responsibility system was first introduced. • Developed a mutually supportive relationship between agriculture and rural manufacturing
Haian Model	Haian City, Jiangsu Province	• Started with eco-agriculture • Emphasized adding value to agricultural products through processing
Daqiuzhuang Model	Tianjin	• Started with structural adjustment with focus on manufacturing growth • Supported by strong collective management
Doudian Model	Beijing	• Started with service industries that supplied food and construction materials to Beijing • Supported by strong collective management
Gengche Model	Suqian City, Jiangsu Province	• Started with handicraft manufacturing • Received investment from households, household corporations, villages, and townships

NOTE: Compiled by the author, based on Wu 1997, pp. 401–407.

Free rural-urban migration and rural industrial development, which characterized the transfer of rural surplus labor in the 1980s, generated negative social and
economic impacts as well. First, the presence of rural surplus labor in cities threat-

ened social stability. Numerous studies and surveys revealed that the rising crime rate in Chinese cities was very much related to rural migration. In Shanghai, rural migrants constituted 31.4 percent of the criminals convicted in 1989. One out of every eight homicides in Shanghai was committed by a migrant. Second, family planning became a difficult undertaking among migrants (SFPPT 1991). A survey conducted by the Capital Planning and Construction Commission revealed that in the Fengtai District in Beijing, the average number of children for migrant women of childbearing age was three (OCPCC 1991). Migrants were free from the control of both the authorities in their original villages and the authorities in their new homes. Third, free rural-urban migration reduced the proportion of male labor in the countryside and consequently resulted in a high proportion of female labor on farms. This situation is not very helpful for promoting the role of women in society (Gao 1994). Fourth, the transfer of surplus labor into rural enterprises generated a large number of low-quality industrial laborers in the rural industries, as surplus laborers were transferred to nonagricultural sectors with little training.

Negative economic impacts include the loss of arable land due to rural industrial growth, pollution, and lack of scale economies in industrial organization. Studies on the use of arable land reveal that in many regions, the development of rural industries was accompanied by rapid encroachment of arable land. In 1994, total arable land was reduced by 71,333 hectares (1.07 million mu), and 63.4 percent of that reduction was caused by rural industrial expansion (*Guangmin Daily*, February 11, 1995).

Environmental pollution was another severe problem brought about by rural industrialization. Experts argued that pollution caused by rural industries was far more severe than that caused by urban industries and called for a shift in the emphasis of pollution control from cities to the countryside (*Guangmin Daily*, January 10, 1995). According to their forecasts, half of China would have polluted water by the year 2000. Coastal areas such as the Beijing-Tianjin region and the Jiangsu-Zhejiang-Shanghai region would be the most polluted areas.

Still another concern about rural industrialization was the lack of scale economies in industrial location. Not only did the scattered locations of rural industries make specialization and effective organization of production impossible, but the lack of concentration of industries also caused the redundancy of service facilities (Chen, Ding, and Chen 1993: 196–197). Both internal and external economies of scale were foregone by dispersed industrial locations.

Rural Surplus Labor Transfer in the 1990s

Characteristics

Agricultural surplus labor transfer in the 1990s had four main characteristics. First, rural industries became less important in absorbing surplus labor than in the 1980s.

The absolute number of employees in rural industries continued to increase in the 1990s, but at a slower pace than in the period 1983–1988. In 1988, there were 57 million workers in rural manufacturing establishments, while in 1991 the number had increased by only 1 million (Figure 5.2). Another 15 million were added between 1991 and 1993, but the increase was partially offset by a reduction in 1994. A study of rural manufacturing establishments revealed that labor requirements were shifting from quantity to quality in these industries. In the period 1978–1988, about 90 percent of the increments of output in rural manufacturing were achieved by increasing input, including the amount of labor (State Statistical Bureau 1996a: 24). As these industries become more capital and technology intensive, their demand for new labor will be low.

Second, the tertiary sector has been playing a more important role in absorbing the surplus labor. This was in the forms of peasants' businesses in small towns and in cities. Figure 5.3 shows the change in the number of employees in the tertiary sector. Two upsurges are discernible. The first was between 1983 and 1985, when the government allowed peasants to settle in towns and cities with their own supply of grain (Han and Wong 1994). It was adjusted and consolidated during the late 1980s. The period 1991–1994 saw another upsurge, which was spurred by the structural demand associated with industrial development. By 1995, about 24.5 million agricultural surplus laborers worked in the tertiary sector.

Third, the primary sector was considered to be the most important sector for the transfer of agricultural surplus labor. The transfer of surplus labor to fishery industries, for example, started in the mid-1980s. On the Shandong and Liaodong peninsulas, nonlocal laborers accounted for about 30 percent of the total labor force in the fishery industries (Shao 1993). In Guangdong province, a large proportion of the farms engaged guest laborers in agricultural production. According to Chinese researchers, the current rate of utilization of low- and medium-quality fields, sea farming areas, and grasslands is only about 70 percent. In addition, a large number of irrigation networks need to be maintained or improved. The result is that the primary sector could absorb 100 million agricultural surplus laborers (Nie and Bao 1996: 108).

Fourth, cross-region and cross-boarder transfers became significant in the 1990s. Cross-region movement of rural surplus labor generated *min gong chao,* discussed in the following section. Cross-boarder surplus labor transfer was mainly in the form of labor export, which, in 1995, reached 3 million people.

Min Gong Chao

In the early 1980s, the number of agricultural laborers who had migrated from their home villages stood at about 2 million. The number increased after the mid-1980s, and in the 1990s, a "flood" of migrating labor accumulated. The alarm bell for this "flood" sounded during the holiday season of the Chinese New Year, when the majority of the migrants sought ways for a return visit with their families for celebrations.

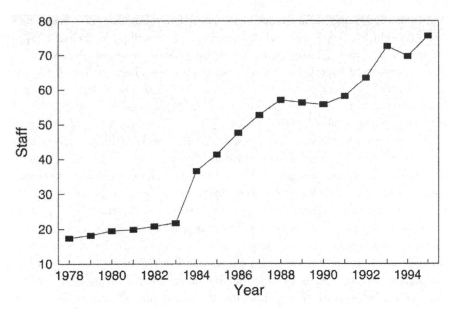

FIGURE 5.2   Staff Employed in Rural Manufacturing Enterprises (in millions)

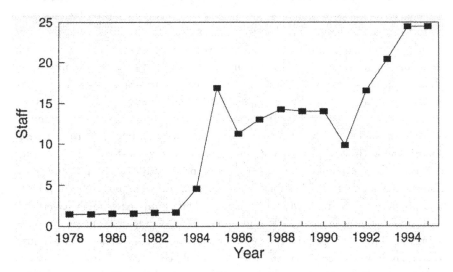

FIGURE 5.3   Staff Employed in Rural Tertiary Enterprises (in millions)

According to researchers at the Policy Research Institute of the Chinese Communist Party Central Committee (CCPCC) and the Ministry of Agriculture, about 65 million migrating laborers formed the *min gong chao* (tide of peasant workers) in 1993. This consisted of approximately 14 percent of the rural labor force (Guo 1995).

In mid-1994, a survey was conducted in seventy-five villages selected from eleven provinces, which uncovered several characteristics of migrating labor. First, a large proportion of migrating labor (36.2 percent) had crossed provincial borders, while 30.7 percent of the migrant labor had moved out of their villages or townships but settled within the same county, and 33.1 percent had crossed the county boundary but settled within the province. Second, about 72.9 percent of the migrating labor had entered cities, while only 20.8 percent had settled in the countryside.[2] Of the 72.9 percent who had migrated to cities, 45.1 percent had settled in medium and small cities, while 27.8 percent had settled in large cities. Third, long-term migration, rather than seasonal migration, made up the majority of the migrating labor, with 53.2 percent of the migrating labor staying longer than ten months. Seasonal migration accounted for about 39.8 percent. Fourth, free migration, rather than government-organized settlement, was the majority. About 39.6 percent of the migrating labor followed village folks who had migrated earlier, and 31.5 percent left their home villages for individual exploration. Another 31.5 percent migrated with the guidance and help of relatives and friends. Only 5.2 percent migrated under the arrangement by government organizations or through employment agents. Fifth, rural enterprises and various nonstate establishments hired the majority of the migrant workers. Only 11.2 percent were working for the state sector.

Regional variations in *min gong chao* were examined in the same survey. Based on the three-region division of China, that is, the eastern, central, and western regions, the survey results indicated that the eastern region was less active than the other two regions in labor migration. The central and western regions had larger volumes of migration, longer migrating distances, and longer times of stay for the migrating laborers than in the eastern region. This was probably due to different levels of economic development in the three regions. Rapid economic growth in the eastern region facilitated the transfer of surplus labor within the same region, without much cross-region migration. In the central and western regions, however, the relatively slow economic growth did little to absorb agricultural surplus labor.

Some researchers argue that *min gong chao* presents the third opportunity in China's rural development. The first opportunity was the introduction of the household responsibility system, which dismantled rural communes and increased the productivity of agriculture. The second opportunity was the development of rural enterprises, which changed the economic structure of the rural area through industrialization. It was believed that *min gong chao* would contribute to realizing market allocation of human resources, which would be a breakthrough in the transition of the planned economy to a market-oriented economy (Wang 1994). Indeed, migrant labor was viewed as a major catalyst for economic development in some central and western provinces. The governments of Sichuan, Guizhou, and Hunan, for example, have considered *min gong chao* an opportunity to upgrade the quality of their rural labor force and to become more competitive in the emerging market economy. Moreover, the capital accumulated by migrating laborers is an important source of investment, and the ideas brought back by migrants to their home villages have become major forces of change in local economies.

Main Approaches Affecting the Transfer

Agricultural surplus labor was transferred through multiple channels in the 1990s. This section analyzes quantitatively the main channels of surplus labor transfer.

It is hypothesized that the amount of labor transferred is dependent on the forces of surplus labor generation and absorption. Surplus labor generation is a function of rural productivity, peasants' attitudes toward migration, and urban attraction to potential migrants. Surplus labor absorption is a function of rural industrial development, nonindustrial growth, urbanization, labor export, and migration. In this chapter, only forces of absorption are examined. The following functional relationship is postulated,

$$SLT = f(INDSC, NISC, URBANC, RLSSA, FLOAT, PRIME)$$

where SLT is surplus labor transfer, measured by the change in the proportion of agricultural labor in the total rural labor force, INDSC is the percentage change of the labor force in the rural manufacturing sector, NISC is the percentage change of the labor force in the rural nonmanufacturing sectors, URBANC is the percentage change of the urban population, RLSSA is the proportion of agricultural surplus labor sent abroad, FLOAT is the proportion of migrating labor in the total rural labor force, and PRIME is the ability of the primary sector to absorb surplus labor.

Data for these variables are drawn from the *China Statistical Yearbook 1991* and *China Statistical Yearbook 1996*, which provide year-end data for 1990 and 1995 respectively. A 13 by 30 data matrix was initially constructed. The 13 variables were consequently transferred into seven indices that represent the dependent and independent variables. The 30 provinces form the analysis units.

By applying a stepwise regression technique, four independent variables entered the equation. They are RLSSA, URBANC, PRIME, and NISC (Table 5.3). Together, these four independent variables explain 50 percent of the total variance of the dependent variable, SLT. Since the forces of surplus labor generation are excluded from the equation, the 50 percent explanation can be considered to be high.

Among the four independent variables, RLSSA ranked first and explained 18.2 percent of the total variance. URBANC ranked second and added 11.2 percent of the explaining power. PRIME ranked third and contributed 11 percent of the explaining power. NISC ranked last as an independent variable. The results suggest that at the provincial level, the export of labor, urban transition of surplus labor, agricultural absorption (through in-depth development), and nonmanufacturing establishments in rural areas were the main factors influencing rural surplus labor transfer. Rural manufacturing development and migrating labor, or *min gong chao*,

TABLE 5.3   Results of the Stepwise Regression Analysis

SLT = 3.35 + 0.80 URBANC − 0.09 PRIME + 2.33 RLSSA + 0.02 NISC
    (3.006) *   (2.236) **   (-2.403) **   (3.002) *   (1.997) **

F = 5.76 *

NOTE: Numbers within brackets are the t values.
      * Significant at p = 0.01
    ** Significant at p = 0.05

were not significant factors. This may be due to the fact that rural manufacturing was concentrated in the coastal provinces, and the origins of *min gong chao* were a few provinces in the central and western regions rather than widely spread among all provinces.

Summary and Conclusions

China's sizable rural population and limited arable land have made agricultural surplus labor an inherent part of national development. Agricultural surplus labor was not visible under the command economic system, as it was disguised by massive political and economic projects and hidden unemployment. The rural reform dismantled the communes and eliminated the egalitarian distribution system in the countryside, which consequently brought agricultural surplus labor into the forefront of economic transition. Since 1978, agricultural surplus labor has been rapidly accumulating. There are now about 200 million surplus laborers in the countryside and cities. It is critical for the government to find proper ways to accommodate the surplus labor in order to maintain social and political stability and to further economic progress.

Policies on agricultural surplus labor transfer in the 1980s emphasized containment, which tried to prevent free rural-urban migration. This migration was perceived to overburden the already poor urban infrastructure and threaten urban industrialization. Two policies, namely, the Urban Policy of 1980 and *li tu bu li xiang*, were advocated to control agricultural surplus labor and curb migration to large industrial cities. The result was the mobilization of local initiatives in developing rural enterprises. Rural enterprises, particularly the rural manufacturing sector, were the major absorbers of agricultural surplus labor in the 1980s.

Since urban reform started in 1984, the blocking force on agricultural surplus labor transfer has been weakened by the growth of nonstate enterprises and policy changes that have allowed peasants to settle in small cities and towns. In addition, rural industries were facing competition from urban industries, which made it desirable to reduce their labor input while adjusting toward capital- and technology-intensive production. This led to a new phase in agricultural surplus labor transfer in the 1990s. Although rural manufacturing and cross-region migration were significant ways of transferring agricultural surplus labor in some provinces, statistical analyses reveal that labor export to the international market, the transition of agricultural surplus labor to urban nonagricultural activities, renewed development of the primary sector, and the development of the rural tertiary sector were common forces that contributed to agricultural surplus labor transfer in the 1990s.

Both positive and negative impacts of agricultural surplus labor transfer are discernible in the Chinese experience. On the positive side, it can be noted that the pushing and blocking forces motivated local governments to generate various models of rural industrial growth. Rural industrialization has become the main approach toward local economic development. *Min gong chao* has become a link between developed regions and backward areas. Migrants bring ideas and capital back to their hometowns and spur local development. Rural industrialization and *min gong chao* are the forces behind the second and third leaps forward in changing China's rural landscape, the first leap forward being the introduction of the household responsibility system.

Negative impacts brought about by agricultural surplus labor transfer include social problems in both cities and the countryside and economic issues specific to rural industrialization. The increase in crime rates and the failure of family planning among the migrating population have been significant social problems in cities, while in the countryside, out-migration of the male labor force has posed a constraint on the role of women in society. The scattered distribution of rural industries has caused serious reductions in arable land and widespread environmental pollution. Dispersed rural industries make it difficult to specialize and thus lack both internal and external economies of scale. Moreover, rural industries compete with modern urban industries for raw materials and markets and thus stress the modern economic sector.

Agricultural surplus labor transfer in China presents a unique model specific to Chinese economic reform. In this model, agricultural surplus labor was first contained in the countryside and then gradually released to cities. Rural industries helped to generate a reservoir for absorbing the surplus labor, while urban reform opened employment opportunities in time to accommodate the surplus labor. The transfer of surplus labor from agricultural to nonagricultural sectors and from the countryside to cities has provided great opportunities for stimulating economic development. If properly guided by appropriate policies, rural industrialization can be used as a check on rural-urban migration and, at the same time, as a complement to modern industries. Similarly, interregional migration can diffuse development from the coastal regions to the inland provinces. There are both challenges and opportunities in formulating innovative policies for agricultural surplus labor transfer.

Notes

1. In addition to the 175 million rural surplus laborers, or rural unemployed, there were more than 100 million peasants working in rural enterprises. Even if half of the labor force in the rural enterprises

had been redundant or hidden unemployed, there still would have been 50 million to add to the 175 million unemployed. This would indicate that there were at least 225 million agricultural surplus laborers.

2. The place of settlement of the remainder, about 6.2 percent of the migrating labor, could not be determined in the survey.

References

Chen, J. 1995. On rural surplus labor and its coping strategies. *People's Daily (Renmin ribao)*, 28 January 1995.

Chen, S. X., J. P. Ding, and N. X. Chen. 1993. *Chinese Industry in the 1990s (90 nian dai zhong guo gong ye)*. Beijing: China Management Press.

Deng, X. 1987. Expedite reforms. In *Hold to the National Policy of Reforms and Opening (Zhong gong zhong yang wen xian yan jiu shi)*. Beijing: People's Press.

Gao, X. 1994. Current trends in rural labor transfer and feminization. *Sociological Research (Shehuixue yanjiu)* 2: 83–90.

Gu, S. 1993. *Theory and Practice of Industrialization and Urbanization*. Wuhan: Wuhan University Press.

———. 1994. Solutions for rural surplus labor in China. *Social Sciences in China (Zhongguo shehui kexue)* 5: 59–66.

Guangmin Daily, January 10, 1995.

Guangmin Daily, February 11, 1995.

Guo, S. T. 1995. Waves of peasant workers: An analysis of rural labor migration. In *The Status and Management of China's Population Migration (Zhongguo ren kou liu dong tai shi yu guan li)*, ed. D. S. Ji and Q. Shao, pp. 196–202. Beijing: China Renkou Press.

Han, S. S. 1994. *Controlled Urbanization in China, 1949–1989*. Unpublished Ph.D. dissertation, Department of Geography, Simon Fraser University.

Han, S. S., and S. T. Wong. 1994. The influence of Chinese reform and pre-reform policies on urban growth in the 1980s. *Urban Geography* 15 (6): 537–564.

Ji, D., and Q. Shao, eds. 1995. *The Status and Management of China's Population Migration (Zhongguo renkou liudong taishi yu guanli)*. Beijing: China Renkou Press.

Li, M. B., and X. Hu, eds. 1991. *The Impact of Floating Population on the Growth of Large Cities and Coping Strategies*. Beijing: Economics Daily Press.

Li, X. 1993. On the two approaches to rural surplus labor transfer. *Population Front (Renkou zhanxian)* 2: 5–9.

Liao Wang, August 5, 1997.

Ma, Q. 1992. Some thoughts on the status of China's rural labor. *Population Journal (Renkou Xuekan)* 4: 22–26.

Nan Fang Daily, December 30, 1955.

Nie, Z., and Y. Bao. 1996. A study of China's rural surplus labor transfer. In *The Rural Problem at the Mid-Stage of China's Industrialization (Woguo gongye hua zhongqi nongcun jingji wenti yanjiu)*, ed. Nie Zhengbang, Wang Jian, and Wu Anan. Beijing: China Planning Press.

Office of Capital Planning and Construction Commission (OCPCC). 1991. Floating population and coping strategies in Beijing. In *The Impact of Floating Population on the Growth of Large Cities and Coping Strategies*, ed. M. B. Li and X. Hu. Beijing: Economics Daily Press.

People's Daily, January 13, 1998.

Shanghai Floating Population Project Team (SFPPT). 1991. A report on the status and management strategies of floating population in Shanghai. In *The Impact of Floating*

Population on the Growth of Large Cities and Coping Strategies, ed. M. B. Li and X. Hu. Beijing: Economics Daily Press.

Shao, Q. 1993. The issue of migrant labor in the fishing industry of the two islands and a gulf in the north. In *The Status and Management of China's Population Migration*, ed. D. Ji and Q. Shao, pp. 130–141. Beijing: China Renkou Press.

State Council Research Office Project Team. 1994. *Policy and Practice of Small Town Development*. Beijing: China Statistics Press.

State Statistical Bureau. 1986. *Labor Force Survey of 100 Villages in China, 1978–1986*. Beijing: China Statistics Press.

————. 1996a. *Rural Statistical Yearbook of China, 1996*. Beijing: China Statistics Press.

————. 1996b. *Statistical Yearbook of China, 1996*. Beijing: China Statistics Press.

Wang, X. G. 1995. Rural economic restructuring and its model selection. In *The Status and Management of China's Population Migration*, ed. D. Ji and Q. Shao, pp. 130–141. Beijing: China Renkou Press.

Wu, T. 1997. *On the Industrialization of Agriculture in China (Zhongguo nongcun gongyehua lun)*. Shanghai: Shanghai People's Press.

Yang, X. 1993. *On the Strategies of China's Rural Labor Transfer*. Nanchang: Jiangxi People's Press.

Yang, Y. 1994. A new interpretation of rural surplus labor. *Economics Information (Jingjixue qinbao)* 3: 20.

Zhou, T. 1993. Problems of serious labor redundancy in agriculture. *Population Journal (Renkou xuekan)* 2: 8–14.

Zhu, F. 1995. *Town Construction and Development (Jizhen jianshe yu fazhan)*. Chengdu: Sichuan University Press.

6

Policies and Spatial Changes of Industrial Development

Dadao Lu

The People's Republic of China (PRC) began to carry out massive natural resources exploitation and industrial construction shortly after its establishment in 1949. From then on, in less than half a century, it has set up a huge industrial system, which includes nearly all industrial sectors and covers most parts of its territory, although there was only a very weak and unequally distributed industrial framework when the civil war ended. This remains an unmatched achievement in the country's history of development. China has become one of the world's leaders in industrial production, particularly in the production of energy, raw and processed materials, and finished goods, such as televisions and refrigerators, and it exports its industrial products to more than one hundred countries.

China's industrial development reflects its distinctive natural, economic, social, and political characteristics, although it has also absorbed the successful experiences of other industrialized countries. China's industrial development has been characterized by the following features:

1: In its early phase, industrial development was oriented toward the domestic market and based mainly on domestic resources. However, it has been gradually internationalized since the early 1980s. Now its industrial production is becoming more and more globalized in terms of both foreign direct investment and import/export.

2: In order to quickly consolidate national power, key domestic resources (particularly capital and technological resources) were initially focused on long-term production-oriented heavy industry. Since the reform, the emphasis of industrial development has been focused on the production of consumer goods and, more recently, on the high-tech industry.

3: The mechanism of a socialist planned economy played a determinant role in shaping the industrial landscape for nearly forty years. During this period, the focus on heavy industry and the location of plants in proximity to raw materials were major trends of development in most provinces. However, with the open-door and reform policies and the gradual transition toward a socialist market system, market forces, both domestic and external, are now guiding industrial development and reshaping industrial structure and its spatial patterns.

4: The central government made great efforts to equally distribute regional development in the 1950s, 1960s, and the first half of the 1970s. Although the effect is somewhat visible, the regional disparity of industrial development remains as a result of the great regional variation in locational advantages, historical precedent, and natural conditions and resources.

These features constitute a framework for analyzing and understanding the industrial development in China since 1949. This chapter examines in detail China's policies regarding industrial development and its consequent spatial patterns. I begin with a short review of China's pre-1949 development and then proceed into the modern period. China's central government and local governments normally establish a socioeconomic development plan every five years, which is called a Five-Year Plan (FYP, hereafter). Except for a three-year readjustment period between the Second and Third FYPs, the FYPs are consecutive. Therefore, the following analysis of the post-1949 period is organized chronologically using the FYPs.

A Brief Review of Pre-1949 Industrial Development

The modern industrialization of China began soon after the Opium War (1839–1842), when Western capitalist economic forces invaded and drove the original rural economy to bankruptcy. The major industrial sectors at that time included ship building and repairing, munitions, textile manufacturing, and food processing. These early industries were mainly located in the treaty port cities along the southeast coast and the Yangtze River (70 percent were concentrated in the southeast coastal cities). However, development occurred very slowly until the end of the nineteenth century. From 1895 to the beginning of the Second Sino-Japanese War in 1936, modern Chinese industry expanded at a brisk pace, particularly during World War I. The key industrial sectors in this period were coal mining, iron and steel production, non-ferrous metal production, and textile manufacturing. In regard to industrial structure, heavy industries developed much faster than light industries. By the early 1930s, the Yangtze River Delta had developed into one of two major industrial areas in China at that time, of which Shanghai was the core city and light industry and textile manufacturing were the main sectors. The second major industrial area was central Liaoning, with coal mining and iron and steel production as its main industrial sectors. Tianjin and Qingdao also became important centers of textile production. Contrary to the rapid growth of the coastal region, the vast inland regions were nearly devoid of industrial development. Only a few provinces had metal-ore mining.

From the beginning of the Second Sino-Japanese War to the establishment of the People's Republic of China, China suffered more than ten years of war. The continuous warfare severely destroyed most of the industrial production facilities and resulted in the serious decline of industrial production, particularly in the southeast coastal region and in central China. In 1949, the total value of industrial output decreased to half of what it had been in 1936. A few new developments in this period were the construction by the Japanese of coal mines and metal-ore mines, iron and steel plants, nonferrous metal-smelting plants, and alkali plants in Northeast China and North China for the purpose of pillaging natural resources from China. In short, the new People's Republic of China and its leaders had a very poor foundation for industrial development.

Industrial Policy and Spatial Patterns Before the Reform

The Recovery and the First FYP (1949–1957): An Effective Beginning

After the establishment of the People's Republic of China, the first important task confronting the new nation was to recover from the damage done during the war and to plan new economic development. A peaceful international environment would have been helpful. However, the Korean War broke out on June 25, 1950. As the Chinese went to North Korea's rescue, the United States punished China through an economic blockade, and its armed forces prevented communist China from recovering Taiwan. Moreover, with the Potsdam Declaration of 1945, the world had been divided into two groups—the capitalist camp of the West led by the United States and the communist camp of the East centered on the Soviet Union. China was the frontier of the communist camp and was menaced by potential conflict. The outbreak of the Korean War increased the possibility of further military confrontation. Under these intense geopolitical pressures, China decided to consolidate political and economic power by giving priority to the development of heavy industry (the production of producer goods, e.g., steel, coal, cement, etc.) in the First FYP.

Recovery and Industrial Policy.   In the eight years of the recovery and the First FYP, the annual growth rate of industrial production was as high as 17.8 percent.

This is partially because the original industrial output was very low. Nonetheless, the main reason for that rapid growth should be attributed to an effective strategy of industrial construction and resource management by the government as well as to domestic political stability.

The strategy of giving priority to heavy industry was clearly reflected in the structure of industrial investment in the First FYP. Eight-nine percent of the total industrial capital construction investment went to heavy industrial sectors, including metallurgy (with 18.6 percent of total investment), machinery (mainly for military defense), coal mining, power generation, chemicals, forest products, and building materials. In light industry, textile manufacturing and food processing (particularly sugar refining) were the major sectors of investment.

Location Strategy and Spatial Patterns.   As early as the recovery period, the central government made efforts to redistribute the unequal regional development by urging coastal plants to relocate inland in closer proximity to raw materials. This strategy was more clearly expressed in the First FYP: "In order to change the irrationality of the original unequal spatial distribution (of industry), new industrial bases must be set up, but the transformation and extension of old industrial bases and the support of them are prerequisites for building new bases."

Although this strategy was mainly a result of defense concerns—the government wanted to reduce the risks from possible military attacks on the coastal region of Southeast China—its practice actually had a sizable economic impact that shaped a new spatial structure of industrial production. Major areas of industrial construction during the recovery included Northeast China, North China, and East China.[1] For security purposes, some important plants were relocated inland. For example, the Shenyang tire plant and the Dandong paper mill were moved to, respectively, Huadian and Jiamusi in Heilongjiang province. At the same time, large-scale industrial construction began in Fulaerji in Heilongjiang, a small town near the China-Russia border. In the following First FYP, many new industrial projects were launched with the support of the Soviet Union. Nearly 850 large- and medium-scale projects (with an investment of more than RMB 10 million yuan) were planned, of which 156 key projects were financed by loans and directed by technicians from the Soviet Union (they are usually called the "156 projects"). These industrial projects constituted the basic structure of new China's industrial system.

The following analysis of the locations of the 156 projects reveals that the spatial emphasis of industrial capital construction shifted from the coastal region to inland areas in the First FYP. Of the 156 projects, four projects were cancelled and two merged, so that 150 projects were actually built. The total amount of investment in these built projects reached RMB 20 billion yuan in five years. Table 6.1 shows the spatial distribution of the 150 built key projects.

Provinces where more than ten projects were located are Shaanxi (mainly defense industries), Liaoning, Heilongjiang, Shanxi, Henan, and Jilin. Projects located in

TABLE 6.1   Spatial Distribution of the Key Industrial Projects in the First FYP

Region	Number of Projects	Percent of Total
Northeast China	56	37.3
Northwest China	33	22.0
North China	27	18.0
Central & Southern China	18	12.0
Southwest China	11	7.3
East China	5	3.4
Total	150	100

these six provinces account for 70 percent of the total 150 projects. In contrast, twelve provinces, including Shanghai, Jiangsu, Zhejiang, Fujian, Shandong, Tianjin, Guangdong, Guangxi, Guizhou, Qinghai, Ningxia, and Tibet, did not receive a single project. The twelve provinces fall in two categories. One category includes the relatively developed coastal provinces, which had a strong economic foundation and the best internal conditions for industrial construction but a bad location in regard to national defense given the international environment; the other category includes periphery provinces whose conditions were not suitable for large-scale industrial construction at that time. Of the eleven coastal provinces,[2] eight were not on the list for the 150 projects. Only 32 of these key projects were located in coastal provinces. As a result of this location strategy, the development of the traditional industrial areas, particularly the Yangtze River Delta and Tianjin, was restrained, and the growth rate of the coastal region declined. From 1952 to 1957, the annual industrial growth rate was 15.5 percent for the whole country, and as high as 17.8 percent for the inland region, but only 14.4 percent for the coastal region.

Along with a general shift of investment from coastal areas to inland regions in the First FYP, another prominent spatial feature was the concentration of industrial investment in northern provinces, particularly those near the Soviet Union. Of the 150 key projects that were built, 116 were located in northern provinces. Moreover, in the First FYP, the amount of capital construction in the three northern regions (Northeast, Northwest, and North China) was nearly three to four times as much as the investment per capita in other parts of China. This spatial pattern was closely related to the international environment China had to cope with at that time, which was characterized by Sino-Soviet friendship and Sino-U.S. antagonism. The northern regions near the Soviet Union were considered to be safe areas.

The Second, Third, and Early Fourth FYPs (1958–1972): A Tumultuous Period

The Great Leap Forward (GLF).   The reasonable and encouraging development trend of the First FYP did not continue in the following years. The radical "left" ideology provoked by the anti-"right" political struggle in 1957 quickly spread to the area of economic development and changed the industrial development process. In early 1958, the central government, directed by Mao Zedong, drew up a "general line" of socialist construction as well as a policy of "walking on two legs" for agriculture and industry. It further emphasized the importance of heavy industry and urged the development of an independent and comprehensive local industrial system not only for economically coordinated regions but also for provinces that could not support such industrial systems. Under this political and economic impetus began the Second FYP, which was characterized by a crash industrialization program—the Great Leap Forward (GLF).

As a result of this rash development strategy, the central planning force was no longer involved in controlling industrial capital construction investment. According to incomplete statistics, more than 216,000 large-, medium- as well as small-scale industrial projects were under construction in the Second FYP, and the total amount of investment was three times as much as that in the First FYP. In a short time, small-scale industrialization had sprung up all over the vast rural areas. In 1958, most counties and even towns set up small-scale iron and steel plants in response to Mao's call for catching up with Britain and the United States in the production of steel. After that, small-sized rural industries, including electric power, cement, coal mining, chemical fertilizer as well as iron and steel industries (usually called the "Five Small Industries") proliferated in a disorderly fashion. These small rural industries were initiated without conducting any cost-benefit analyses.

At the same time, the strategy of "steel as key link" was carried out.[3] A large quantity of resources (capital and technicians) was used to build two new iron and steel plants in Jiuquan in Gansu province and Panzhihua in Sichuan province, although seven "old" plants—at Anshan, Benxi, Shanghai, Maanshan, Wuhan, Taiyuan, and Chongqing—had already been under large-scale expansion. This increased capital construction investment beyond the government's means, which in turn resulted in a large deficit and a sharp decline in the living standard.

The unexpected economic chaos that ensued forced the central government to resume strong planning power in 1961 and sharply cut investment. Total capital construction investment was reduced to RMB 12.7 billion yuan, from the previous year's 38.9 billion and was further reduced to 7.1 billion in 1962. Hundreds of thousands of industrial projects had to be abandoned. The GLF resulted in a huge economic loss and a decline in the economic growth rate. The annual industrial growth rate averaged only 1.63 percent in the Second FYP, less than one-tenth of that of the First FYP. For the provinces, the greater the increase in investment in the GLF, the greater the decline in their growth rate. Several key provinces involved in industrial capital construction, like Sichuan, even had negative growth. To recover from the GLF disaster and the subsequent famine disasters in 1961 and 1962, three years of readjustment without much industrial investment followed the Second FYP.

The Third Front Strategy and Its Spatial Pattern.   As if it were not enough for China to fall into domestic economic chaos, its geopolitical situation deteriorated in the early 1960s. The ideological conflict with the Soviet Union finally led to the end of Sino-Soviet friendship, and a thirty-year rift began in 1960. At the same time, Taiwan became increasingly belligerent toward Mainland China with the support of the United States. The simultaneous antagonism with the two world powers caused the top leaders of China to make an overly pessimistic judgment that wars might break out at any moment. Accordingly, the central government made two strategic decisions: (1) a large-scale shift of industrial capital construction into the "Third Front" areas,[4] and (2) further emphasis on the development of comprehensive local industrial systems aiming at independent economic survival during wartime. The general aim of this strategy was to develop the inland area, particularly the "Third Front," into a strong rear base with complete industrial structure able to support potential wars against invasions.

According to this strategy, from 1965 on, major areas of industrial capital construction were transferred into the "Third Front," whose core region comprised Sichuan, Shaanxi, and Guizhou provinces as well as the western parts of Henan, Hubei, and Hunan provinces. Key invested sites included Panzhihua (iron and steel), Jiuquan (iron and steel), and Chongqing (iron and steel and machinery). In addition, provinces were required to set up their own "Minor Third Front"—a provincial rear base against potential wars. The location principle in the "Third Front" period embraced "regional dispersal and situation concentration," "far away from big cities" and "hidden in mountains or caves wherever possible," all of which led to severe external diseconomies as well as ineffective investment.

The total investment of capital construction of the whole country in the Third FYP was about RMB 120 billion yuan, 66.8 percent of which went to inland areas. In contrast, the inland share of investment in the First and Second FYPs was only 47.8 percent and 53.9 percent respectively (Table 6.2). The greatest increase of investment among regions occurred in Northwest China and Southwest China. These two regions' share of total investment rose to 35.1 percent in the Third FYP from only 16.9 percent in the First FYP. At the same time, investment in Shanghai, Tianjin, Liaoning, and Jiangsu decreased.

The "Third Front" strategy continued until 1972, and two major waves of inland construction can be distinguished. The first one occurred between 1964 and 1966 with the key area of construction being Sichuan, Hubei, and Gansu, whereas the second one occurred in the period from 1969 to 1972. Although more provinces were involved in the second wave, the key areas consisted of Sichuan, Hubei, Shaanxi, Henan, and Guizhou.

The Late Fourth and Fifth FYPs (1973–1980): The Inception of Opening

Changes in Industrial Policy.   The international political environment changed substantially in 1972. U.S. President Nixon's visit to China and the subsequent Shanghai

TABLE 6.2   Regional Distribution of Capital Construction Investment in China, 1952–1975 (%)

Period	Coastal Region	Inland Region	In the "Third Front"
First FYP (1952–1957)	41.8	47.8	30.6
Second FYP (1957–1962)	42.3	53.9	36.9
Readjustment Period (1962–1965)	39.4	58.0	38.2
Third FYP (1965–1970)	30.9	66.8	52.7
Fourth FYP (1970–1975)	39.4	53.5	41.1
Total (1952–1975)	40.0	55.0	40.0

Joint Communiqué issued on February 28, 1972, broke the hostile encirclement and reduced the threat of potential war with the United States. This geopolitical change, as well as concerns over the delay of economic development caused by the Cultural Revolution, encouraged China's top leaders to open a crack in its closed economic system to the importation of industrial equipment in order to stimulate and promote economic development.

In January 1973, the central government worked out a plan of "increasing equipment imports and enlarging economic exchange," which defined major items to be imported. Guided by this plan, forty-seven sets of key equipment were imported in the 1970s, most of which was processing equipment and power equipment. Among them were thirteen sets of synthetic ammonia–processing equipment with an annual capacity of 300,000 tons, four sets of chemical fiber–producing equipment, three sets of petrochemical processing equipment with ethylene processing being the core, three sets of large generators, a 1.7 meter-wide steel rolling mill, and more than ten sets of coal-mining equipment. These were the key industrial projects underway in the 1970s, which played an important role in improving the irrational industrial structure and enhancing national economic power. Most of them were put into production by 1977.

The industrial capital construction from 1973 to 1980 was planned to facilitate the above-mentioned import projects. The majority of the industrial investment went to the power industry, the chemical industry, and the iron and steel industry in order to alleviate the shortage of raw and processed materials and power supply. In addition, the share of investment to light industry and to the textile manufacturing was also raised to balance the development of light and heavy industries. In the Fifth FYP, capital construction investment in the textile, food processing, and paper making industries increased by 129 percent, 82 percent, and 38 percent, respectively, compared with those of the Fourth FYP. Correspondingly, the share of investment to the metallurgical industry decreased from 17.5 percent in the Fourth FYP to 15.4 percent in the Fifth FYP, and that of the machinery industry decreased from 22.2 percent to 14.5 percent. The sharp decrease in investment in the machinery industry was due to cuts in defense spending.

Spatial Patterns.   The location of industrial development in this period shifted toward the coastal region, particularly the "old" industrial areas. The major characteristics of the spatial pattern are as follows:

1: Most of the new industrial projects were located in traditional industrial areas. Of the forty-seven import projects, twenty-four went to the coastal region, twelve to the middle region, and eleven to the western region. Major invested areas in the coastal region included central Liaoning, the Beijing-Tianjin-Tangshan Triangle, the Yangtze River Delta, and the Shandong peninsula, while key areas in the middle and western regions were the Sichuan Basin and the Hanjiang River Plain (centered on Wuhan). These areas are traditional industrial centers or former major areas of industrial construction after 1949, which had much better infrastructure than other areas and were able to offer the necessary technological support and production cooperation for new projects.

2: In regard to specific locations, most key projects were located on the coast or on the Yangtze River, which facilitated water supply and cheap water transport. For example, of the ten new large oil refinery projects and four refinery expansion projects between 1973 and 1980, six were on the Yangtze River (and this does not include Shanghai). To achieve external economies, most new chemical fertilizer works were also located on the coast or on the Yangtze River in proximity to refineries.

3: The development of industrial districts was another major feature of the spatial pattern in this period, which is totally different from the dispersed pattern in the "Third Front." New plants were planned and organized to constitute an integrated industrial district; those industries with close cooperative ties were platted together to achieve external economies. Industrial districts were planned in large or medium-sized cities or in close proximity to raw materials.

Open-Door Industrial Policy and
Its Spatial Patterns in the 1980s

New Policies and Structural Changes

At the end of the 1970s, great political and economic changes took place in China. In December 1978, the Third Plenum of the Chinese Communist Party's Eleventh Central Committee decided to give priority to economic growth instead of political struggles. In 1979, the central government ratified a "special policy and flexible measures" in Guangdong and Fujian provinces. In 1980, the Standing Committee of the People's Congress approved four special economic zones (SEZs): Shenzhen, Zhuhai, Xiamen, and Shantou. From then on, China entered a new era of industrial development. In early 1984, the central government further decided to give fourteen coastal port cities special economic policies.[5] These cities are called "open cities." Over the next few years, the Pearl River Delta, the Yangtze River Delta, the Xiamen-Quanzhou-Zhangshou Triangle (in southern Fujian), the Shandong peninsula, and the Liaoning peninsula were designated as (economic) open areas.

The purpose of opening these relatively industrialized cities and areas was to encourage international economic and technological exchanges, attract foreign investment and advanced technology to transform traditional industries, and speed up the development of new industries. The open cities were also expected to transfer their imported technologies to inland areas by way of regional cooperation. The result of the new industrial policy was stunning. The annual industrial growth rate in the 1980s was as high as 14 percent, much higher than that of the previous two decades. A large number of industrial enterprises in the coastal region, particularly the open cities, carried out large-scale replacements of equipment and greatly improved their production technology. In addition, new industries, such as electronics and electrical appliances, developed quickly.

From the early 1980s on, the emphasis of industrial development shifted to the energy industry and to consumer goods production. The share of industrial capital construction investment going to the energy sector increased rapidly, from 39.8 percent in the Fifth FYP to 44.0 percent in the Sixth FYP. The key energy project was the construction of the Shanxi Coal Base, whose annual coal production rose to 0.33 billion tons in 1987 from only 0.19 billion tons in 1982 and which exported more than 0.1 billion tons to other regions in 1985. In addition, the construction of three large-scale open-cut coal mines in eastern Inner Mongolia were initiated, and quite a few large-scale power stations were completed, including the Gezhouba hydropower station. These key energy projects played an important role in alleviating the shortage of energy in the coastal region. The investment in light industry also increased significantly, from 12.7 percent of the total industrial capital construction investment in the Fifth FYP to 15.5 percent in the Sixth FYP, which led to a great improvement in the living standard during the 1980s.

Locational Patterns in the Sixth FYP

The Sixth FYP witnessed a locational strategy of "priority to the coastal region." The share of industrial investment of the eleven coastal provinces (Hainan included in Guangdong) rose to 46 percent from 44 percent in the previous FYP. In 1984, Guangdong's share of total capital construction investment was the highest among all China's provinces. By utilizing foreign capital, especially Hong Kong's capital, Guangdong rapidly developed into a strong base for consumer goods manufacturing. Moreover, industrial investment in Shanghai, Fujian, Zhejiang, Jiangsu, and Shandong also increased significantly.

In contrast, the investment share of the inland areas decreased sharply. For example, Sichuan was once the key area of industrial construction, and its share of national capital construction investment was as high as 14.5 percent between 1965 and 1967, but it was reduced to only 4.4 percent in 1984. Nonetheless, a few important industrial projects were constructed inland, including energy projects in Shangxi, Heilongjiang, Inner Mongolia, Anhui, and Henan as well as the Gezhouba hydropower station, and nonferrous metallurgical projects in Shanxi (aluminum), Henan (aluminum), Jiangxi (copper), and Hunan (lead and zinc). In addition, several new industrial projects and expansion projects were underway in central Shaanxi, southern Sichuan, and in Urumqi.

Change of Locational Strategy in the Seventh FYP

In the Seventh FYP, the strategy of "priority to the coastal region" further developed into a "three regions" strategy. The dual division of "coastal region" and "inland region" was too simplistic and inadequate as the framework for industrial location, because China is a vast country with enormous variations. Premier Zhao Ziyang first used the terms "coastal region," "middle region," and "western region" in as early as 1982. Guided by his ideas, the State Planning Committee formulated a "three regions" locational strategy in the Seventh FYP. The guiding ideology of this new strategy was to locate new industrial projects mainly in the coastal and middle regions in the next decade and beyond and to move major areas of industrial construction gradually into the middle region while continuing to promote the rapid development of the coastal region. This was clearly stated in the plan:

> The relationship of economic development between the coastal, middle and western regions must be handled properly. Before the end of this century, the economic development of the coastal region is to be further promoted, but the location of energy industry and raw material industry will mainly be shifted to the middle region. At the same time, the development of the western region will be further planned and prepared actively. The development of the coastal region will be well integrated with that of the middle and western regions so as to support and promote the development of each other.

However, the actual spatial pattern of industrial location did not change much in the Seventh FYP. The coastal region became more important as foreign direct investment flowed mainly into this region. In the middle and western regions, key areas of industrial construction were nearly the same as those in the Sixth FYP.

Changes in the 1990s

In the 1990s, China continued its rapid pace of economic development. Industrial growth surged with substantial improvements in production technology and industrial infrastructure. From 1991 to 1995, the annual industrial growth rate was more than 20 percent on average, and for 1992, 1993 and 1994, it was 24.7, 27.3, and 24.2 percent respectively. In 1995, the total value of industrial output reached RMB 9,198.4 billion yuan (at current prices), more than three times that of 1980. Coal output exceeded 1.36 billion tons, electricity production topped 1007 billion kilowatt hours, steel production exceeded 95 million tons, cement production reached nearly 476 million tons, and nearly 35 million television sets were produced. China has in fact become a leader in industrial production, especially the production of raw and processed materials.

A significant feature of industrial development since 1992 has been the increased significance of market forces, rather than central planning, in guiding the adjustment of industrial structure. First, the development of the energy industry and the raw and processed material industry was further strengthened in order to satisfy the increasing demand caused by rapid economic growth. Key projects included the geological prospecting and exploitation of crude oil in Xinjiang, the Bohai Bay, the mouths of the Yangtze and Pearl Rivers, the southern Huanghai Sea, and Beibuwan Bay; the development of hydropower on the Hongshui River in Guangxi and on the upper and middle reaches of the Yangtze River and the Lancangjiang River; the further expansion of several large iron and steel plants as well as new petrochemical processing, cement, and glass-making plants. Second, the greatly expanding domestic and export markets led to sharp increases in the production of electrical appliances, electronic components, textiles, communication equipment, and personal computers. For example, the output of color television sets doubled, production of washing machines and refrigerators increased by 70 percent, production of cameras increased by nearly ten times, and production of radios increased by 170 percent in the Eighth FYP. Third, the import-substitution strategy promoted the development of automobile manufacturing. In 1995, China manufactured 1.45 million motor vehicles, 22.4 percent of which were cars, an enormous increase over the 8.3 percent five years earlier.

In regard to spatial patterns of industrial growth, a "T-shaped" framework is emerging, in which major industrial growth points are concentrated on the eastern coast and on the Yangtze River (forming a horizontal 'T'). The key characteristics of this framework include the following:

1: In the coastal region, industrial sectors such as electronics, communications equipment, automobile manufacturing, and electrical appliances have developed quickly into new growth points, while the energy, iron and steel, petrochemical, engineering, and shipbuilding industries remain strong. Moreover, the export of manufactured goods has increased by a large measure.

2: In the Yangtze River industrial corridor, industrial sectors like iron and steel, petrochemical processing, automobile manufacturing, and light processing and manufacturing are given priority. This industrial corridor includes three of the major industrial areas in China: the Yangtze River Delta, the Wuhan area in the middle reach, and the Chongqing area in the upper reach.

3: The special economic zones, high-tech parks, and other types of economic and technological development zones have become new growth poles and technological innovation bases in both the coastal region and the Yangtze River industrial corridor.

In addition to the "T-shaped" framework, a few other areas in the western region are becoming more important in the locational map of China's industrial development. These are mainly areas endowed with rich power resources, such as the upper reach of the Yellow River (with Lanzhou being the core city), the Shanxi-Shaanxi-Inner Mongolia Triangle (coal based), and the Hongshuihe River in Southwest China. High-energy-demand industries, like nonferrous metallurgy, are being or will be located in these areas.

In the late Eighth FYP, a new strategy of "coordinated regional development" was formulated by the central government, mainly to alleviate political pressures from the middle and western provinces caused by the increasing disparity of economic development. This may influence the industrial location pattern in the future. However, thus far, few specific measures to fulfill this strategy have been put into practice. The major effort seems to rely on regional economic and technological cooperation. Therefore, it appears that it will be very difficult, if not impossible, to reduce the disparity among regional industrial development efforts or to change significantly its spatial pattern in the near future.

Notes

1. Northeast China includes Liaoning, Jilin, and Heilongjiang provinces. North China includes Beijing, Tianjin, Hebei, Shanxi, and central Inner Mongolia. East China refers to Shanghai, Jiangsu, Shandong, and Zhejiang.
2. Coastal provinces include Liaoning, Hebei, Beijing, Tianjin, Shandong, Jiangsu, Shanghai, Zhejiang, Fujian, Guangdong, and Guangxi.
3. The "steel as key link" strategy called for giving priority to the development of the iron and steel industry and using it as the guiding principle in capital construction.
4. The "First Front" referred to the coastal area and the northern border area, which were presumably susceptible to military attack. The "Third Front" referred to the remote central and western mountainous areas, particularly Sichuan, Shaanxi, and Guizhou provinces, which were considered to be safe from bombing attacks. The "Second Front" is relatively ambiguous in geographical definition and referred to the areas between the "First Front" and the "Third Front."
5. These ports include Dalian, Qinhuangdao, Tianjin, Yantai, Qingdao, Lianyungang, Nantong, Shanghai, Ningbo, Wenzhou, Fuzhou, Guangzhou, Zhanjiang, and Beihai.

7

Changes in China's Space Economy Since the Reform

Max Lu

The economic reform launched by the Chinese government in the late 1970s undoubtedly represents a major change in the People's Republic's economic development strategy. During the reform, market principles and an open-door policy were introduced into China's Soviet-style, centrally planned economic system. Although equity is still considered a goal of economic development, in practice the new development strategy has emphasized efficiency (Fan 1995, 1997; Yang 1990, 1991). To boost efficiency, local and especially provincial governments have been granted more power in decisionmaking. The previous emphasis on military-oriented heavy industry has been readjusted to favor consumer goods production. Private enterprises are allowed to not only compete with the state-owned and collective enterprises but also thrive in the post-reform period, while the state-run sector experiences persistent difficulties, including massive layoffs. In order to attract foreign investment and technological know-how and to gear the economy to the outside world, China established a series of special economic zones (SEZs), open cities, and a variety of open areas, with various preferential policies in place for foreign investors. Though punctuated by events like the 1989 Tiananmen Square incident, the over-two-decade-long reform and opening to the outside world has drastically transformed China's rigid command economy into a vibrant market economy. In the nineteen years since the inception of the reform in 1978, China's GDP grew at an average rate of 9.8 percent per year in constant prices. Following the Asian financial crisis in 1997, which sent many Asian economies reeling, and the recent worldwide economic slowdown, China's economy still expanded at an impressive annual rate of more than 7 percent in the 1998–2001 period. More foreign investment is pouring into the country than ever before. In 2000, China's GDP exceeded for the first time the $1,000 billion landmark. The fast and sustained economic growth has made China the seventh largest economy in the world since 1998 in terms of GDP. Many experts believe China's fast economic growth will continue in the foreseeable future.

Nowhere has the impact of the reform been more significant than in the changes in China's economic structure. Prior to the reform, the Chinese economic structure was relatively simple and dominated by the state-run heavy industry that mainly served the military. It has since been changed to an economy in which consumer goods production is emphasized and the importance of the state-run sector is shrinking. The rapid expansion of the private sector has been most impressive. According to an Asian Development Bank report in early 2002, enterprises that are not state-owned or collectively owned have been generating more than 60 percent of China's GDP. The Chinese government has announced that by the end of 2001, China had 2.03 million registered private enterprises, which employed 27.14 million people. More than 85 percent of China's medium-sized and small enterprises are privately owned. Foreign joint ventures of different kinds numbered 398,900, with a total of $776.5 billion of contracted foreign investment ($409.3 billion actually used). These foreign joint ventures employed 23 million people. The private sector has in fact become the most viable part of the Chinese economy, responsible for the almost uninterrupted fast growth since the late 1970s.

The reform has also profoundly altered China's spatial economic patterns. To maximize the return on investments, China's central government has concentrated its economic development efforts in regions where economic and institutional infrastructure is available or where the main factors of production, especially exploitable natural resources, can be found. Not coincidentally, such regions are also favored by foreign investors (Leung 1990). With a combination of relatively good infrastructure, readily available agricultural and industrial raw materials, a substantial skilled labor pool, and large markets, the coastal area has become the favored place for both domestic and foreign investment and has consequently enjoyed one of the world's fastest-growing economies. Not only the physical landscape but also the social and cultural fabric in this part of China have undergone significant transformations. The especially remarkable economic performance of South China's Guangdong province, which is adjacent to Hong Kong and contains three of the country's five special economic zones, has prompted many to call this region the fifth "Asian Tiger," following South Korea, Taiwan, Hong Kong, and Singapore.

In contrast, economic development in the vast interior of the country has lagged behind. Little foreign investment has found its way there. There are few skyscrapers or luxury hotels, few Western fashions or fast-food outlets. The majority of China's 65 million people living in poverty in the mid-1990s were found in the inland provinces (Lu et al. 1997). Many farmers in those provinces have flocked to the booming coastal cities in search of employment opportunities and become the so-called "floating population." Although the interior provinces have always been less developed than their coastal counterparts, many fear that the economic reform of the past two decades has left these provinces even further behind (Wei 1994).

Yet the widening coast-inland disparity in economic development is only one facet of China's changing space economy. The new development strategy and the revenue-sharing scheme between the central and local governments have resulted in a different set of forces shaping regional economic structures and the ebb and flow of regional economies. In their pursuit of fast economic growth, regions have competed both for foreign investment and for preferential policies (Yang 1991). They have not only emulated the successful development measures adopted by the SEZs and some other coastal localities but also tried to develop industries in which they have comparative advantages. The net result of these practices appears to be increasing regional economic specialization. Furthermore, the differentiated growth driven mainly by the lopsided distribution of foreign and domestic investment, particularly the rise of South China, may have eroded the dominant role of the Yangtze River Delta in China's space economy. The principal axis of the Chinese space economy may have been shifting from one of east-west orientation to one of north-south orientation.

The purpose of this chapter is to analyze the major changes in China's space economy since the initiation of reform. Although much research has focused on recent changes in China's regional inequalities, little attention has been paid to other dimensions of China's changing spatial economic pattern (Wei 1999). This chapter will fill the void in the literature on Chinese regional development by focusing on other changes in China's spatial economic pattern in the post-reform period, particularly the trend toward regional specialization and the axial shift in dominant regions of economic development, though a brief discussion on regional disparities is also presented. The next section of this chapter reviews the features of regional development prior to the reform, which provides a point of departure for the later discussion. This is followed by a discussion of the regional dimensions of China's economic reform and the widening gap between the coast and inland areas. After elucidating the forces that shape regional specialization, the chapter analyzes the trend toward specialization. The final section of the chapter discusses the axial shift in China's space economy.

Regional Economic Development Prior to the Reform

To facilitate the discussion of the changes in China's space economy since the reform, it is worth reviewing the situation prior to it. China is a very complex and pluralistic society in many respects. This complexity is reflected not only in the diversity of its culture but also in the unevenness of its economic development. China's economy was traditionally heavily biased toward the coastal area, as it was the center of foreign influence and development in the nearly half century preceding the founding of the

People's Republic in 1949. The main concentrations of industrial capacity were found in the lower Yangtze River Basin centered on Shanghai and southern Jiangsu province, in the North China Plain around the treaty port city of Tianjin, and in northeastern China, especially Liaoning province (Xie and Costa 1991). Along the coast, several foreign colonies and treaty ports, such as Dalian, Yantai, Qingdao, and Shanghai, were established during the nineteenth century. The Japanese were very active in Northeast China (Manchuria) during the eight years (1937–1945) of the Second Sino-Japanese War and had developed that region as a heavy industrial and material support base for its home islands (Field 1986; Wu 1979). By contrast, the vast interior of China was largely underdeveloped and devoid of modern industry.

Since coming into power in 1949, the communist government has been in a race with time to achieve national economic development and to raise the living standard of the Chinese people. It modeled its socialist system after that of the Soviet Union, which inspired the Chinese revolution and was China's only major ally of the time. Soviet experts went to China to help build a socialist state. The Soviet-style socialist system is such that the central government controls all aspects of the national economy, and all major decisions regarding economic development strategy, investment, and allocation of economic activities are the outcome of direct bureaucratic control, or the "visible hand." The Chinese government set for itself the task of modernizing China by means of industrialization. Regional interests were sacrificed in favor of pursuing national goals. Provincial and local governments handed over their revenues to the central government to be redistributed. For the most part, the government implemented policies that promoted macroeconomic growth and often included unrealistically high economic targets. China had little trade with countries other than the Soviet Union. In fact, the Western countries had placed a trade embargo on China following the communist takeover. After the bitter split with the Soviet Union in the late 1950s, China's economy developed in self-imposed isolation for almost two decades.

China's economic development strategy during the pre-reform period was largely influenced by the potential threats the Chinese leadership perceived from both the Soviet Union to the north and from the United States and Taiwan, which China considered to be a renegade province, to the east. Defense naturally became Chinese leaders' overriding concern. The allocation of public investment and government revenues bore clear imprints of this military obsession. In order to increase its war readiness, China focused on developing heavy and military industries, such as metallurgy, machinery building, coal mining, power generation, and petrochemicals, often at the expense of light industry and consumer goods production (Zhao and Gu 1995). Geographically, China shifted its focus from the coastal area to the inland. The government allocated a disproportionately large amount of financial and human resources to remote, largely inaccessible mountainous areas in the central and western parts of the country between 1952 and 1978 (Cheng and Zhang 1998; Yang 1990). The ill-fated "Third Front" construction campaign of the 1960s and 1970s epitomized this military-oriented strategy. The campaign promoted dispersed, mountainous, and hidden locations for industrial development. The ten interior provinces of Sichuan, Guizhou, Shaanxi, Gansu, Qinghai, Ningxia, Yunnan, Hubei, Hunan, and Shanxi were designated as the "Major Third Front," which was to be the national focus of industrial development. Remote mountainous areas in individual provinces were considered to be the "Minor Third Front," also to be favored for investments (Zhao and Gu 1995). Many manufacturing plants were relocated from coastal cities like Shanghai to inland areas, and new ones were also established there. Based on official Chinese sources, the total investment in various Third Front projects amounted to about 700 billion yuan (US $200 billion). During the same period of time, investment in coastal provinces stagnated. In the frontline province of Fujian, which faces Taiwan across a strait, virtually no investment was made.

A number of new industrial bases grew out of the Third Front projects in the interior provinces. The Panzhihua Iron and Steel Works in Sichuan province and the Second Automotive Works in Xiangfan, Hubei, are among the well-known projects. Although economic growth improved in the interior, it slowed down in the coastal provinces. The government's goal of reducing the polarized pattern of development was partially fulfilled, though huge economic losses were incurred. Between 1952 and 1978, the share of gross industrial output value in the coastal area exhibited a modest decline, while the interior registered a strong increase (Field 1986; Pannell 1988; World Bank 1985). The interior's total population increased by 45.2 percent during that period, while the coastal population increased by 40.7 percent if the three cities of Shanghai, Beijing, and Tianjin are excluded from consideration. The coastal share of urban population declined by 10 percent to the same level of all the interior provinces combined. But the coast, especially the areas centered on Shenyang in the northeast, Tianjin in the north, and Shanghai in the east remained the country's most important economic regions. Shanghai was particularly prominent as China's industrial and commerce center, setting the national pace with its manufacturing capabilities, universities, and entertainment industry. The level of development generally declined from the Yangtze Delta area (Shanghai) to the inland, giving rise to an east-west orientation.

Along with the westward shift in industrial development, China promoted the establishment of self-sufficient, comprehensive regional economic systems. The logic behind this strategy was that in case some parts of the country came under foreign attack, other regions would continue to function. The whole country was divided into seven major regions with about five to six provinces each. Within each region an independently functional economic system with substantial steel and grain production capacities was to be established. In practice this policy prompted all regions, large and small, to attempt to obtain various manufacturing capabilities, which resulted in inefficient, redundant industrial developments all over the country. Economic decisions were often made without regard to local resources or existing industrial bases. The government particularly encouraged local areas to develop small-scale iron and steel mills, coal and iron ore mining, cement kilns, electric power facilities, and chemical fertilizer factories, or the so-called "Five Small Industries." Such a development strategy inevitably led to homogenization of economic structures between regions.

The highly centralized economic system and the defense-motivated development strategy prior to the reform were extremely susceptible to inefficiency due to the absence of market mechanisms and counterproductive locational choices (Rothenberg 1987; Lakshmanan and Hua 1987). Under the centralized system, production units (firms) were only engaged in the production process. Production, material support, and product purchase were all centrally planned and vertically integrated. There were virtually no horizontal linkages between production units in the sense of genuine market contracts. Poor performance did not force exit from production. No incentive structure was in place to encourage productivity. The poor management left China's economy in a complete shambles. It was to stave off economic collapse and social upheaval as well as to revive the stagnant economy that the economic reform and the open-door policy were introduced in the late 1970s.

Economic Reform and Widening Regional Disparities

Economic restructuring in China followed shortly after the death of Mao Zedong and the removal of the "Gang of Four," led by Mao's widow, from power in 1976. Chinese leaders revived the drive to four modernizations (agriculture, industry, defense, science and technology), first put forward more than a decade earlier by Premier Zhou Enlai. But the government continued to pursue macroeconomic growth and set for itself the goal of quadrupling the country's total 1980 GNP (451.76 billion *yuan*) by the end of the century. Development of heavy industry was still considered to be the key to this goal. A number of assembly lines were introduced from Japan and other Western countries to upgrade China's outdated manufacturing facilities. The most noticeable change was that most of the new manufacturing projects were placed along the coast or in areas with readily exploitable natural resources. The Baoshan Iron and Steel Works, developed near Shanghai at that time, became China's single largest project in terms of investment.

The economic reform enacted in 1978 marked a real departure from the Soviet-style orthodox command economy, though it was initially a program to stop the country's economy from reeling without abandoning socialism or the Communist Party's exclusive grip on power. Following the great success in rural reforms in the late 1970s and early 1980s, market mechanisms were also introduced in urban areas to tackle both the chronic inefficiency plaguing industry and the severe shortage of life necessities and consumer goods. In order to generate the highest returns on its investment, the central government focused its development effort on the coastal provinces in the hope that the wealth created there would seep into the more backward regions in the interior. The coastal area, which consists of the twelve coastal provinces, comprises less than one-eighth of China's territory but 40.8 percent of China's 1.24 billion people (1997 figure). The central government has responded to the need for additional economic and social infrastructure in this region by constructing new ports, airports, highways, and communication facilities. Certain areas on the coast were designated as zones with special investment incentives and more liberal policies. They were allowed to experiment with market mechanisms and reinvest their own tax revenues (Huan 1986; Wang and Bradbury 1986). Those areas include the five special economic zones (SEZs) of Shenzhen, Zhuhai, Shantou, Xiamen, and the island province of Hainan, fourteen open cities along the coast from Dalian in the north to Beihai in the south, and the open regions of the Yangtze River Delta, the Zhujiang (Pearl River) Delta, and the Minnan (the southern Fujian province) Delta (Figure 7.1). Although various other cities and areas located in the inland were also allowed to use investment incentives for foreign investors, the coast was clearly favored by the central government as the focus of economic development and, hence, investment. After several years of discussion, the central government formally endorsed a coast-oriented development strategy in 1988.

The shifts in development strategy and regional focus since the reform have resulted in significant changes in China's space economy. The most politically charged change may be the exacerbation of regional disparities in development levels, especially those between the coastal and the inland provinces. Many studies have documented the changes in regional inequality (Barnard and Shenkar 1990; Chan 1992; Chu 1988; Fan 1992, 1995; Fleisher and Chen 1997; Fujita and Hu 2001; Goodman 1989; Lee 1995; Lu and Wang 2002; Pannell 1988; Wei and Ma 1996; Wu 1979; Xie and Dutt 1990; Ying 1999; Zhao and Gu 1995). The general consensus seems to be that for the country as a whole interprovincial disparities have diminished, while the interregional gaps among the eastern, central, and western zones (see Figure 7.1 for the delimitation of these zones) have increased. Some scholars have questioned the divergence view on the ground that interprovincial disparities have declined and downplay the concern over the unevenness of development. But what concerns people in China and may potentially threaten social stability has always been the interregional gap between the east and west. The fact that this gap has indeed been widening indicates that the concern is legitimate.

The increase in regional inequality is accompanied by a sharply rising disparity in personal incomes. The value of the Gini coefficient—a common measure of income disparity—has gradually increased from about 0.3 in 1980 to 0.458 in 2000. By international standards, China has become a very polarized country in terms of income distribution.

It is worth pointing out that the increase in China's regional disparities is not caused by some regions growing while others stagnate or decline. Every region has made economic progress since the reform. The widening disparities are the outcome

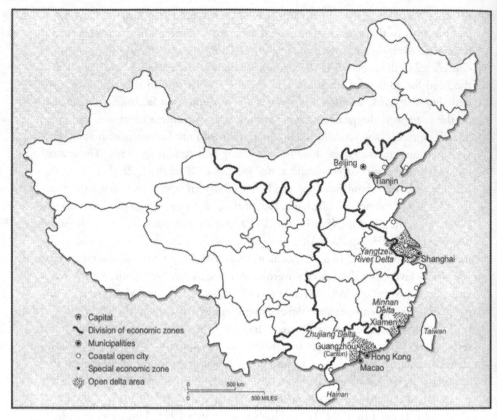

FIGURE 7.1   China's Economic Zones, Selected Open Cities and Regions

of the drastically differentiated rates of economic growth among the regions. The coastal area has maintained a much faster rate of economic growth than has the rest of the country. Guangdong province, in particular, enjoyed a remarkable 14.2 percent annual rate of GDP growth during 1978–1997. Coastal provinces have consequently gained more economic strength than have the interior ones. For example, from 1978 to 1997, the coastal provinces increased their share of industrial output in the national total from 59.9 percent to 65.4 percent, while the interior provinces' combined share dropped from 40.1 percent to 34.6 percent (State Statistics Bureau 1998). Disparity between urban and rural areas and between affluent and poor rural areas is even larger (World Bank 1985). In 2000, rural per capita income amounted to only 35.7 percent of urban residents' per capita income. Rural per capita income in the eastern zone was 3,649.4 yuan, 1,215.4 yuan above the national average, whereas the same figures for the central and western zones were only 2,170.6 yuan and 1,605.8 yuan. The ratio of personal incomes of peasants between the richest province (Shanghai) and the poorest one (Gansu in 1978, Guizhou in 2000) was

2.78 in 1978 but increased to 5.20 in 2000 (5,914.9 yuan in Shanghai, and 1,136.4 yuan in Guizhou).

The widening gap has revived the strains between the interior and the coastal provinces. Local authorities in the vast hinterland watched the coastal boom with growing discontent. The inland governments want more investment, higher prices for their products (mainly raw materials), and freedom to develop and experiment in unfettered capitalism. They demand a bigger share of central investment funds. The burgeoning regionalism has often led to conflicts and tension between regions. To enforce their demands, many inland regions have set up trade barriers, hoarded raw materials in scarce supply, and diverted state-distributed goods to the free market. The "battle" over raw materials among different regions has erupted a number of times (Lee 1998; Yang 1990). For example, Guangdong province relies on other provinces for industrial and other scarce materials and thus has often come under fire for its aggressive buying. Poor provinces object to Guangdong's using its financial clout to bid up prices and deprive them of raw materials and even food. In 1988, Guangdong's neighboring Hunan province, suffering periodic pork and grain shortages despite its surplus production, set up its own customs patrols to stop farmers from selling their pigs for the higher prices available in Guangdong. Other provinces have complained that Guangdong may be stocking up on much-needed supplies at the expense of others and have even asked the central government to intervene. Guangdong, on the other hand, sees the conflict as a clash between their open, competitive economy and the "ossified," inefficient system of the inland.

Since the reform started, the Chinese government has tried several times to redress the striking disparity in regional development. For example, it frequently called for more domestic and foreign investment in the western part of the country and encouraged coastal provinces to forge economic partnerships with the impoverished but resource-rich interior provinces. But these measures achieved very little success. The real hope for the vast interior may lie in the new strategy adopted by the Chinese government in 1999, called "Western Big Development" (xibu da kefa), which has been touted as a top national priority for the next decade. This new initiative targets twelve western provinces and autonomous regions, which together cover 71.4 percent of China's area but account for only 28.1 percent of the country's population and 17.2 percent of the GDP. The region contains China's most important oil, natural gas, and mineral resources and has over 80 percent of the nation's hydroelectric resources, 90 percent of its grassland, and 40 percent of its arable land. In promoting this strategy, the Chinese government seems to be counting on the economic development of the western provinces to boost domestic demand for new products, stem the widespread environmental degradation in the region, ease ethnic and religious tensions, and achieve regionally balanced growth. As a concrete step for carrying out the new strategy, China's State Council issued several policy measures in December 2000. The central government has also committed a significant amount of investment to improve the infrastructure and the environment in western provinces. There are also

signs that foreign businesses and international lending agencies are paying more
attention to China's interior. The Asian Development Bank, for example, announced
in April 2002 that it will allocate 70 percent of its investment in China to the interior provinces and only 30 percent to other parts of the country, reversing its past
practice. But development of China's western region also faces several major challenges, such as its imposing size, harsh environment, and disparate culture. Only time
will tell whether this new strategy succeeds in bringing interior provinces much
improvement in their lot.

The Forces That Shape Regional Economic Structures

The reform and decentralization have also changed the configuration of the forces
that shape China's regional economic structures and the ebb and flow of regional
economies. Two sets of forces are at work. The first set of forces facilitates regional
economic specialization. For example, the new revenue-sharing policy, which allows
local governments to retain a higher portion of their revenues, and greater decisionmaking power and economic autonomy have provided the incentives and possibilities for local governments to pursue rapid economic growth. But the limited amount
of financial resources available means that each region has to support those industries in which it has a comparative advantage over other regions. In the 1980s, many
local governments hired consultants to identify their pillar industries and formulate
development strategies. Consequently, regional economies started to focus on local
strengths and move toward specialization. Enhancing specialization is the uneven
geographic distribution of foreign as well as domestic investment. Foreign investment tends to not only concentrate in the major economic centers, especially in
national and regional capitals and economically more developed areas, but also favor
certain industrial sectors, hence accentuating specialization in regional economic
development (Leung 1990, 1993).

The second set of forces, on the other hand, works against specialization and
tends to breed homogenization of regional economic structures. Aside from the
legacy of the pre-reform development strategy that stressed regional self-sufficiency,
the new development approach and the central-local government benefit-sharing
structure have given rise to a very important, though hitherto seldom discussed,
phenomenon in China's regional development—what may be called "regional imitation behavior." Since the reform, regions in China have been competing for foreign investment as they strive to increase economic growth. Local authorities all over
the country have been pursuing unhindered expansion even while the central government was trying to curb overheated growth through the use of macroeconomic
controls. But three decades of a planned economy had produced a whole generation
of government officials who know little about economics, much less market-
oriented economics. In the uncharted waters of reform, everyone "crossed the river

by feeling the stones," as stated in the official Chinese euphemism for not really knowing what to do. The central government's firm control over ideology, on the other hand, makes implementing innovative approaches to economic development politically risky. Under these circumstances, imitating others becomes the safest way to carry out reform. What has been happening in China is that once one place achieves some success with a new measure of revving up economic growth or attracting investment, everybody else clamors to emulate that success. The two decades of reform are replete with examples of emulation in selecting both industrial sectors and reform measures.

The development of China's automobile industry in the reform period may be the most spectacular example of regional emulation. China's automobile industry started in the 1950s. Although China did not reach a capacity of 1 million vehicles per year until 1995, it nevertheless had some 126 automobile companies located in twenty-seven provinces (all provinces except for Gansu, Ningxia, and Tibet) and more than 3,000 parts-producing companies in the early 1990s, perhaps more than any major automobile-producing country has. Even after the reorganization of the early 1990s aimed to increase the industry's competitiveness by forming enterprise groups, there were still ten major automobile production groups in different provinces. This dispersed distribution is the result of the regional self-sufficiency strategy in the pre-reform stage (Sit and Liu 1997).

After China decided to shift the focus of automobile manufacturing to passenger cars in 1987, many provinces and cities rushed to develop their own automobile industry. Twenty-four of the twenty-seven provinces with automobile-manufacturing capabilities designated the auto industry as the key sector of their economy. Lacking research and development capabilities, almost all of these provinces turned to foreign carmakers to form joint ventures. By 1996, eight passenger carmakers appeared in China with a combined capacity of 780,000 cars. About 91 percent of the domestic market share belongs to three major automakers. The First Automotive Works (FAW), located in Changchun of northeastern China's Jilin province, formed a joint venture with Volkswagen in 1991. A merger brought Audi into the venture. FAW-VW produces both Audi and Jetta sedans with a capacity of 230,000 cars. The Second Automotive Works (the Dongfeng Group), built between 1967 and 1978 under the influence of the self-reliance strategy, established a joint venture with French carmaker PSA Peugeot Citroën in Shiyan, Hubei province, which produces the Citroëns, marketed as Shennong Fukong, with a capacity of 60,000. Shanghai's Dazhong Group is a pioneer in introducing foreign automobile technology and is currently also the largest manufacturer. Its joint venture with Volkswagen formed in the early 1980s can produce 300,000 cars (marketed as Santanas) a year. Among the smaller players, Beijing and Chrysler have been making Cherokee Jeeps. Tianjin and Japan's Daihatsu jointly produce compact cars called Xiali (Daihatsu), once widely used as taxicabs in Beijing. Japan's Suzuki and Fuji have also formed joint ventures in Chongqing—a municipality since 1997—and Guizhou province respectively.

Guangzhou—the capital of South China's Guangdong province—had no producers of automobiles until it created a joint venture with France's Peugeot, but that cooperation ended in 1997 with heavy financial losses.

China's auto industry landscape continues to evolve as the world's major automakers compete for the potentially significant Chinese market. With the addition of a $1.5 billion joint venture between General Motors and Shanghai Automotive Industry Corporation (SAIC) in the city's Pudong Development Zone, Shanghai is shaping up as China's automobile capital. Shanghai-GM started to produce Buick sedans and wagons in 1999. Determined not to be left out of the auto industry bandwagon, Guangzhou established a joint venture with Japan's Honda Motor, with its first Accord rolling off the line on July 1, 1998. Guangzhou Honda's current capacity is 30,000 Accords a year, but in early 2002, talks were under way to expand its capacity to 300,000 cars annually. In January 2000, Guangzhou and Japan's Isuzu signed an agreement to jointly produce Isuzu buses. Guangzhou is hoping the two joint ventures will make the city one of China's most important automobile producing bases. In November 1999, Toyota and Tianjin Automobile Company received permission to jointly produce passenger cars. BMW's October 2001 announcement that it will form a partnership with China's Huachen Group to produce the BMW 3 series cars was one of the top ten auto industry news events in China for that year. In May 2002, SAIC, GM, and the former Liuzhou Wuling Automotive Company, in Liuzhou, Guangxi province, agreed to set up a three-way alliance to develop China's largest automobile production base in Liuzhou, with an annual capacity of at least 300,000 vehicles (mainly minivans and minitrucks).

Imitation in industrial development has led to both serious overcapacity and fierce competition. According to China's third industrial census, conducted in 1996, less than 60 percent of production capacity is used in making more than half of China's 900 major products. The capacity is 40 percent redundant for textile products, 70 percent for air conditioners, and 75 percent for electronics products. Mountains of oversupply have plagued bicycles, appliances such as washers and refrigerators, color TVs, and more recently VCDs (video CDs) and DVDs. The car glut touched off a price war in early 1998. Shanghai's Dazhong not only ramped up production in recent years but also significantly slashed the prices for its Santanas in hopes of grabbing a larger market share. The price war prompted the state to set minimum prices in 1998 in order to protect the car industry in other parts of the country, particularly the state-owned enterprises (Roberts 1998). China's accession into the World Trade Organization in November 2001 has drastically reduced tariffs on imported automobiles while at the same time increasing import quotas. This will make it much easier for foreign-made automobiles to enter the Chinese market. Because the majority of the automakers are small and unlikely to be competitive, it is only a matter of time before they fade away or are merged with the major players in the near future.

Imitation behavior has also been common in choosing reform measures. Economic reform in China has been a gradual process, with some coastal areas allowed to experiment with capitalism. Other regions have often lobbied the central government for the permission to do the same. Shortly after the central government approved the establishment of the four special economic zones (SEZs) in Guangdong and Fujian provinces in 1980, many other regions also set up a variety of economic and technological development zones with preferential policies for attracting investment and promoting economic growth. The imitation has been so widespread that a December 1998 news report showed that there are about 4,210 economic development zones nationwide, with 3,082 zones established without the required authorization from higher-level governments.

The situation has been similar with the high and new technology development zones (parks) and free trade zones (FTZs). To encourage technological transfer and promote the growth of high technology, China established its first high and new technology development park in Northwest Beijing's Zhongguancun, a university and research district, in 1988. The hype about the potential of high-tech development prompted many city governments to lobby for a similar zone in their jurisdiction. In less than three years, 27 national high-technology development parks (zones) sprouted up across China (Chen 1992). The number had increased to 53 national zones, 61 provincial zones, and numerous local government-sponsored zones by the mid-1990s (Chen and Li 1998; also see Chapter 9 by Shuguang Wang in this volume). Most of the high-tech zones are located in the economic and technological development zones along the coast. These zones are supposed to play an important role in the commercialization and industrialization of new technologies such as microelectronics, new materials, laser, new energy sources, and bioengineering. Experiences in developed countries indicate that a successful high-tech development zone must be supported by substantial local R&D capabilities. Such capabilities are lacking in most of China's high-tech zones. It is not surprising that many high-tech zones have not played the role they were designed to play.

The first free trade zone in China was established in the Waigaoqiao area of Shanghai's Pudong Development Zone in September 1990 following the central government's decision to develop Pudong (the eastern suburban area of Shanghai). In the free trade zone, foreign companies enjoy policies that are more favorable than those in the SEZs. They may set up trading firms, warehouses, and export production enterprises that are exempt from import and export duties and other taxes if their products are sold outside of China. Before long, many coastal provinces and port cities were lobbying the central government for a free trade zone of their own. The central government ended up approving fourteen more FTZs, with one or more FTZs in most coastal provinces. Similarly, after China's first stock market was opened in Shanghai in November 1990, followed by another in Shenzhen the following year, many cities also wanted to have a stock market. Some opened their own stock exchange without the required approval from the People's Bank of China (China's central bank). Had the central government not kept tight control over the operation of stock exchanges, China would most likely have dozens of them by now.

Local authorities' attempts to develop industries with regional comparative advantages, the highly uneven geographical distribution of domestic and foreign investment, and the emulation behavior in regional economic policymaking are the three keys to understanding China's regional economic structures and their specialization or homogenization.

The Trend Toward Regional Specialization

To assess the degree to which regional specialization has occurred after the reform, two dissimilarity measures can be calculated to show how provincial economies have differed from that of the rest of the country. Different economic variables may be used in the calculation of the measures. This study uses the employment structure defined by China's State Statistical Bureau in its statistical yearbooks. The sectoral breakdown of employment varies slightly over time in the Chinese statistics, but the following sectors are usually listed: 1) farming, forestry, animal husbandry, and fisheries; 2) industry/manufacturing; 3) geological prospecting and water conservancy; 4) construction; 5) transportation and communication; 6) commerce; 7) real estate, social services; 8) health care, sports, and social welfare; 9) education, culture, and arts; 10) scientific research and technical services; 11) government and social organizations. It would have been better if the categories were less aggregated, but they do reflect the structural features of the provincial economies.

The first dissimilarity measure is based on the so-called expected information measure proposed by Theil (1972):

$$DS\,1_t = \sum_{i=1}^{n} E_{China,t,i} \; \ln \; (E_{China,t,i} \; overE \; _{Province,t,i}),$$

where, $E_{Province,\,t,\,i}$ denotes the proportion of employment in sector i for a particular province at time t; $E_{China,t,i}$ is similarly defined for the rest of China; and n is the number of sectoral classifications of the economy. The second dissimilarity measure has the form of the Chi-square goodness-of-fit measure used to examine the closeness of an empirical distribution to the Chi-square distribution. The dissimilarity measure is calculated by

$$DS\,2_t = \sum_{i=1}^{n} \frac{(E_{Province,t,i} - E_{China,t,i})^2}{E_{China,t,i}},$$

where the variables are defined as in the first measure.

The two dissimilarity measures have common characteristics; that is, as the proportions $E_{Province}$ and E_{China} become more unequal, or the local economy differs more from the economy of the rest of the country, the dissimilarity measures become larger. If

$E_{Province}$ and E_{China} are equal, the dissimilarity measures approach zero—the lower limit for both measures. The two dissimilarity measures provide a summary characterization of the disparity in employment structure between each province and the rest of China. A small value for either measure indicates that the employment structure of a province is not much different from that of the rest of the country, and hence little specialization exists in that province in comparison to the economy of the rest of the country, whereas large dissimilarity values reflect substantial specialization within provincial economies.

Because the purpose of this research is to determine whether provincial economies have become more specialized since the reform, the two dissimilarity measures are calculated for both 1986 and 1995. The data are from the Statistical Yearbook of China (State Statistical Bureau 1987, 1998). The results are listed in Table 7.1 and plotted in Figures 7.2 and 7.3. The table shows that the dissimilarity scores for the majority of the provinces are very small and showed little change between 1986 and 1995. This means that the employment structure of most provinces is very similar to that of the rest of the country. The level of regional specialization is in general very low and has not changed much since the reform. But the two figures show that the dissimilarity values experienced a small upward shift between 1986 and 1995. The level of dissimilarity for 1995 tends to be slightly larger than the 1986 level. If the values had not changed, the dots representing them would have fallen approximately on the line in the two charts. This shift is significant because it indicates a trend toward regional specialization. The changes in dissimilarity value for Shanghai, Beijing, and Helongjiang are particularly pronounced, indicating that these provinces' economies have made bigger strides toward specialization. The relatively large dissimilarity scores for Xizang (Tibet) and Xinjiang in both 1986 and 1995 reflect the fact that both have had a different economic structure than the rest of the country. While Xinjiang has been moving closer to the rest of the country, Xizang has increased its distance over the 1986–1995 period.

An Axial Shift in China's Space Economy

The new development strategy implemented in China since the reform not only has led to widening coast-inland disparities and increased regional specialization but may have also altered the roles different regions play in the national economy. The foremost change may be the rise of South China as a new engine of growth and center of trade (Xu and Li 1990). The return of Hong Kong and Macao to Chinese sovereignty in 1997 and 1999 respectively and the increasing integration of the economies of Hong Kong and Guangdong province make South China arguably one of the most important regions in China. In fact, much evidence indicates that the principal axis of the Chinese space economy may be shifting from one of east-west orientation to one of north-south orientation. This section presents some of the evidence that supports this argument, but more research is needed to confirm the shift.

TABLE 7.1   Dissimilarity Measures of the Provincial Economies

	Dissimilarity Measure 1		Dissimilarity Measure 2	
	1986	1995	1986	1995
Beijing	0.118	0.204	0.423	0.639
Tianjin	0.064	0.091	0.081	0.115
Hebei	0.020	0.024	0.039	0.040
Shanxi	0.040	0.078	0.050	0.188
Inner Mongolia	0.018	0.048	0.043	0.120
Liaoning	0.034	0.031	0.060	0.053
Jilin	0.010	0.015	0.017	0.035
Heilongjiang	0.047	0.122	0.124	0.444
Shanghai	0.069	0.838	0.106	0.242
Jiangsu	0.016	0.042	0.028	0.070
Zhejiang	0.027	0.089	0.039	0.085
Anhui	0.009	0.012	0.017	0.018
Fujian	0.008	0.031	0.016	0.047
Jiangxi	0.015	0.016	0.036	0.030
Shandong	0.035	0.043	0.047	0.059
Henan	0.028	0.033	0.043	0.043
Hubei	0.008	0.037	0.019	0.054
Hunan	0.010	0.013	0.019	0.027
Guangdong	0.039	0.062	0.099	0.071
Guangxi	0.026	0.040	0.055	0.085
Hainan		0.421		2.581
Sichuan	0.029	0.031	0.037	0.037
Guizhou	0.035	1.040	0.072	0.085
Yunnan	0.044	0.060	0.099	0.127
Tibet	0.400	0.575	0.858	0.990
Shaanxi	0.034	0.027	0.057	0.044
Gansu	0.014	0.013	0.032	0.029
Qinghai	0.073	0.085	0.276	0.195
Ningxia	0.038	0.044	0.108	0.114
Xinjiang	0.313	0.279	1.648	1.827

SOURCE: *Statistical Yearbook of China,* 1987, 1996.

The pre-reform pattern of China's regional economic development was characterized by an east-west difference in well-being. From the coast to the inland, the level of economic development declined sharply. The Yangtze River Delta centered on Shanghai has long assumed the position of dominance in China's space economy. It is the focus of the Chinese economy and the densest, richest, and most important region. With the Yangtze River and its tributaries draining the vast productive interior of the country and the position assigned to Shanghai by the central government as the center

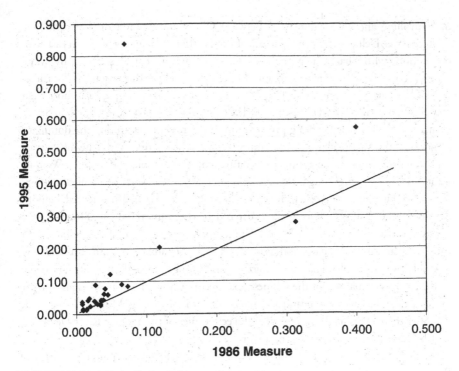

FIGURE 7.2   Dissimilarity Measure 1

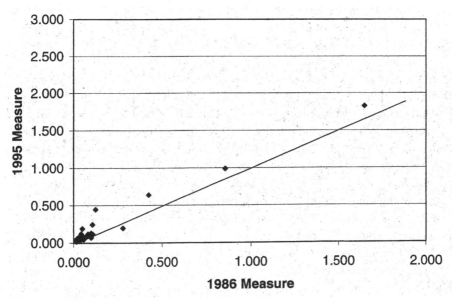

FIGURE 7.3   Dissimilarity Measure 2

of controlling and channeling flows between China and the rest of the world, the Yangtze River Delta set the principal axis of flow within the country along an east-west orientation, with Shanghai at the most critical part of that axis. In many ways, Shanghai was the economic juggernaut of China and had the capacity to set the direction of the national economy. Before 1978, Shanghai alone accounted for more than one-eighth of China's gross industrial product. But the sustained rapid growth in South China, especially in Guangdong province, since the reform has eroded the dominant position of the Yangtze River Delta and Shanghai. Guangdong is playing an increasingly important role in the national economy, eclipsing the role of Shanghai. With its return to China, Hong Kong is becoming, more than ever, a center of channeling flows between China and the world. The completion of the north-south oriented Beijing-Kowloon Railway in 1996 greatly expanded the hinterland of Hong Kong, Guangzhou, and Shenzhen (Ye 1998). The railroad links booming South China with the central and northern parts of the country and has already been redirecting the flow of goods away from Shanghai to Guangzhou and Hong Kong. All these are effectively shifting the principal axis of the Chinese space economy to one of north-south orientation. The economy of the Pearl River Delta, where Hong Kong, Macao, Guangzhou, and Shenzhen are located, is larger than the economies of the Yangtze River Delta and North China's Beijing-Tianjin-Tangshan region combined.

Guangdong province is leading China in virtually every material measure. Benefiting from the preferential treatment by the central government, such as low fiscal commitments, the setting up of special economic zones, greater autonomy in foreign trade, and permission to retain a high proportion of foreign-exchange earnings, Guangdong maintained a 14.2 percent average annual growth in its GDP in the two decades after the reform, much higher than the national average of 9.8 percent. It is among the fastest growing provinces in China. From 1978 to 1997, the share of its GDP in the national total increased from 5.1 percent to 9.8 percent, while Shanghai's share dropped from 7.5 percent to 4.5 percent; the share of its revenue increased from 3.5 percent to 6.2 percent, while Shanghai's shrank to 3.8 percent from 14.9 percent. In 1997, Guangdong received $11.7 billion in foreign investment, compared to Shanghai's $4.2 billion. In 2001, Guangdong's GDP exceeded one trillion yuan, becoming the largest provincial economy in China. The per capita GDPs of Shenzhen and Guangzhou, at $5,222 and $4,586 respectively in 2001, were the highest among China's major cities (excluding Hong Kong).

Many people from other provinces are migrating to Guangdong to seek their fortunes. "Going south" has become synonymous with getting rich among the general public. Scarce commodities as well as money from all over China also flow to this most commercialized province because of the higher prices or better return on investment there. Contrary to the perception that Guangdong's manufacturing is dominated by low-skilled, labor-intensive sectors, the province has China's largest high-technology and new-technology industries, which have been growing at 33 percent in recent years. Guangdong accounted for 41.6 percent of all Chinese exports in

1997, nearly 4.5 times as much as China's eighteen central and western provinces combined. It is of special importance in China's new economic development strategy. With three of China's five special economic zones, including the most successful one—Shenzhen, being in Guangdong and its location bordering Hong Kong, it is the main threshold for the open door of the nation. Though Guangdong's economic growth has slowed in the wake of the Asian financial meltdown, its growth rate continues to lead the country. Guangdong's ambition is to be Asia's fifth "Tiger" after Hong Kong, Taiwan, Singapore, and South Korea. In fact, the province has boasted that by 2010 its per capita GNP will catch up with South Korea's.

Conclusion

Since the reform, China has undergone a dramatic economic transformation. It has shifted from an inward-looking, self-reliant, and self-(in)sufficient, centrally planned economy to a robust, export-oriented, fast-industrializing economy. The rapid economic growth has significantly altered regional economic structures and the spatial development pattern. Coastal provinces like Guangdong have experienced much faster rates of growth and often also have more say over how their economies are run than do their inland counterparts. As a result, the coast-inland disparities have widened substantially. The increasing gap has revived regionalism and resulted in regional conflict. It has indeed become a sensitive political issue in China. The slowly developing inland provinces have been fighting for a share in the coast's prosperity. The central government seems to have given in to their demands recently by promising to significantly increase investment in the inland. The stake in this new government policy is high both politically and economically. The delicate task facing the Chinese government in the years ahead will be to continue an aggressive macroeconomic policy but at the same time distribute the growth to interior provinces without undermining the areas most responsible for the generation of national wealth. Notwithstanding the ongoing discussion in China on achieving coordinated development among the coastal and interior provinces, the bias toward coastal development is unlikely to change in the foreseeable future.

Ever since the reform, the various regions of China have been competing with each other for more liberal policies from the central government, a bigger share of state investment as well as more investment from the outside world. They are becoming more interdependent as the country's economy grows to a higher level. But at the same time, interregional relationships have grown more complex, with burgeoning regionalism and regional tension. To achieve the fastest possible growth, local authorities have been looking for industries with comparative advantages to support and emulating successful reform efforts. In this process, regional economies are constantly being reorganized to increase competitiveness. The analysis in this chapter shows that all these have pushed regional economies slowly toward specialization. The trend is likely to continue in the future.

The rise of south China, particularly Guangdong province, may be one of the most remarkable stories in China's regional development since the reform. The integrated economies of Guangdong, Hong Kong, and Macao are making south China one of the most important regions of China and indeed the world. The principal axis of China's space economy may have been shifting from one of east-west orientation to one of south-north orientation. The new Beijing-Kowloon Railway has opened up a vast hinterland for south China and redirects the flow of goods to the south, which will only serve to accelerate the shift.

One note of caution is that the fortunes of China's regions remain subject to shifts in political winds. After the pro-democracy movement in 1989, China's central government shifted its development focus away from the South and toward central and northern China. The major sign of this changing policy was the approval of Shanghai's ambitious plan for rejuvenating its economy by developing its eastern suburban area (Pudong) into an international economic, commercial, and financial center. Shanghai and its surrounding provinces have benefited from special policies that enable them to maintain a growth rate slightly higher than the national average. This has reinforced the pre-eminence of the Shanghai region in China. But it is unlikely to reverse the axial shift that is under way in the Chinese space economy. After more than two decades of rapid development and an aggressive policy that promotes Guangdong–Hong Kong–Macao integration, Guangdong already has enough capability to grow on its own.

References

In addition to the references cited below, the research reported in this chapter benefited from numerous news reports that appeared both in the print media and on the Internet (www.people.com.cn; www.wenxuecity.com; www2.chinesenewsnet.com; www.sohu.com).

Barnard, Mark, and Oded Shenkar. 1990. Variations in the Economic Development of China's Provinces: An Exploratory Look. *GeoJournal* 21(1–2): 177–192.

Chan, Kam Wing. 1992. Economic Growth Strategy and Urbanization Policies in China, 1949–1982. *International Journal of Urban and Regional Research* 16(2): 275–305.

Chen, Hanxin. 1992. The Current Situation of the Hi- and New-Tech Industry and the Distribution of the Developing Areas in China. Paper presented at the 27th International Geographical Congress, August 9–14, 1992, Washington, D.C.

Chen, Hanxin, and Xinfeng Li. 1998. Distribution and Prospects of High-Tech Development Zones in China. Paper presented at the International Conference on "China and the World in the 21st Century," August, Hong Kong.

Cheng, Joseph Y. S., and Mujin Zhang. 1998. An Analysis of Regional Differences in China and the Delayed Development of the Central and Western Regions. *Issues & Studies* 34: 35–58.

Chu, David. K. Y. 1988. Some Analysis of Recent Chinese Provincial Data. *The Professional Geographer* 40: 19–32.

Fan, C. Cindy. 1992. Foreign Trade and Regional Development in China. *Geographical Analysis* 24: 240–256.

_____. 1995. Of Belts and Ladders: State Policy and Uneven Regional Development in Post-Mao China, *Annals of the Association of American Geographers* 85(3): 421–449.

_____. 1997. Uneven Development and Beyond: Regional Development Theory in Post-Mao China. *International Journal of Urban and Regional Research* 22: 425–442.

Field, R. M. 1986. "China: The Changing Structure of Industry." In *China's Economy: Looks Toward the Year 2000*. Vol. 1, 505–547. U.S. Congress Joint Economic Committee. Washington, D.C..

Fleisher, Belton M., and Jian Chen. 1997. The Coast-Noncoast Income Gap, Productivity, and Regional Economic Policy in China. *Journal of Comparative Economics* 25: 220–236.

Fujita, M., and D. Hu. 2001. Regional Disparity in China, 1985–1994: The Effects of Globalization and Economic Liberalization, *Annals of Regional Science* 35: 3–37.

Goodman, D. S. G. 1989. *China's Regional Development*. London: Routledge for Royal Institute of International Affairs.

Huan, G. 1986. China's Open Door Policy, 1978–1984. *Journal of International Affairs* 39: 1–18.

Lakshmanan, T. R., and Chang-i Hua. 1987. Regional Disparities in China. *International Regional Science Review* 11: 97–104.

Lee, Jongchul. 1995. Regional Income Inequality Variations in China. *Journal of Economic Development* 20 (2): 99–118.

Lee, Pak K. 1998. Local Economic Protectionism in China's Economic Reform. *Development Policy Review* 16: 281–303.

Leung, Chi Kin. 1990. Locational Characteristics of Foreign Equity Joint Venture Investment in China, 1975–1985. *Professional Geographer* 42: 403–421.

_____. 1993. Personal Contacts, Subcontracting Linkages, and Development in the Hong Kong–Zhujiang Delta Region. *Annals of the Association of American Geographers* 83: 272–302.

Lu, Dadao, et al. 1997. *Regional Development Report of China: 1997*. Beijing: Shangwu Yinsuguan (Commercial Press).

Lu, Max, and Enru Wang. 2002. Forging Ahead and Falling Behind: Changing Regional Inequalities in Post-reform China. *Growth and Change* 33: 42–71.

Pannell, Clifton W. 1988. Regional Shifts in China's Industrial Output. *Professional Geographer* 40: 19–32.

Roberts, Dexter. 1998. So Much for Competition: Beijing Slaps on Price Controls to Stanch Deflation. *Business Week*, November 30: 56, 58.

Rothenberg, J. 1987. Space, International Economic Relations, and Structural Reform in China. *International Regional Science Review* 11: 5–22.

Sit, V. F. S., and W. D. Liu. 1997. The Restructuring and Internationalization of the Chinese Automobile Industry After the Adoption of Open-Door Policy. *Dili Yanjiu* (Geographical Research) 16: 1–11.

State Statistical Bureau. 1987, 1996. *Zhongguo Tongji Nianjian 1987, 1996* (Statistical Yearbook of China). Beijing: Zhongguo Tongji Chubanshe.

_____. 1998. *Chengjiu Huihuang de 20 Nian* (Twenty Years of Remarkable Achievements). Beijing: Zhongguo Tongji Chubanshe.

Theil, Henri. 1972. *Statistical Decomposition Analysis with Applications in the Social and Administrative Sciences*. Amsterdam: North-Holland.

Wang, Jici, and John H. Bradbury. 1986. The Changing Industrial Geography of the Chinese Special Economic Zones. *Economic Geography* 62: 307–320.

Wei, Houkai. 1994. On the East-West Disparities and Speeding Up Development of the West. *Guizhou Shehui Kexue* (Guizhou Social Science) 5: 2–8.

Wei, Y. D. 1999. Regional Inequality in China. *Progress in Human Geography* 23: 49–59.

Wei, Yehua, and Laurence J. C. Ma. 1996. Changing Patterns of Spatial Inequalities in China, 1952–1990. *Third World Planning Review* 18 (2): 177–191.

World Bank. 1985. *China: Long-Term Development Issues and Options.* Washington, D.C.: World Bank.

Wu, C. T. 1979. *Development Strategies and Spatial Inequality in the People's Republic of China.* Nagoya, Japan: United Nations Centre for Regional Development.

Xie, Yichun, and Frank J. Costa. 1991. The Impact of Economic Reforms on the Urban Economy of the People's Republic of China. *Professional Geography* 43: 318–335.

Xie, Yichun, and Ashok K. Dutt. 1990. Regional Investment Effectiveness and Development Levels in China. *GeoJournal* 20: 393–407.

Xu, Xueqiang, and Siming Li. 1990. China's Open Door Policy and Urbanization in the Pearl River Delta Region. *International Journal of Urban and Regional Research* 14: 49–69.

Yang, Dali. 1990. Patterns of China's Regional Development Strategy. *China Quarterly* 122: 230–257.

_____. 1991. Reforms, Resources, and Regional Cleavages: The Political Economy of Coast-Interior Relations in Mainland China. *Issues & Studies* (September):43–69.

Ye, Shunzan. 1998. Hong Kong and the Development of the Areas Along the Newly Built-up Beijing-Kowloon Railway. *Dili Xuebo* (Acta Geographica Sinica) 53: 157–165.

Ying, L. G. 1999. China's Changing Regional Disparities During the Reform Period. *Economic Geography* 75: 59–70.

Zhao, Simon X., and C. Gu. 1995. A Policy Review on Spatial Strategy of Regional Development in China 1953–1992. *China Report* 31 (3): 385–410.

8

Sustainable Development in the Yangtze Delta Area

Mei-e Ren

The Yangtze Delta Economic Region includes the municipality of Shanghai and fourteen provincial cities in Jiangsu and Zhejiang provinces, namely, Nanjing, Zhenjiang, Changzhou, Wuxi, Suzhou, Yangzhou, Taizhou, and Nantong in Jiangsu, and Hangzhou, Jiaxing, Huzhou, Shaoxing, Ningbo, and Zhoushan in Zhejiang (see Figure 8.1).[1] This region has an area of 38,429 square miles, nearly 60 percent of which is hilly and mountainous. In the western part of Hangzhou and around the reservoir of Xinan Jiang lies a mountainous area that has peaks over 3,280 feet in elevation. If defined geomorphologically by the 16-foot contour, the delta would have an area of only about 15,444 square miles, with its apex near Zhenjiang. The delta is low-lying, swampy, dotted with numerous small lakes and ponds and crisscrossed by a dense network of rivers and canals, with more than one mile of waterway per square mile of land. At the center of the delta is the Taihu Lake, which, with an area of 864 square miles, is one of the largest freshwater lakes in China. The Suzhou-Wuxi-Changzhou area, with over 30 percent of its total area covered with water, is the famous "Water Country" of China. Suzhou's old city proper is well known in the world as the "Venice of the Orient."

This region is the most productive agricultural area in China and also one of the country's most industrialized regions. It forms the economic backbone of China. For one thousand years, Suzhou and Hangzhou were known as "Heaven on Earth," as boasted in a Chinese proverb: "Up in the sky is the paradise, down on earth there are Suzhou and Hangzhou." Shanghai and the Suzhou-Wuxi-Changzhou corridor have recently been regarded by some economists as the "economic core" of the Yangtze Delta.

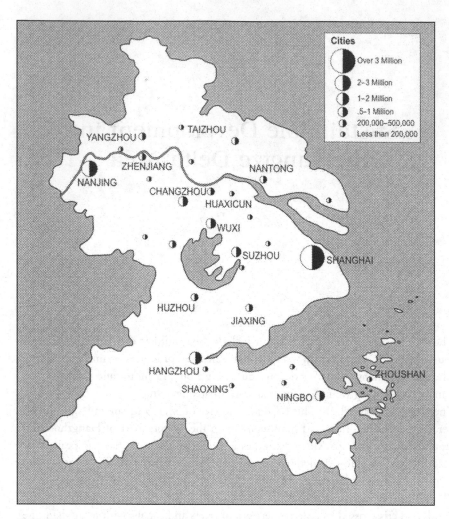

FIGURE 8.1   Major Cities and Towns in the Yangtze River Delta
(after She 1997)

The reason for including the hilly and mountainous area in the Yangtze Delta Economic Region is mainly socioeconomic. Traditionally, the Yangtze Delta is roughly delimited by the Shanghai-Nanjing-Hangzhou triangle. Ningbo, largely mountainous, lies far beyond the delta. However, since the opening of the port city of Shanghai in 1842, large numbers of Ningbo natives have flocked to Shanghai and become businessmen, bankers, and so on. It is estimated that by 1949, nearly one-third of Shanghai's population had its roots in Ningbo. Owing to the close historic and recent economic ties, Ningbo, including Zhoushan, a provincial city formerly under Ningbo's jurisdiction, was included in the Yangtze Delta Economic Region. The region is delimited by administrative units, with provincial cities as basic units for statistical convenience. In the rest of this chapter, the Yangtze Delta or the Delta will be used to denote the Yangtze Delta Economic Region.

Economic Characteristics of the Yangtze Delta

The Yangtze Delta is one of the most important economic regions in China. With approximately 1 percent of China's territory and 6 percent of the population, it contributes 15.4 percent of the national GDP and 21.8 percent of the national revenue (Table 8.1). The combined GDP of Suzhou, Wuxi, and Changzhou, which totals 203.4 billion RMB yuan, almost equals that of Shanghai. More than one-half of the GDP of the whole delta comes from the small area formed by Shanghai, Suzhou, Wuxi, and Changzhou. This area also has the densest cluster of cities and towns. From Changzhou-Wuxi-Suzhou to Shanghai, urban built-up areas spread almost continuously along the Shanghai–Nanjing Railway. The Suzhou-Wuxi-Changzhou area also contains China's wealthiest county, its wealthiest town, and its wealthiest village.

Xishan, formerly a suburb of Wuxi, now produces a GDP of 21.2 billion yuan (RMB), almost equal to that of the Wuxi city proper (21.57 billion yuan). Besides its phenomenal growth in regard to industry, Xishan is also a national model for agricultural modernization. Similarly, Wujin, formerly a suburb of Changzhou, now has a GDP of 16.37 billion yuan, exceeding its mother city, Changzhou city proper (11.54 billion yuan).

Shengze, in the city of Wujiang, has been known for centuries as China's silk textile capital. An old saying boasts that the town "produces 10,000 feet of silk everyday to clothe people of the whole country." Today, with large, modern silk-weaving factories in a garden-like setting and a huge silk textile market worth more than 10 billion yuan every year, Shengze seems more prosperous than Wujiang city proper and has been named the wealthiest town in China. Although the town currently has a population of only 110,000, a very small city by Chinese standards, it is an economic giant. Its annual agricultural and industrial output is valued at 3.57 billion yuan. There are a number of towns like Shengze in the Suzhou-Wuxi-Changzhou area.

China's richest village is Huaxi in the city of Jiangyin. Owing to its outstanding agricultural development in the 1970s, it was widely known as a "model village" of the country. However, Huaxi has now become a modern industrialized community, with villa-like houses, a magnificent conference hall, a theater, and all kinds of modern conveniences. Sales of its industrial goods reached 1.62 billion yuan in 1995, and its tourist industry generates an income of more than 0.3 billion yuan a year. Recently, Huaxi established two new associated villages in the remote areas of Northwest China and Northeast China: Huaxi village in Ningxia and Huaxi village in Zhaodong (Heilongjiang province). In addition to agriculture, factories have been set up in Ningxia to produce wool garments and other products. The branch in Zhaodong has even acquired a state-owned factory. Thus in only a few years, Huaxi

TABLE 8.1   GDP of Cities in the Yangtze River Delta, 1995 (in billion yuan)

City	GDP	City	GDP	City	GDP
Shanghai	246.257	Nanjing	57.646	Angzhou	29.920
Suzhou	90.311	Nantong	46.657	Zhenjiang	28.586
Hangzhou	76.201	Shaoxing	41.121	Taizhou	26.975
Wuxi	76.111	Changzhou	36.970	Huzhou	22.731
Ningbo	60.926	Jiaxing	32.128	Zhoushan	7.349

NOTE: US$1 ≈ 8.2 RMB yuan.

has helped to lift 6,000 local people in Ningxia out of poverty and has more than doubled the annual per capita income of villagers in Zhaodong Huaxi village.[2]

There are 322 towns in the Suzhou-Wuxi-Changzhou area, averaging one town every 54 square kilometers (1997). Many of the small towns are economic giants, making this region a model of rural urbanization in China. Recently, a far-reaching program of readjustment of villages was implemented in southern Jiangsu, through which inhabitants and rural factories are moved to new towns in order to protect and reclaim farmland. This program will farther enhance urbanization of rural areas in Suzhou, Wuxi, and Changzhou.

In the twenty-first century, the Yangtze Delta will be at the forefront of the globalization of China's economy and the march to the highly competitive world market. Several major problems, however, must be addressed to ensure sustainable development of the Delta. They include 1) construction of the Shanghai International Shipping Center, 2) the increase in population, the loss of farmland, and the readjustment of agriculture, 3) industrial restructuring, and 4) environmental problems. The rest of this chapter discusses these issues.

The Shanghai International Shipping Center

China is making a strenuous effort to develop Shanghai into an international economic, financial, and trade center. The Shanghai International Shipping Center (SISC) is an indispensable part of this goal. SISC will provide Shanghai, the Yangtze Delta, and the whole Yangtze Valley with an important window on the world and will facilitate the entry into the international market.

SISC, following examples of Kobe-Osaka and Yokohama-Tokyo in Japan and Los Angeles-Long Beach in the United States, will be a port association or port group instead of a single Shanghai port. It will consist of three ports: Shanghai as the central port and Beilun in Ningbo and Taicang in Jiangsu as secondary ports. The office of the Shanghai International Shipping Center Port Association was formally established in Shanghai on September 29, 1997. Owing to the great importance of containers in maritime transport, emphasis will be placed on the development of international container hub-ports.

Three important factors affect the location and development of a container hub-port. First, the port must meet certain physical conditions: It should have a water depth of 43 feet–49 feet at the channel and alongside the berth, weak winds and wave energy, and enough flat land adjacent to the berth for handling, storing, and transporting containers. Second, there must be an extensive well-developed hinterland. Third, there must be good transportation facilities for both inland and ocean transport. In the delta area, owing to usually short hauls (less than 200 miles), expressways, instead of railways and inland waterways, are the most important means of inland transport for containers to and from the port. The availability of direct intercontinental shipping lines with frequent and regular sailing is a dominant factor in the development of a container hub-port. An interesting example is the Yantian port in Shenzhen. The port was opened in late 1994, and with the subsequent opening of direct shipping lines with regular sailing to Europe, its throughput has rapidly increased to 450,000 TEUs in 1996 (Wang 1997).[3] It is evident that ports with direct intercontinental container-shipping lines and frequent sailing will have a great advantage over other ports without these facilities. In this context, a comprehensive evaluation of Shanghai, Beilun and Taicang ports may be summarized as follows.

Physical Conditions.   Waigaoqiao and Taicang are river ports on the Yangtze River, with the latter only thirty-seven miles upstream from the former. Historically, both ports have been handicapped by a sandy shoal at the Yangtze River mouth, which limits the water depth of the present shipping channel to only 23 feet. However, improvements aimed at deepening this crucial shipping channel started at the end of 1997. If everything goes smoothly, it is expected that the shipping channel will be deepened to 41 feet by 2007, and then both Waigaoqiao and Taicang will be accessible to third- and fourth-generation container ships. Taicang port has more than twelve miles of continuous deepwater riverfront available for development. The central and best part, which is six miles long, has recently been allocated to China Ocean Shipping Corporation (COSCO) and will accommodate twenty new berths for third- and fourth-generation container ships. Available deepwater frontage is limited at Waigaoqiao, however, and at the adjacent Wuhaogou port area.

In Taicang port, there is a wide plain behind the waterfront, which is suitable for establishing a new development zone—COSCO International City. Waigaoqiao and Wuhaogou, on the other hand, are already situated in the Pudong Development Zone and also partly in the Pudong Free-Trade Zone, but current land prices are rather high, and flat land available for port development is limited.

Beilun is a seaport. Its physical setting is similar to the Victoria harbor of Hong Kong. Flushed by a strong tidal current with a speed of approximately 3.3 feet per second, the port has a water depth of more than 49 feet. A mountainous island to the east (on the seaward side) shelters the port from strong winds, and therefore wind and wave energy in the harbor is weak, enabling container berths to operate almost throughout the whole year. Thus, on the basis of physical conditions alone, Beilun is the best port in the Shanghai International Shipping Center.

Developed Hinterland.   Shanghai and Taicang ports have an extensive and highly developed hinterland. The area adjacent to Taicang port, including Suzhou, Wuxi, and Changzhou, has a total GDP almost equal to that of Shanghai and has a large volume of foreign trade. Indeed, Taicang port has the advantage of cheap inland container transport between the port and its direct hinterland. A recent study by the province of Jiangsu showed that the direct cost savings of inland container transport will be 0.87 billion yuan a year when Taicang handles 2.4 million TEUs in 2010. In contrast, Beilun has only a limited hinterland.

Transportation Facilities.   Shanghai, Beilun, and Taicang all have good express-ways or high-grade highways. However, Shanghai container ports are located on the east side of the Huangpu River (Pudong), whereas a large part of the industrial com-plex is on the west side of the Huangpu River. With the rapid increase in popula-tion and industrial development in Pudong, the existing tunnels and bridges across the Huangpu River are already overcrowded. Unless measures are taken quickly, the transport of a large number of containers across the Huangpu River may lead to a bottleneck, limiting the development of Shanghai container ports.

As mentioned earlier, direct intercontinental container-shipping lines and frequent sailing are of prime importance to the success of a container hub-port. Taicang's con-tainer berths, together with COSCO International City, will be constructed and man-aged jointly by COSCO and the Jiangsu provincial government. Each party is contributing half of the total investment of the project. COSCO is the fourth largest ocean container carrier in the world. Its participation in the Taicang container port will facilitate the rapid opening of numerous intercontinental container-shipping lines and frequent sailings between Taicang and the leading ports of the world. Recently, Shanghai and Beilun have made great efforts to recruit additional direct container-shipping lines and to increase sailing frequency. However, the direct participation of COSCO in the Taicang port gives Taicang a unique advantage over the two other ports. It is expected that by 2010, a segment of the intercontinental containers com-ing out of other parts of South Jiangsu will use Taicang as an outlet to take advantage of its numerous intercontinental shipping lines and its efficient sailing schedule. For the same reason, many of the intercontinental containers coming out of North Jiangsu will also be attracted to Taicang port (via the Jiangyin Yangtze Bridge).

Political Feasibility.   Finally, because the Shanghai International Shipping Center is a state project involving Shanghai, Jiangsu, and Zhejiang, the local interests of all three parties concerned must be well coordinated and reasonably satisfied so that resources can be fully mobilized for the construction of the center. Furthermore, full agreement and earnest commitment might result in an earlier opening date. In view of the fierce competition for container cargo among the existing international container hub-ports in the Far East, this is in China's vital interest. In fact, some international organizations have recently listed "political feasibility" as an essential condition for major projects on sustainable development (involving several nations or regions/provinces) (ICSU 1997).

Overall, the present scheme for the establishment of the SISC is the best one because not only is it the most technically and economically rational plan but it is also politically feasible.

Population Growth, Diminishing Farmland, and Agricultural Readjustment

The greatest challenge to the sustainable development of agriculture in the Delta is increasing population and decreasing cultivated land. From 1975 to 1985, total population increased by 10 percent. Between 1990 and 1995, natural increase added 1.7 million to the region's population, averaging about 340,000 people per year. The population pressure has been made more serious by the influx of large numbers of migrant workers, estimated at 5 million people, from adjacent regions and provinces. In cities, there is a floating population of about 3 million, composed of mostly private business owners. In some highly developed regions, migrant workers constitute a high proportion of the local population. For example, there are about a half million migrant workers in Suzhou, which is nearly 10 percent of the total local population. Huaxi village in Wuxi has only 1,479 local residents (1994) but employs more than 1,000 migrant workers. These workers usually are not registered in the official population statistics, but they consume local food and housing and enjoy all the local conveniences. This greatly aggravates pressure on the land.

On the other hand, cultivated land has been decreasing at an alarming rate since 1956. From 1990 to 1995, cultivated land decreased about 180,000 hectares, averaging 36,000 hectares per year. The rapid decrease in cultivated land is attributed to several factors.

First, industrialization has consumed large amounts of cultivated land. It is estimated that in Suzhou, from the late 1970s to the late 1980s, 40,000 hectares of cultivated land were lost to industrial development. Second, the expansion of urban built-up areas and the construction of highways, roads, and other urban infrastructure has devoured land that used to be devoted to agriculture. Third, the establishment of economic development zones has led to the further diminishment of arable land. In the 1990s, there was an economic development zone craze; every county established one or more state, provincial, or local economic development zones. These zones enclose large tracts of good arable land, which is not fully used. For example, Suzhou has five state and ten provincial economic development zones. In Suzhou city proper alone are the Singapore Industrial Park and the Hi-Tech Development Zone (APEC Industrial Park). The planned area of the former is 27 square miles and that of the latter is 20 square miles. Both are much larger than old Suzhou city proper. In fact, two new cities are being built, each close to the size of Nanjing city proper.[4] The ambitious and unrealistic enclosure of arable land for economic development zones results in a great waste of a precious resource. Moreover, it is clear that rapid economic growth (GDP is increasing at a double-digit rate) will further decrease cultivated land in the Delta.

The consequence of increasing population and decreasing cultivated land is already apparent. From 1990 to 1995, per capita cultivated land shrank from 0.0486 hectare to 0.0453 hectare, about 0.008 hectare below the lowest limit (0.053 hectare) set by the United Nations. If the migrant population is taken into consideration, the situation is even more serious. To meet these challenges, a sweeping readjustment of agriculture must take place.

Grain Production

Traditionally, the Yangtze Delta has been a grain surplus region. The Grand Canal, constructed in 605 A.D., was chiefly designed for shipping surplus grain from the Yangtze Delta and the Middle Yangtze Basin to the capitals of the Chinese empires in North China, such as Luoyang, Xian, Kaifeng, and Peking. However, since 1984, the Delta has changed from a grain surplus region to a grain deficit region. It is estimated that about 4 million tons of grain (chiefly feed corn) has to be imported every year, and Shanghai (with about 10 million people) is totally dependent on imported grain from other parts of China and abroad.

Although the Yangtze Delta is a high-yield area, with about one-third of the paddy fields producing 15 tons per hectare of grain (rice and wheat) a year, this increase in grain production is more than offset by the great loss of cultivated land, especially the loss of grain acreage. From 1990 to 1995, grain acreage decreased by 6.9 million hectares. Therefore, annual grain production has remained virtually unchanged in the past decade, and per capita grain production is decreasing.

The grain deficit in the Yangtze Delta is becoming more serious with the beginning of the new century, as population increases by 1.7 million people and the grain deficit reaches 6.5 million tons. Nonetheless, there is the potential to significantly increase grain production. About 40 percent of total cultivated land, or about 1.35 million hectares, suffers from low yield. By applying existing agricultural techniques, the yield of these fields could be greatly increased. In addition, the rate of double-cropping could be increased significantly by adopting interplanting techniques, practicing appropriate crop rotation systems, and so forth. For example, through the use of an annual rotation system in 1996, which consisted of two dry crops (including corn) and one paddy rice crop, the rate of double-cropping in cultivated land in Nantong increased 238 percent, a 60-percentage-point increase from the previous level. In addition, the amount of chemical nitrogen currently applied to paddy fields is twice the amount actually needed, which results in land degradation. Through a more efficient use of chemical fertilizers, and by increasing the use of organic fertilizers, the yield of paddy fields can be raised significantly.

Regarding grain deficiency, it must be acknowledged that with the rapid growth of the economy and the increase in the nonagricultural population, one should not expect the Yangtze Delta to be totally self-sufficient in grain.[5] The importation of a reasonable proportion of grain is unavoidable.

The Market Economy

An important factor in the decrease in grain production is the market economy. Due to the high costs of land and labor in the Yangtze Delta, traditional grain farming has been replaced by the production of high-grade agricultural products with high economic returns. Therefore, large areas of rice paddies have been converted to fishponds producing high-value fish and special aquatic products such as crabs and eels, to vegetable gardens producing special produce for both domestic and export markets, and to flower gardens producing flowers for city markets.

Silk and cotton are the two major traditional cash crops produced in the Yangtze Delta, and the market economy has had significant effects on these two crops as well. Mulberry cultivation, silkworm raising, and cotton growing are all very labor-intensive activities. Owing to the higher costs of labor and land in the Yangtze Delta, mulberry and cotton production has moved from South Jiangsu to North Jiangsu, where both labor and land are cheaper. In 1994, three-fourths of the total production of mulberry in Jiangsu province took place in North Jiangsu; only 17 percent remains in the Suzhou-Wuxi-Changzhou corridor. For the same reason, mulberry cultivation in the famous fishpond-mulberry dike ecosystem in the Pearl River Delta has now shifted to the hilly regions of northern Guangdong, and mulberry trees in the area have been replaced by vegetables, flowers, and sugar cane. Similarly, cotton production in South Jiangsu and in the suburban counties of Shanghai, Ningbo, and Shaoxing has decreased. The remarkable change in the geographical distribution of mulberry and cotton production has had a great impact on the location of the silk and cotton textile industries.

Another important factor in the sustainable development of agriculture is the effort to increase the incomes of farmers. Various measures have been taken to reach this goal. The most notable are the encouragement of larger farms and the integration of agricultural production, food processing, and marketing under a single administration. However, a number of problems involving ownership, capital, and the division of dividends need to be resolved.

Restructuring Industry

The Yangtze Delta is a highly industrialized region in China. The value of its industrial output accounts for about 20 percent of the country's total. The region's key industries, such as synthetic fibers, textiles, and electronics, account for a large proportion of their respective industries in the country as a whole, and the value of their products make up 30–45 percent of the country's total (She 1997, p. 110).

A great majority of the large enterprises are state-owned. They are usually handicapped by several shortcomings.

Small-scale production.   Take the automobile industry for an example. Shanghai Volkswagen Corporation, capable of producing 300,000 passenger cars a year, is the largest car manufacturer not only in the Yangtze Delta but also in China. Its actual production in 1996 was about 200,000 cars, only a fraction (less than 10 percent) of Toyota, GM, or Ford's annual production. As a result, Shanghai Volkswagen's cost of production is high, and its products are not competitive in the world market. In Jiangsu and Zhejiang provinces, automobile makers usually produce fewer than 100,000 cars a year, and their labor efficiency is very low. It is estimated that the efficiency of a worker in the Jiangsu automobile industry is less than 10 percent of that of developed countries. In fact, it is only through protective tariffs that the automobile industry in the Delta is able to make a profit every year. When China joins the World Trade Organization and protective tariffs are eliminated or lowered, automobile manufacturers in the Delta may find their products uncompetitive even in the domestic market, and the whole industry may be in the red.

Outdated technology.   The technology and equipment of Shanghai Volkswagen were imported from Germany's Volkswagens in the 1980s. They are now considerably outdated. Moreover, its key technology is still in the hands of the Germans. China's ability to improve technology and develop new car models is rather weak.

Inability to produce high-grade new products.   State enterprises spend a large sum of money to import equipment from abroad but often pay little attention to improving foreign technology. Moreover, they usually invest little in research and development, which is often only less than 1 percent of their annual output value (much lower than that in developed countries). Consequently, they often lack the ability to produce high-quality new products, and their products are unable to compete in the world market.

Most rural industry is in the red.   Rural industry is a very important part of the industrial sector in the Yangtze Delta. In the early 1980s, a boom in rural industry in South Jiangsu (indeed in the whole Delta) was widely regarded as a landmark in rapid economic growth of the region. But rural industry is seriously handicapped by its small scale, a lack of investment, and old equipment. Its products are unable to meet the changing demands of the market. Consequently, a large part of rural industry is now in the red.

From the above discussion, it is clear that a radical restructuring of industry is urgently needed. Recently, the central government has paid great attention to this. It is expected that through mergers and acquisitions and by restructuring, the situation could be improved. A wave of mergers among large state enterprises as well as small rural industries has already begun. The four largest state petrol-chemical enterprises around Nanjing—Yangtze Petrol-chemical, Jinling Petrol-chemical, Yizheng Chemical Fibers, and Nanjing Chemicals—have recently merged into a giant corporation, China Eastern United Petrochemical Group, which has 145,700 employees and 45 billion yuan in assets. This is the largest merger not only of state enterprises in the Yangtze Delta but perhaps also in the whole country. It is expected that through restructuring and mergers, state enterprises can return to profitability. However, most state enterprises are now overburdened with too many employees, often unproductive ones. To lay off large numbers of these employees would threaten the security of Chinese society. However, several large state enterprises have been successfully restructured and have become highly competitive. A notable example is the Little Swan Company in Wuxi. By updating technology and adopting modern management, it has not only acquired a large share of the domestic washer market but also exported its products to Southeast Asia, South America, and the Middle East (Clifford et al. 1997).

Similarly, some rural enterprises have emerged as rural giants by introducing new technology and improving management. For example, Hongdou Group in Wuxi is now China's top rural enterprise in the garment and knitwear sector, and Longshan Group in Dantu county is China's largest rural enterprise engaged in food processing. Longshan Eel Corporation, specializing in culturing eels, making eel feeds, and processing eels, now has about 40,000 employees and more than fifty branches in China and abroad, including one in the United States. Its sales of baked eel account for more than 30 percent of the Chinese market and 25 percent of the Japanese market.

The cotton textile industry used to be the most important industry in the region, but owing to its outdated equipment and inefficient management, it has been continuously losing money in recent years. Today, the cotton textile industry is not regarded as a pillar industry in the region, but because of the millions of workers employed by the industry, it is still very important in the Delta. With profound changes in the geographical distribution of cotton-producing areas, a radical relocation and restructuring of the cotton textile industry has taken place. In Shanghai, since 1992, 164 factories were closed, 220,000 workers (about 40 percent of the total) were laid off, and 300,000 spindles were moved to the interior province of Xinjiang—the largest cotton producer in China today. Similarly, a large number of ordinary cotton spindles and looms have been moved from South Jiangsu to North Jiangsu, a major cotton-producing region of the province. Between 1984 and 1995, in terms of the number of spindles and production of plain clothing and low-end cotton textiles, South Jiangsu has dropped from 52 percent of the province's total to 43 percent, while North Jiangsu increased its share. Also about 20 percent of cotton spindles, or approximately 1 million, in Jiangsu are too old and must be scrapped. The small number of cotton spindles and looms that are left in South Jiangsu have been completely updated to produce high-grade cotton products. It is expected that through these readjustments, the efficiency of the cotton textile industry will gradually improve and the industry will again become profitable.

A similar change in the silk textile industry has also occurred. Although Suzhou, Huzhou, and Hangzhou are still famous centers for the production of fine silk textiles, a significant part of raw silk and low-end silk fabrics are now manufactured in North Jiangsu. Recently, the government of Suzhou altered its annual production plan from the manufacture of 66 million feet of ordinary silk clothes to 16.4 million feet of high-grade products mainly for export in an effort to recover lost foreign markets. In conjunction with this change, twelve large state silk spinning, weaving, and dying factories in Wujiang have recently merged to form a huge corporation. All of these measures are designed to revive the struggling traditional silk weaving industry.

Environmental Problems

The rapid economic growth in the Yangtze Delta has led to the serious deterioration of the environment. The plain of the Delta is very densely populated, and consequently its environmental capacity is rather limited. The environmental deterioration of the delta has received the serious attention of the Chinese people and their government as well as the international community. This section discusses the major pollution problems.

Water Pollution

This is the most serious pollution problem of the region. It includes three components.

Inland Water Pollution.   Contrary to the common belief that the Yangtze Delta is rich in water resources, it is now troubled by a shortage of water. This does not mean that the quantity of water is insufficient but rather that the quality of water is poor and therefore undrinkable and unsuitable for household use.

The delta around Taihu is crisscrossed by innumerable canals and waterways, the most important being the Grand Canal. The South Jiangsu section of the Grand Canal, 129.4 miles long, has been completely renovated recently and is now navigable for ships and barges up to 500 tons. It is the busiest inland waterway in the country. More than 100 million tons of cargo are transported by the canal every year, exceeding the volume transported by the Yangtze River. Transported goods are chiefly bulky commodities, such as coal, grain, and building materials.

Taking advantage of the cheap waterway transport, many cities, towns, and factories are located along the canal. They discharge a large amount of industrial and household sewage water into the canal, often without any treatment. Consequently, the water in the Grand Canal is heavily polluted, becoming dark and malodorous nearly six months out of the year. Another source of pollution is excessive application of chemical fertilizers and pesticides in cultivated fields, which are washed into inland waterways.

The shore waters of Taihu Lake along the large urban centers of Suzhou and Wuxi are also seriously polluted, as can be vividly seen from satellite imagery. During the summer months, this pollution often triggers entropication and has caused great harm to Wuxi's water supply and to local people's health.

Pollution of the Yangtze River.   For the most part, the Yangtze River is relatively unpolluted. However, there are many large cities and several petrochemical, chemical, and iron and steel factories along the lower reaches of the river, and waters near these cities and factories are seriously polluted.

Pollution of the Adjacent Sea.   Because the Yangtze River discharges large amounts of polluted water into the ocean, the seawater adjacent to the Yangtze Delta is also polluted. Shanghai, with an estimated discharge of 2.7 billion tons of sewage water per year by 2010, is the largest source of pollution. The sewage water is discharged by the Huangpu and Suzhou Rivers through underground pipes to the sea. The Suzhou River, a large tributary of the Huangpu River, has long been notorious for its dark and malodorous water throughout the year. The sea has a strong diluting force, but since the ocean current flows toward the southeast, pollutants from Shanghai and the Yangtze are carried to the Zhoushan Islands, the largest ocean fishing ground in China. Toxic pollutants will gradually accumulate in the fish and other seafood, affecting the health of a large number of people.

Air Pollution

Air pollution mainly comes from coal-fired power plants. Along the lower Yangtze River between Nanjing and Shanghai, there is a large power plant (capable of producing more than 1 million kilowatts) about every nineteen miles. These large thermal power plants, together with many small ones, emit large amounts of sulfur dioxide and particulate matter and are the main sources of air pollution in the Yangtze Delta, especially along the lower Yangtze River. Now, in large cities including Shanghai, the concentration of particulates is over the permissible limit. Sulfur dioxide is the source of acid rain in the Delta. Acidity of rain is particularly high in some areas in the fall and winter. About 100 million tons of coal is needed to fuel the plants, which must come from North China. The long-distance haul of huge volumes of coal adds pressure on the rail and sea transport systems, and the shipping of large amounts of coal in waterways is an important source of water pollution.

The Rising Sea Level and Land Subsidence

There are already many publications dealing with the rise in the sea level. In delta areas, the hazard of inundation due to a rise in the sea level is aggravated by land subsidence (Milliman and Haq 1996). The Yangtze Delta is a notable example (Ren 1994; Ren and Milliman 1996). Chiefly due to the overpumping of groundwater, Shanghai's rate of land subsidence exceeds the rate of eustatic rise of the sea level and threatens the safety of its business center, "the Bund," where solid dikes have been built to protect it from storm surges. The dikes are high and massive. Their lower level is used to house travel services and American fast-food restaurants, like Kentucky Fried Chicken, while their upper level provides a wide promenade with an excellent view of the Huangpu River. The dike around the Bund has become a popular tourist spot in Shanghai, perhaps unique among the world's large cities.

A little inland from the sea is the low and swampy deltaic plain around Taihu Lake with an elevation below the flood level of both Taihu and the Yangtze River. Strong dikes and other engineering works have been built to protect the region from the inundation of floodwater. The overpumping of groundwater has resulted in large areas of land subsidence, which is especially serious in the urban areas of Suzhou, Wuxi, and Changzhou. Recently, various measures have been taken to combat excessive groundwater withdrawal, and land subsidence is now under control. However, over the wide rural area, especially in industrialized towns like Shengze, land subsidence remains unchecked, and towns have had to build strong dikes to protect themselves from the risk of floodwater inundation. Land subsidence is also a serious environmental risk in the region as a whole. Recently, some bridges on the Shanghai-Nanjing expressway were damaged due to land subsidence.

The serious water and air pollution in the Delta, particularly in the lower Yangtze valley, is a threat to sustainable development. Fortunately, the environmental awareness of the people and local officials has greatly increased. The Chinese people agree that they cannot afford "pollution first, combating pollution later." The rapid economic growth of the Delta must be coordinated with pollution management and control. Many encouraging examples have emerged.

Yangzhong, a county-level city, is on a sandy island in the Yangtze River. The city has made efforts to coordinate its economic development with population growth, resource development, and environmental protection and has established a model eco-city. Population growth has been rationally controlled, with a natural growth rate of −0.34 percent and a birth rate of 3.1 percent achieved in 1997. The loss of cultivated land has been offset by newly reclaimed land and through the rearrangement of housing and factories. Another example is Tengtou village in Fenghua county near Ningbo, where an eco-agricultural system has been effectively developed. This village was designated by the United Nations Environment Programme as one of the world's 500 best agro-eco villages in 1993.

Measures have also been taken to control pollution in large urban cities. For example, the main source of water pollution in the urban area of Suzhou, a large paper factory, was closed down, despite its high profits. Furthermore, Taizhou city authorities have refused to grant permits to foreign companies to build factories that would have caused serious pollution. Another illuminating case is that the people of Wujiang city (a county-level city of Suzhou) have successfully produced shrimp free of pollutants in the local canals and lakes and proudly serve fresh raw shrimp as a delicacy in local restaurants. This illustrates that through the strenuous efforts of the local government and people, inland waters can be kept clean.

On the other hand, one should not underestimate the enormous difficulty of coping with China's serious environmental problems. To illustrate this point, one need only review the history of air pollution control efforts in the lower Yangtze valley. Owing to their heavy air pollution, it has been suggested that no more coal-fired power plants should be built in the lower Yangtze valley. Instead, they should be located on the coast of North Jiangsu to prevent further deterioration of air quality along the lower Yangtze River. However, because of cost/benefit considerations and the urgent need for more power, many large power plants will nonetheless be built there and existing power plants will be enlarged. All these power plants are coal-fired. It was estimated in 1998 that expansion of the existing power plants alone would add 2.25 million kilowatts of new capacity, which would require an additional 5.6 million tons of coal as fuel. Moreover, large new coal-fired power plants are planned in Taicang and Changshu for the development needs of Suzhou. This example illustrates that controlling air pollution in the lower Yangtze valley will be tough, and its prospect in the near future looks rather gloomy.

Conclusion

As the twenty-first century begins, the Yangtze Delta has an unparalleled opportunity for sustainable development but also faces serious challenges. The favorable location, the opening of the Pudong Special Economic Zone, the establishment of the Shanghai International Shipping Center, and the rich resources of sci-tech personnel and skilled workers make Shanghai and the Yangtze Delta the gateway to the world market. Its unparalleled opportunity is matched only by the Pearl River Delta in the south, but the later lacks a rich and extensive hinterland like the Yangtze River Valley. For sustainable development, Shanghai and the Yangtze Delta must place a special emphasis on coordinated development of the two regions and make concrete plans to foster mutual benefits.

The world economy is changing very rapidly, and the world market is highly competitive. The Yangtze Delta faces a large number of challenges, with the most important being time and vision.

"Time is money." Major construction projects must be completed in time; otherwise good opportunities may be lost forever. The Shanghai International Shipping Center is a case in point. An international container hub-port must be established in the Yangtze Delta by 2007; otherwise transshipment of intercontinental containers from the Delta region will be divided by the existing large container hub-ports around mainland China. The construction of an international container hub-port in Shanghai depends solely on the success of deepening the shipping canal at the Yangtze River mouth to 41 feet by 2007. If the dredging cannot attain that depth or cannot be completed by 2007, then the whole SISC project will fail to achieve its goal. Therefore, it is important that the Beilun port be quickly developed in order to avoid losing the market to other Asian ports. Likewise, speed is of the utmost importance in restructuring existing industries, whether for the purpose of reducing large annual losses or for creating products designed for the world market.

Serious water and air pollution is the greatest challenge to sustainable development in the Delta. Because pollution comes from many sources in different districts (counties and cities), comprehensive and coordinated efforts by all districts and economic sectors are necessary to effectively combat pollution. Political vision is urgently needed so that the short-term interests of local enterprises do not compromise sustainable development and the long-term interests of the broader delta region. It is hoped that the cases cited in the preceding section will provide examples for others and that citizens and government officials will unite to work for a common cause—sustainable development.

Notes

In preparing this chapter, I have freely used information in publications of the local governments in the Yangtze Delta and data in the interim reports of colleagues working in the Delta. To all these organizations and people I am deeply indebted. However, the conclusions and opinions in this article are my own, and I alone am responsible for any error in the text.

I am also grateful to Zhang Yongzhan and Yang Baoguo, of the State Pilot Laboratory of Coastal & Island Exploitation, Nanjing University, for their help in the preparation of the manuscript.

1. There are three levels of cities in China. The first includes cities under the state's direct administration. Only four cities belong to this category: Beijing, Shanghai, Tianjin, and Chongqing. The second level contains provincial cities that have a number of xians (counties) under their administration. Many counties have been renamed to cities, which are called county-level cities.

2. At present, China still has 60 million people in poverty. Poverty eradication is a key issue in the sustainable development of the whole country.

3. TEU is an international measure for container cargo, standing for a 20-foot unit equivalent or a 20-foot-long cargo container. General cargo is shipped in steel containers of various sizes, ranging from 20-foot-long boxes to those over 40 feet long.

4. The area of Nanjing city proper inside the ancient city walls is 21.68 square miles.

5. For example, in 1995, among 550,000 rural workers in Jiangyin, 450,000 (or about 80%) were nonagricultural workers.

References

International Council of Scientific Unions (ICSU). 1997. *Science International* 64: 12.

Clifford, M. L., et al. 1997. China: Can It Really Reform Its Economy? *Business Week,* September 29:116–124.

Milliman, J. D., and B. U. Haq, eds. 1996. *Sea Level Rise and Coastal Subsidence.* Dordrecht, Netherlands: Kluwer Academic Publishers.

Ren, M. 1994. Relative Sea Level Rise in China and Its Socio-Economic Implications, *Marine Geodesy* 17:37–44.

Ren, M., and J. D. Milliman. 1996. Effect of Sea Level Rise and Human Activity on the Yangtze Delta, China. In *Sea Level Rise and Coastal Subsidence*, ed. J. D. Milliman and B. U. Haq, pp. 205–214. Dordrecht, Netherlands: Kluwer Academic Publishers.

She, Z. 1997. *The Water-Land Resources and Regional Management in the Yangtze River Delta.* Hefei, China: Chinese Science and Technology University Press.

Wang, J. 1997. Hong Kong: One Container Load Center for Two Economies. *Jinji Deli (Economic Geography)* 17(3): 33–39.

9

From Special Economic Zones to Special Technological Zones

Shuguang Wang

It has been almost a quarter century since China opened its doors to the outside world and embarked on an economic reform. An important part of the reform has been the establishment of various development zones at different times, including Special Economic Zones (SEZs), Economic and Technological Development Zones (ETDZs), and High and New Technology Industry Development Zones (HNTIDZs). The sequence of their establishments reflects important policy adjustments in the course of economic reform. While SEZs and ETDZs are well known, HNTIDZs represent a new form of economic reorganization and investment opportunities and have not received sufficient attention from geographers outside China. This chapter examines the shifting emphasis in establishment of development zones and related policy changes in China, with a special focus on the characteristics and performances of the HNTIDZs.

SEZs and ETDZs: Forerunners of the Open Policies

Special Economic Zones were the earliest to be established in China. In August 1980, the Chinese Parliament approved four SEZs: Shenzhen, Zhuhai, Shantou, and Xiamen. In April 1988, the fifth SEZ, also the largest, was established in the newly created Hainan province. All five SEZs are located in South China for two reasons. First, South China is the ancestral home of many overseas Chinese who had a desire to invest in the opened China. Second, South China is close to Hong Kong, Taiwan, Macao, and Southeast Asian countries, which were the main sources of foreign investment in China in the early 1980s (Wang and Bradbury 1986).

The SEZs were designed to be forerunners of China's open-door policies, to attract foreign investment and technology, expand exports, and train personnel with expertise in international trade (*Economic Areas in China* 1993). Incentives for foreign investment include preferential tax policies and a high level of administrative autonomy. In the first four years, the central government had no regulations for corporate income tax or import tariffs for SEZ enterprises. It simply endorsed the interim policies formulated by the provincial government of Guangdong (Ma and Fang 1995). Not until November 1984 did the central government work out and promulgate the first set of state regulations (State Council 1984). The central government invested heavily in the SEZs to improve infrastructures.

With tax incentives and financial assistance from the central government, SEZs developed rapidly, which greatly stimulated local and regional economic expansion in South China, especially in Guangdong province, where three of the five SEZs are located. This prompted other provinces to ask for approval of SEZs in their territories. Instead of approving more SEZs, the central government decided to create a new set of development zones: Economic and Technological Development Zones. Between 1984 and 1986, fourteen ETDZs were approved and established, all in seaboard cities.[1] The ETDZs were granted basically the same tax and tariff incentives as those enjoyed by SEZs (State Council 1984), but they had less administrative autonomy. For example, SEZs were relatively independent administrative divisions, whereas ETDZs were under the direct leadership and jurisdiction of local municipal governments to carry out certain special policies (*Economic Areas in China* 1993). Furthermore, the central government provided only limited capital investment to ETDZs for construction of infrastructure. Except for low-interest loans from state banks, it became a local responsibility to find capital for that purpose.

In general, both SEZs and ETDZs were established to create localized environments with favorable conditions that would attract foreign investment, technology, and equipment. Most preferential policies for SEZs and ETDZs were offered to foreign investors only; those applying to domestic investors were very limited. The policies therefore were criticized by Chinese enterprises as promoting unfair competition. Also, the preferential policies were described as "areal investment policies," meaning that all foreign investment made in these zones, whether in high- or low-technology sectors, would be eligible for the preferential policies.

Establishment of Special Technological Zones—HNTIDZs

China's experiment with SEZs and ETDZs took place at a time when the world economy had just started a major restructuring. In the early 1980s, major industrialized countries were making strategic plans for research and development of high technologies in the interest of increasing their competitiveness in the world economy. At one level, OECD governments took collective efforts to deal with the essential issue of how the resources of science and technology could help to solve the economic difficulties caused by the oil crisis of the 1970s and the subsequent economic recession. Agreeing that accelerated scientific and technological progress held the key to the future economic well-being of their countries, the OECD leaders worked in collaboration to devise science and technology policies for the 1980s (OECD 1980). At another level, individual OECD member countries launched their own race against one another (Bylinsky 1986). For example, the Reagan administration announced the "Strategic Defense Initiative," which aimed at boosting American high-tech industries through heavy defense spending and enormous military procurement (Markusen, Hall, and Glasmeier 1986). The Japanese government initiated its "Visions for the 1980s" (Tasuno 1986), in order to boost Japan's innovative capacity. Not wanting to be left behind the United States and Japan, nineteen European countries jointly launched the "Eureka" program—an idea first proposed by President François Mitterrand of France (Dickson 1988)—to promote cooperative R&D and forge industrial alliances among these European countries and to bring their new technologies to market. By the mid-1980s, a three-way race among the United States, Japan, and Western Europe was well under way (Peterson 1990).

After several years of experimenting with SEZs and ETDZs, the Chinese government realized that for both political and economic reasons, Western industrialized countries would not transfer their state-of-the-art technologies to China; most foreign investment went to labor-intensive, assembly-type manufacturing with little on-site R&D. As a result, the SEZs and the ETDZs were evolving more like "export processing zones" than what they had been designed for, and the industries in these zones made low value-added products that were not competitive enough to enter foreign markets in large quantities.

Concerned with China's future, four Chinese scientists submitted a letter to the Central Committee of the Chinese Communist Party on March 3, 1986, in which they strongly suggested that China join the international race for high-technology development.[2] Two days later, the letter reached Deng Xiaoping, who promptly issued the following instruction: "This matter should be decided with no delay." Under Deng's instruction, the State Council called together 124 senior Chinese scientists to draft China's strategic plan for the development of science and technology. The strategic plan was officially named the "High-Tech Research and Development Program of China" (State Science and Technology Commission 1992), but since both the letter of the four scientists and Deng's instruction occurred in March 1986, this plan came to be known as the "863 Plan." The core of the plan was to monitor high-tech research in the world and, more importantly, to develop high technologies and commercialize them by Chinese scientists. Because China was still a developing country and was unable to fund high-technology research on a broad scale, only seven areas were selected as priorities.[3] The State Science and Technology Commission was delegated the authority to implement the strategic plan.

To nurture China's own high-tech industries, the State Science and Technology Commission designed the "Torch Program," with the main task of establishing High and New Technology Industry Development Zones in selected cities (State Science and Technology Commission 1992). The HNTIDZs were meant to be China's special technological zones, where the achievements of scientific research by Chinese scientists could be transformed into competitive commodities and a new generation of entrepreneurs and competent managers could be bred. As new technologies and management mechanisms matured, they would then be disseminated to other regions of the country and used to transform and revitalize China's traditional industries, especially the state-owned large and medium-sized factories. This process is signified by the program's name, which connotes that single torches can start prairie fires. To implement the program, a special government agency known as the Office of the Torch Program was created under the auspices of the State Science and Technology Commission.

Almost at the same time as the Torch Program was drafted and approved, the State Council designated China's first national experimental HNTIDZ—the Beijing Experimental Zone for Development of New Technology Industries (this name is still being used today)—and encouraged other cities to prepare for development of their own HNTIDZs. Beijing was chosen as the site for the experiment because it, as the national capital, has the best-qualified labor force and the highest concentration of advanced infrastructure in the country.[4] With the designation of the Beijing HNTIDZ in 1988, the State Council also ratified the criteria for determining high- and new-technology enterprises and the interim preferential policies, both formulated by the Beijing Municipal Government. Only those enterprises that met the criteria were entitled to preferential policies, which included tax exemption or reduction and tariff exemption or reduction for import of materials and export of finished products (Beijing Municipal Government 1988).

After about three years of experimentation in Beijing, the State Council in March 1991 issued its "Document No. 12 of 1991," in which twenty-six zones in other Chinese cities were selected and designated as national HNTIDZs (Zhang 1993). At the same time, the State Council issued national preferential policies and criteria for determining high-tech enterprises, which would apply to all HNTIDZs (State Science and Technology Commission 1991; State Administration of Taxation 1991). The Chinese government defined high and new technologies as being related to the following eleven fields: microelectronics and electronic information; space science and aerospace; photo electronics and photo-mechanical-electronic integration; life science and bioengineering; material science and new materials; energy science, new energy technology and efficient energy consumption; ecological science and environmental protection; earth science and marine engineering; science of fundamental matter and radiation; medical science and biomedical engineering; and other new engineering technologies that can be applied to the existing traditional industries in China (State Science and Technology Commission 1992). In principle, only those enterprises that engage in the research and development of the above fields are permitted to enter the HNTIDZs, but the last field on the list is quite vague and can be interpreted as many different things. In addition, HNTIDZ enterprises must meet specific requirements for employee qualifications and R&D spending. For example, 30 percent or more of the employees of high-tech enterprises should have post-secondary degrees, at least 10 percent of employees should be engaged in R&D activities, and at least 3 percent of their corporate income should be spent on R&D activity.

One year later, in November 1992, the State Council designated another twenty-five HNTIDZs in its "Document No. 169 of 1992." By then, the total number of state-designated HNTIDZs had reached fifty-two (including the Beijing zone), and the allocation of China's special technological zones was completed (Table 9.1).

The designation of fifty-one HNTIDZs in two years triggered a bandwagon effect in China. Many provinces hastily announced the creation of more such zones (the number reached 79 in 1993), with the hope that these province-designated technology zones would later be elevated to national HNTIDZs. In 1993, however, the State Council had to issue its "Document No. 33 of 1993" to cool down this HNTIDZ fever. The number of national HNTIDZs was fixed at fifty-two, but the fifty-two national HNTIDZs will be evaluated regularly for performance; those that perform poorly will be eliminated from the national team and be replaced by selected provincial HNTIDZs.

In general, the HNTIDZs have three major differences from the SEZs and ETDZs that had been previously established (Table 9.2).

1: Geographically, HNTIDZs are more widely distributed in the country than the other two types of development zones, which are limited to coastal provinces.[5] As shown in Table 9.1, HNTIDZs are located in twenty-eight of the thirty-one provinces, autonomous regions, and municipalities in mainland China.[6] The three provinces that do not have a HNTIDZ are Ningxia, Qinghai, and Xizang (Tibet). The widespread distribution aims to reduce the regional economic disparities that had worsened after the establishment of the SEZs and ETDZs and to alleviate the problem of population migration from the interior to the eastern coastal regions. Most HNTIDZs are located in large metropolitan areas or in provincial capitals. All provinces and autonomous regions with HNTIDZs have one in their capital city, except for Inner Mongolia.

2: Unlike the preferential policies for SEZs and ETDZs that apply mainly to foreign investors, those for HNTIDZs apply equally to both domestic and foreign investors, so that domestic enterprises have a fair chance to compete with foreign-invested businesses. These policies are called "areal sector policies," as opposed to the "areal investment policies" for SEZs and ETDZs. This means that only enterprises in the government-prescribed

TABLE 9.1   Geographical Distribution of China's 52 HNTIDZs

Province	HNTIDZ	Subtotal
Coastal provinces (28)		
Fujian	Fuzhou, Xiamen	2
Guangdong	Foshan, Guangzhou, Huizhou, Shenzhen, Zhongshan, Zhuhai	6
Guangxi[a]	Guilin, Nanning	2
Hainan	Haikou	1
Hebei	Baoding, Shijiazhuan	2
Jiangsu	Changzhou, Nanjing, Suzhou, Wuxi	4
Liaoning	Anshan, Dalian, Shenyang	3
Shandong	Jinan, Qingdao, Weifang, Weihai, Zibo	5
Shanghai[b]	Shanghai	1
Tianjin[b]	Tianjin	1
Zhejiang	Hangzhou	1
Interior provinces (24)		
Anhui	Hefei	1
Beijing[b]	Beijing	1
Chongqing[b]	Chongqing	1
Gansu	Lanzhou	1
Guizhou	Guiyang	1
Heilongjiang	Daqing, Harbin,	2
Henan	Luoyang, Zhengzhou	2
Hubei	Wuhan, Xiangfan	2
Hunan	Changsha, Zhuzhou	2
Inner Mongolia[a]	Baotou	1
Jiangxi	Nanchang	1
Jilin	Changchun, Jilin	2
Shaanxi	Baoji, Xian	2
Shanxi	Taiyuan	1
Sichuan	Chengdu, Mianyang	2
Xinjiang[a]	Urumqi	1
Yunnan	Kunming	1

[a]Autonomous regions
[b]Municipalities equivalent to provinces

TABLE 9.2   Differences Among SEZs, ETDZs, and HNTIDZs

	SEZ	ETDZ	HNTIDZ
Geographical distribution	In coastal provinces	In coastal provinces[a]	In both coastal and interior provinces
Type of preferential policy	Areal investment policies; apply mainly to overseas investors	Areal investment policies; apply mainly to overseas investors	Areal sector policies; apply to both overseas and domestic investors
Favorableness of preferential policy and government support	Preferential policies most favorable; with large sums of capital investment from central government	Preferential policies less favorable; with limited capital investment from central government	Preferential policies least favorable; with virtually no capital investment from central government

SOURCES: State Council 1984; State Administration of Taxation 1991; State Science and Technology Commission1991; personal interviews with officials of Office of the Torch Program and HNTIDZ administrators 1995.
[a]Until 1992, all ETDZs were in coastal provinces, but between 1992 and 1994, 18 more ETDZs were established, many of which are in interior provinces.

high-tech fields (or sectors) are permitted to enter the HNTIDZs and are eligible for preferential policies; investment in low-tech sectors, whether domestic or foreign, is not allowed to enter, nor are low-tech sectors eligible for preferential policies.

3: Preferential policies for HNTIDZs are much less favorable than those for SEZs and ETDZs. For example, for all SEZ and ETDZ enterprises with a contract life of ten years or longer (except those engaged in service trade), a two-year tax exemption commencing with the first profit-making year is allowed, followed by a 7.5 percent preferential income tax rate in the following three years. But for newly established HNTIDZ enterprises, corporate income tax is exempted in the first two years of operation (for newly established Chinese-foreign joint ventures in HNTIDZs, corporate income tax is exempted in the first two profit-making years); after that, a 15 percent preferential income tax applies with no exception (normal enterprise income tax rate is 30 percent). In addition, whereas SEZs and ETDZs were given the autonomy to grant extra tax incentives to foreign investors, HNTIDZs were denied the same autonomy. Moreover, the central government provides little in the way of capital investment to the HNTIDZs for construction of infrastructure, unlike the healthy financial support of SEZs. The only state support for HNTIDZs, in addition to preferential policies, is low-interest bank loans. Even bank loans are very limited because state banks are reluctant to finance high-technology enterprises, especially start-up firms, due to their high-risk nature, as heavy investment in R&D does not necessarily guarantee successful products.

HNTIDZ Structure

High-tech industry development zones have existed in many industrialized countries since the early 1950s under the general name of "technopoles." The term technopole has its origin in French, meaning planned development. Castells and Hall (1994:1, 8–9) appropriated this term into the English language to refer to various deliberate attempts by government, in association with universities and private companies, to plan and promote technologically innovative industrial production within one concentrated area. Alternatively, technopoles can be interpreted as growth poles driven by high-technology industries, whose function is to generate the basic materials of the information economy. Castells and Hall (1994) distinguished three types of technopoles: industrial complexes of high-tech firms; technology parks; and science cities.

Industrial complexes of high-tech firms are spontaneously formed geographical agglomerations of R&D facilities and related manufacturing establishments. Silicon

Valley and Boston's Route 128 in the United States, and the M4 Corridor and the Cambridge Phenomenon in the United Kingdom, are all examples. Strictly speaking, these industrial complexes were not deliberately planned; Castells and Hall included them because government and universities indeed played a crucial role in their development.

Technology parks, alternatively called science parks or research parks, are deliberately planned high-technology business areas. Operationally, they are a group of research organizations and businesses devoted to the development of scientifically proven concepts from the laboratory stage to the factory production stage. Physically, they are a group of small to medium-sized office and laboratory-type buildings in a high-quality landscaped setting (Worthington 1982). The first technology park in the world was alleged to be Stanford Research Park created in 1951 (Rogers and Larsen 1984), which attracted, and spun off, numerous high-tech firms in the surrounding region and subsequently led to the formation of the famous Silicon Valley.

The concept of science city, also known as "technopolis," originated in Japan. In 1980, the Ministry of International Trade and Industry (MITI) of Japan advocated the technopolis concept that suggested the creation of cities in which scientific research, high-tech industry, and high-quality living are all brought together in an organized relationship (Bloom and Asano 1981; Nishioka and Takeuchi 1987; Fujita 1988; Glasmeier 1988; Edgington 1989; Masser 1990). Other examples of science cities are Russia's Akademgorodok and South Korea's Taedok (Castells and Hall 1994).

Although China's HNTIDZs were modeled on the general concept of technopole, they are not an exact duplicate of any of the three types of technopoles described above. The HNTIDZs are larger in area (ranging from 10 to 100 square kilometers) than typical technology parks in North America but smaller than science cities, as they are only a designated area within a city. Being government-planned developments, they are also different from such industrial complexes of high-tech firms as Silicon Valley and Boston's Route 128. China's HNTIDZs are composed of two portions: a "policy area" and a "new development area." The policy area is located in the built-up part of the host city, where universities and research institutes cluster. However, this area usually has little land available for new development, and it often includes many non-high-technology enterprises that existed before the HNTIDZ was designated. The new development area is located either near the edge of the city or in suburban counties. In new development areas, all occupants are required to meet state criteria for high-tech firms. These areas are planned also to provide convenient and comfortable living facilities to attract scientists and engineers. The policy and new development areas may be adjacent to each other, or they may be spatially separated, resulting in a "multiple-site" HNTIDZ. Examples of the latter include the HNTIDZs in Shanghai, Hangzhou, Tianjin, and Dalian (Gu 1995). In recent years, the Beijing HNTIDZ designated two new development areas in its suburbs: one in Fengtai District, the other in Changping County. The disadvantage of multiple-site zones is the high cost of providing infrastructure and services to constituent enterprises, as this reduces the scale economies that are usually associated with spatial agglomeration. Despite the differences from the technopoles in other countries, Chinese planners do have a long-term development sequence in mind: that is, to start from intracity HNTIDZs, gradually expand HNTIDZs to form science cities, and eventually build more science cities to form regional techno-belts.

The management system of the HNTIDZs also has Chinese characteristics. In North America, technology parks are often managed by their developers or financial sponsors. When a park is collectively owned by all occupants, a board of directors and an executive committee consisting of representatives of the occupants would manage it.[7] In the United Kingdom, the majority of technology parks either employ a full-time manager or rely on a university's industrial liaison officer to interact with tenant firms; the single manager is usually supported by a secretary and a property service officer (Monck et al. 1988). In Japan, prefecture councils administer the technopolises.

The management system in China's HNTIDZs is much more complex. Each HNTIDZ has an elaborate Administrative Commission, which is given a unique position in the government hierarchy. Each commission is granted municipal powers over development of the respective HNTIDZ. In structure, the Administrative Commission is like a miniature municipal council, aiming at consolidating responsibilities and increasing operation efficiency (see Table 9.3). Its power ranges from admitting enterprises into the zone to planning and financing construction of infrastructure, leasing land, collecting taxes, and dealing with foreign trade and cooperation for its constituent enterprises. In one area of responsibility, however, the Administrative Commission's autonomy is rather limited: It is prohibited from initiating any tax incentives that are additional to those specified in the national policies, regardless of specific local circumstances. This effectively eliminated an important means of competition among the HNTIDZs.

Besides the key departments listed in Table 9.3, most HNTIDZs have a Pioneering Service Centre, equivalent to incubators in other countries (Allen 1985; Cooper 1985; Cooper et al. 1985; Smilor and Gill 1986). Unlike the key departments that perform administrative functions, the Pioneering Service Centre is a profit-making agency providing commercial services. In addition to the normal types of services (such as secretarial, logistic, and facility support), some Pioneering Service Centres provide limited venture capital to start-up firms. When venture capital is provided, the Pioneering Service Centres are also involved in enterprise management and production decisionmaking.

In China's HNTIDZs, six types of enterprise ownership coexist, more than in any country's technopoles. The state-owned and collectively owned, which are a heritage of the centrally planned economy, are the two dominant types of enterprises in HNTIDZs. Although collectively owned enterprises have outnumbered the state-owned, they are much smaller on average and employ less than half as many people as the state-owned. By state regulations, a collectively owned enterprise can raise its

TABLE 9.3   Key Departments Under the HNTIDZ Administrative Committee

Key Department	Responsibilities
Liaison and Public Relations Department	Liaison, coordination, promotion, file keeping, and reception of visitors
Labor and Personnel Department	Labor affairs and personnel training
Planning and Development Department	Planning and construction of infrastructure and management of fixed assets
Enterprise Development and Management Department	Determining and admitting high and new technology enterprises and project approval, contract registration, and product exhibition
Finance, Audit, and Tax Bureau	Monitoring enterprises' financial management, auditing their accounts, and collecting taxes
Office of Foreign Affairs	Invitation and reception of foreign visitors and approval of Chinese staff members going abroad
Department of International Trade and Foreign Investment	Approving foreign investment and import/export of production materials and finished products
Strategic Development Research Center	Conducting policy research and evaluation and making recommendations to administrators

SOURCE: Compiled from various HNTIDZ Investment Guides.

initial capital either from eight or more individuals or from state-owned factories or other public institutions. Within the first three years of operation, the collectively owned enterprises must return their initial investment, plus 15 percent interest, to the original investors. Once principal and interest are paid, the entire accumulation of capital and property belongs to all employees of the enterprise, and the original investors cannot claim sole ownership (Beijing Experimental Zone for Development of New Technology Industries 1994). Private enterprises can be owned by one or more individuals who contributed to the initial investment, but they are not bound by the state regulations governing collectively owned enterprises. This type of enterprise was allowed in China only after the economic reform. In the HNTIDZs, private enterprises are the smallest in terms of both their numbers and their size.

In general, the enterprises in HNTIDZs have all suffered from a shortage of capital, which has hampered their growth and expansion. In 1992, the State Science and Technology Commission and the State Commission of Economic System Reform authorized experimentation with joint-stock enterprises—a typical form of production organization in capitalist economies. The purpose of introducing this type of enterprise into socialist China was to raise capital from shareholders and to decentralize investment risk. In joint-stock enterprises, ownership and management are separate. Managers are responsible to the shareholders but are under the supervision of a board of directors selected by shareholders. This has been recognized in China as a more effective management mechanism than state-owned and collectively owned enterprises. In 1993, only 5.8 percent of all HNTIDZ enterprises were of the joint-stock type; in 1995, they increased to 14.5 percent.

The reliance on China's indigenous R&D capability by no means excludes foreign participation. Foreign participation is welcome for two reasons. First, it brings in much needed capital for the fast development of China's HNTIDZs. Second, foreign investment is often accompanied by at least partial technology transfer, which helps to further increase China's indigenous technological capability. China now uses its lucrative market for exchange of foreign investment and high technology. That is, only if foreign investors are willing to invest in high-tech sectors and agree to transfer some high technologies are they allowed to sell their products in China. Attracted by preferential policies and the Chinese market, foreign-backed enterprises have developed rapidly in China's HNTIDZs. Some of them are owned solely by foreign investors (including those from Hong Kong, Taiwan, and Macau); others are joint ventures with Chinese partners. By the end of 1995, 1,937 such enterprises had been established, which accounted for 15 percent of all HNTIDZ enterprises in the country and employed 158,884 people.

Analysis of HNTIDZ Performance

This analysis is based on data published by China's State Science and Technology Commission (SSTC), which since 1992 has been publishing a mini-statistical yearbook for HNTIDZs. Because there is at least a one-year time lag in releasing the statistics by SSTC, the most recent data available at the time of analysis were those for 1995. But even with data for only four years, some important patterns in HNTIDZ development are identifiable.

Between 1992 and 1995, China's HNTIDZs grew steadily. By the end of 1995, a total of 12,864 high-tech firms had been admitted into the fifty-two zones, which provided nearly 1 million jobs and yielded an annual income of 152.89 billion Chinese yuan—equivalent to 2.5 percent of the national GDP in that year. Of the 152.89 billion yuan, 49 percent was from sales of self-made products, especially products from the microelectronics and electronic information industry. Seventeen percent of the 12,598 high-tech products manufactured by the HNTIDZ enterprises were successfully exported. Products from the sectors of new energy technology and environmental protection, though less significant in number than electronics products, also had much higher-than-average export rates. These represent remarkable achievements in a period of only four years.

Despite the overall steady growth, many HNTIDZs are still at low levels of development. By the end of 1995, thirty HNTIDZs still had less than 30 percent of their employees with post-secondary education; twenty-three HNTIDZs had less than 10 percent of their staff engaged in R&D activities; twenty-nine HNTIDZs spent less than 3 percent of their total income on R&D. This implies that nearly half of the fifty-two HNTIDZs must have a considerable number of registrants that do not meet the state criteria for high-tech enterprises. Furthermore, only 10 percent of the 12,598 products have met international standards (Office of the Torch Program 1996).

For various reasons, neither all fifty-two HNTIDZs nor all six types of enterprises are expected to perform equally well. Three hypotheses are proposed regarding their varying performance. The first hypothesis is that HNTIDZs in coastal provinces in general perform better than those in the interior. This hypothesis is derived from the common belief that coastal provinces in China are economically more developed than interior ones. They have higher levels of urbanization and better-established industrial bases, all favorable conditions for the development of HNTIDZs at relatively low cost. But this does not mean that all cities in coastal provinces have more favorable conditions for development of high-tech industries than cities in interior provinces. The status of being provincial capitals may matter. In China, provincial capitals are usually given priorities in receiving government investment in infrastructure, such as international airports, major land transport terminals, and communication facilities. They often have relatively high concentrations of intellectual resources, since major universities and research institutions tend to be located at provincial capitals, and provincial capitals also have better developed business services, especially banks, than other cities. This suggests that capital cities in interior provinces may possess better conditions for the development of high-tech industries than do many coastal cities. The second hypothesis therefore states that the HNTIDZs located in provincial capitals will perform better than those in other cities. The third hypothesis postulates that new forms of enterprises, including both joint-stock and foreign-invested ones, perform better than the traditional state-owned and collectively owned enterprises. Joint-stock enterprises represent an effective way of raising capital for the development of high-risk industries. They should therefore perform better than the cash strapped state-owned and collectively owned enterprises. Foreign-invested enterprises may perform better because they have ready access to capital as well as advanced production and management techniques.

Total corporate income is usually used to measure industrial performance, but this indicator is considered inappropriate in the Chinese context. HNTIDZs and their enterprises calculate their total corporate income by summing up three types of earnings: (1) income from sales of self-made high-tech products; (2) income from technology transfer and technical services; and (3) income from retail or wholesale of products made by outside enterprises or imported from overseas. Because the last part of income does not reflect internal R&D capacities, this study uses only the first two types of earnings to measure the performance of the HNTIDZs and their enterprises. The sum of these two types of earnings is referred as "technology-generated income," which in 1995 accounted for 76.2 percent of the total corporate income of the 52 HNTIDZs.

The ranking of the 52 HNTIDZs by their technology-generated income leads to two observations. First, of the 28 HNTIDZs located in coastal provinces, only 15 (or 53 percent) are ranked among the top 26 (i.e., the top half of all HNTIDZs). This indicates that HNTIDZs in coastal provinces in general did only slightly better than those in interior provinces, so there is no strong support for Hypothesis 1.

In fact, 10 of these 15 HNTIDZs are located in only four of the eleven coastal provinces: Shandong (3),Guangdong (3), Liaoning (2), and Jiangsu (2). Second, of the 27 HNTIDZs located in provincial capitals, 17 (or 63 percent) are ranked among the top 26. Eleven inland-based HNTIDZs are ranked among the top 26. Eight of them, or 73 percent, are in provincial capitals, whereas only 60 percent of the 15 coastal province-based HNTIDZs in the top 26 are in provincial capitals. This provides support for the second hypothesis. Even when both locational attributes, which are not mutually exclusive, are considered simultaneously, only 9 (or 53 percent) of the 17 HNTIDZs that are located in capital cities and ranked among the top 26 are in coastal provinces—not a significant majority. Observations about Hypotheses 1 and 2 are supported further by chi-square statistics: For the association between HNTIDZ rank and location of coastal province, $x^2 = 0.31$ with p = 0.58, whereas for the association between HNTIDZ rank and the status of provincial capital, $x^2 = 3.77$ with p = 0.05. With the above statistics, Hypothesis 1 is actually rejected. To conclude, HNTIDZs in coastal provinces as a whole do not perform much better than those in interior provinces, but the majority of the HNTIDZs in provincial capitals exhibited above-average performance. This implies that capital cities in interior provinces compete well with coastal cities in developing high-tech industries and should not be overlooked by domestic or foreign investors as good investment locations.

Notably, the three privileged municipalities of Beijing, Shanghai, and Tianjin ranked among the highest. The Beijing HNTIDZ has many of the country's most successful high-tech enterprises, such as the nationally renowned Stone Group, Legend Group, Peking University Founder Group Corporation, Beijing Kehai High-Tech Group Company, and Hope Computer Company, all electronics-based companies.

Differences in performance exist among the six types of enterprises. In terms of overall technology-generated income, the state-owned enterprises ranked the highest, followed by the foreign-invested and joint-stock enterprises; but this is mainly because the former type had many more enterprises than the latter two types. When measured on a per employee basis, the foreign-invested and joint-stock enterprises outperformed all other types; by the same measure, even the privately owned enterprises performed better than the state-owned and collectively owned enterprises. This implies that the new types of enterprises have the highest productivity, or they produce the highest value-added products, and the older types are much less efficient. Hypothesis 3 is therefore strongly supported. To find out whether internal or external factors are more important to enterprise performance, a correlation analysis at the HNTIDZ level was conducted, in which technology-generated income was correlated both to HNTIDZ investment in infrastructure and to R&D spending.[8] Whereas R&D spending was found to be highly correlated with technology-generated income (r = 0.73 with p = 0.00), investment in infrastructure showed no correlation at all (r = 0.05 with p = 0.72). The ranks of per employee technology-generated income are largely correlated with the ranks of per employee R&D spending. This indicates that construction of infrastructure, which is the responsibility of the HNTIDZ administrative commissions and external to individual enterprises, has been slow, and its positive impacts on enterprise performance have not yet shown up, whereas the effects of R&D spending, which is direct input made by individual enterprises, are quickly reflected in enterprise performance. Apparently, the difference in labor productivity among the six types of enterprise is attributed to the varying levels of per capita R&D spending. The difference in itself points to a new direction of economic reform in the future: Transforming state-owned and collectively owned enterprises to the new forms of production organization will be economically rewarding. Not surprisingly, the Fifteenth National Congress of the Communist Party of China in 1997 formally adopted a resolution to speed up the transformation of state-owned industrial operations into joint-stock enterprises.

Future Directions and Challenges

Through the establishment of various development zones in the past two decades, China has gained valuable experiences in participating in the new global economy. The Chinese government has frequently adjusted the country's economic reform and open policies. In April 1990, the central government announced the creation of Pudong Economic Development Zone in Shanghai. Although the Pudong Economic Development Zone is similar to the five SEZs in all aspects, it was not named a special economic zone. In fact, there will probably be no new SEZs in China in the future. Between 1992 and 1994, eighteen more ETDZs were established in China, but the new ETDZs were no longer limited to seaboard cities; many of them were designated in interior provinces.[9]

In 1995, when the preferential policies for SEZs and ETDZs expired, the Chinese government made two major changes. First, most preferential policies would not be renewed, except for the preferential across-the-board 15 percent corporate income tax. The exemption from tariffs on imported materials and equipment used for producing exports would be phased out over 1995–2000; at the same time, the overall tariff on imports has been reduced from 36 percent to 23 percent due to pressure from the World Trade Organization (*People's Daily*, April 29, 1996:2). Second, investments in labor-intensive and low-tech manufacturing should no longer be approved in SEZs and ETDZs; they should be encouraged to move to interior provinces to create jobs and help stimulate economic development there. In other words, SEZs and ETDZs must reposition themselves in the face of increased globalization of the world economy. With these policy changes, the SEZs and ETDZs are expected to play similar roles to those of HNTIDZs in the future, and the differences between the types of development zones will diminish.

Future development of China's HNTIDZs faces at least two major challenges. First, their development will be impeded by a shortage of capital, which may be aggravated by inadequate government support and lack of venture capital. As in many other countries, the state explicitly intervenes in technology development in China; but the Chinese government has not really formed relationships with HNTIDZ enterprises, including the state-owned enterprises, by providing R&D grants and making large-scale procurement of domestically made high-tech products, a common form of government support for high-tech industries in many other countries. Although tax reduction and tariff exemption indeed helped many enterprises to get a foothold in the special technological zones, the state support in most HNTIDZs is inadequate. The severe lack of capital has prevented many HNTIDZs from completing the basic infrastructure needed to attract potential investors and entrepreneurs, either domestic or foreign. Their main source of income has been from leasing land and buildings in their zones. The shortage of funding is compounded by the lack of venture capital—a typical capitalist form of high-risk business financing (Liles 1974; Tyebjee and Bruno 1984a,b; Florida and Kenney 1988a,b,c). Until now, venture capital has been very limited in China; an important part of the "Silicon Valley culture" is therefore missing. While innovative industries must be nurtured by capital, new enterprises without pre-existing reputations have had great difficulty obtaining loans from state banks. Even though the Chinese government has encouraged establishment of venture capital, the government itself is reluctant to divert resources away from the state-owned factories— the backbone of the national economy—to finance high-risk ventures. As a result, many infant firms with business potential are suffering from "malnutrition." Even large companies have to rely on the retailing of someone else's products to accumulate development capital. This is why some HNTIDZs, including the Beijing HNTIDZ, earn a relatively high proportion of their total income from retailing activities.

The second challenge is the rapid phasing-out of preferential policies. In fact, the preferential policies from the beginning were meant to be interim supports. They were offered to attract high-tech firms to the HNTIDZs, which were expected to become self-sustainable in three to five years. After that, preferential policies were to be gradually withdrawn. In spite of the request by HNTIDZs that national preferential policies be continued and be broadened, the central government recently began to narrow the policy scope, with the intention of eventually withdrawing most of the preferential treatments. The state tax system was reformed in 1994 with at least three changes affecting HNTIDZ enterprises: (1) a new tax was levied, which is calculated on the basis of added values of finished products; (2) HNTIDZ enterprises are no longer exempt from construction tax or from tax on investment in fixed assets; and (3) exemption from and reduction of corporate income tax for newly established firms is now strictly limited to the first five years, with no possibility for extension (Beijing Experimental Zone for Development of New Technology Industries 1995). The first change hit HNTIDZ enterprises especially hard because high-tech companies produce mostly high value-added products, and much of the profit needs to be retained to support R&D of new products and to offset the losses incurred from unsuccessful R&D activities. In April 1996, the exemption from import tariffs was canceled, and normal tariffs were restored on the import of all materials, parts, equipment, and instruments, even though they are used to make products for export (*People's Daily*, April 1, 1996:2).

Few countries in the world maintain permanent preferential policies, but China's withdrawal is thought to be much faster than expected. These policy changes, which apply to both domestic- and foreign-invested ventures, have stemmed from two important considerations. First, China has been negotiating with the World Trade Organization to become a full member, and the prospect is promising. Once accepted, China will be obliged to bring down most trade barriers and remove many forms of government protection of its industries. In the future, HNTIDZs must not rely on preferential policies but need to find their own means of raising capital. Second, the Chinese government wants to discourage quick profit-makers that rely on imported equipment and materials and on cheap Chinese labor for assembly of high-tech products and to encourage those producers that use Chinese-made equipment and domestic materials to manufacture export products, and accordingly, the exemption from export tariffs is unchanged.

China's HNTIDZs are still in their infancy, and their role in the country's economic development cannot yet be clearly stated. Because of financial constraints and the high-risk nature of high-technology development, there is no guarantee that all of the fifty-two HNTIDZs will succeed. The experiences of other countries have shown that it takes considerable time to build technopoles that are able to yield significant results. For example, it took Silicon Valley twenty years to lay the basic foundations; Japan planned to spend twenty to thirty years to complete its technopolises. China would need at least the same amount of time to build its HNTIDZs. Even if only half of the HNTIDZs succeed, they will make invaluable contributions to the transformation of the national economy. Their success will also contribute to an accelerated shift of the center of world economic development to Pacific Asia, where Japan has already generated a strong gravity force.

Notes

1. The seaboard cites with ETDZs are Fuzhou, Guangzhou, Zhanjiang, Qinhuangdao, Lianyungang, Nantong, Dalian, Qingdao, Yantai, Tianjin, Ningbo, and Shanghai (three ETDZs were established in Shanghai).

2. The four Chinese scientists are Wang Daheng, Wang Ganchang, Yang Jiachi, and Chen Fangyun.

3. The seven priority areas are biotechnology, information technology, automation technology, energy technology, advanced material technology, space technology, and laser technology.

4. The Beijing HNTIDZ is located in the northwest quadrant of the city. This is a built-up area of 100 square kilometers with an unusually high concentration of intellectual resources; it includes more than 50 universities and 130 research institutes, with about 300,000 people engaged in scientific research and development of new technologies. Around 1985, a number of researchers and professors from this area broke through the old system and pioneered some small-scale enterprises to commercialize the laboratory achievements of their scientific research. Most of these enterprises were engaged in production of electronics and were concentrated on one street in Zhongguancun (a local community). The street hence got the nickname of "Zhongguancun Electronics Street." The goal of the pioneers was rather ambitious: to build this street into China's Silicon Valley. When the Chinese government decided to establish the

first high-tech industry development zone in 1988, the area centered on Zhongguancun Electronics Street became the first and only candidate.

5. All SEZs are in coastal provinces. Until 1992, all ETDZs were also in coastal provinces, but between 1992 and 1994, eighteen more ETDZs were established, many of which are in interior provinces.

6. Mainland China is divided into 22 provinces, 5 autonomous regions, and 4 municipalities directly under the central government, all having equal political status. The 4 municipalities are Beijing, Shanghai, Tianjin, and Chongqing, but Chongqing did not become such a municipality until early 1997.

7. An example of the first type is Saskatchewan's Innovation Place in Canada, where the University of Saskatchewan leased a parcel of land to Saskatchewan Economic Development Corporation, which in turn developed and has been operating and maintaining the technology park. Sheridan Park in Mississauga, Canada is an example of the second type.

8. The correlation analysis was conducted for the 52 HNTIDZs instead of the 6 types of enterprise in order to make use of a much larger sample size for more reliable results. Since both investment in infrastructure and R&D spending fluctuated from year to year, and their effects are long-term and cumulative, their three-year totals (1993, 1994, and 1995) were used for the correlation analysis. The 1992 data were excluded due to their incompleteness. Investment in infrastructure refers to expenditures on such things as land, offices, industrial buildings, utility lines, and residences for HNTIDZ workers.

9. The 18 new ETDZs are in Yingkou, Weihai, Wenzhou, Fuqing, Zhangzhou, Kunshan, Wuhan, Wuhu, Hangzhou, Chongqing, Changchun, Harbin, Shenyang, Huizhou, Xiaoshan, Fanyu, Beijing and Urumqi (Source: *Progress of Economic and Technological Development Zones,* 1995).

References

Allen, D. N. 1985. An Entrepreneurial Marriage: Incubators and Startups. In *Frontiers of Entrepreneurship Research,* ed. J. A. Hornaday, E. B. Shils, J. A. Timmons, and K. H. Vesper. Wellesley, Mass.: Centre for Entrepreneurial Studies, Babson College.

Beijing Experimental Zone for Development of New Technology Industries. 1994. *Beijing HNTIDZ Annual Report, 1993–1994.* Beijing.

_____. 1995. *Beijing HNTIDZ Annual Report, 1995.* Beijing.

Beijing Municipal Government. 1988. *Interim Regulations on Relevant Policies for the Beijing HNTIDZ.* Beijing.

Bloom, J. L., and S. Asano. 1981. Tsucuba Science City: Japan Tries Planned Innovation. *Science* 212:1239–1247.

Bylinsky, G. 1986. The High Tech Race: Who's Ahead? *Fortune* 114(8): 26–44.

Castells, M., and P. Hall. 1994. *Technopoles of the World: The Making of 21st Century Industrial Complexes.* London and New York: Routledge.

Cooper, A. C. 1985. The Role of Incubator Organizations in the Founding of Growth-Oriented Firms. *Journal of Business Venturing* 1:75–86.

Cooper, A. C., W. C. Dunkelberg, and R. S. Furuta. 1985. Incubator Organizations: Background and Founding Characteristics. In *Frontiers of Entrepreneurship Research 1985,* ed. J. A. Hornaday, E. B. Shils, J. A. Timmons, and K. H. Vesper. Wellesley, Mass.: Centre for Entrepreneurial Studies, Babson College.

Dickson, D. 1988. Ureka. *Technology Review* 91(6):26–33.

Economic Areas in China. 1993. Beijing: New Star Press.

Edgington, D. W. 1989. New Strategies for Technology Development in Japanese Cities and Regions. *Town Planning Review* 60:1–27.

Florida, R., and M. Kenney. 1988a. Venture Capital, High Technology, and Regional Development. *Regional Studies* 22:33–48.

_____. 1988b. Venture Capital-Financed Innovation and Technological Change in the USA. *Research Policy* 17:119–137.

_____. 1988c. Venture Capital and High Technology Entrepreneurship. *Journal of Business Venturing* 3:301–319.

Fujita, K. 1988. The Technopolis: High Technology and Regional Development in Japan. *International Journal of Urban and Regional Research* 12:556–594.

Glasmeier, A. K. 1988. The Japanese Technopolis Program: High-Tech Development Strategy or Industrial Policy in Disguise? *International Journal of Urban and Regional Research* 12:268–284.

Gu, C. L. 1995. A Preliminary Investigation of China's High-Tech Parks. Beijing: Institute of Geography, Chinese Academy of Science. (An internally circulated research paper)

Liles, P. R. 1974. *New Business Ventures and the Entrepreneur.* Homewood, Ill.: Richard D. Irwin.

Ma, H., and Z. Fang. 1995. *The Present and Prospect of Regional Economic Development in China.* Beijing: Economic Management Press.

Markusen, A. R., P. Hall, and A. Glasmeier. 1986. *High-Tech America.* Boston: George Allen and Unwin.

Masser, I. 1990. Technology and Regional Development Policy: A Review of Japan's Technopolis Programme. *Regional Studies* 24:41–53.

Monck, C. S. P., R. B. Poter, P. R. Quintas, D. J. Storey, and P. Wynarczyk. 1988. *Science Parks and the Growth of High Technology Firms.* London: Croom Helm.

Nishioka, H., and A. Takeuchi. 1987. The Development of High Technology in Japan. In *The Development of High Technology Industries: An International Survey,* ed. M. J. Breheny and R. W. McQuaid. London: Croom Helm.

OECD. 1980. *Science and Technology Policy for the 1980s.* Paris: Organization for Economic Co-operation and Development.

Office of the Torch Program. 1993. *1992 Statistics on HNTIDZs.* Beijing: State Science and Technology Commission.

_____. 1994. *1993 Statistics on HNTIDZs.* Beijing: State Science and Technology Commission.

_____. 1995. *1994 Statistics on HNTIDZs.* Beijing: State Science and Technology Commission.

_____. 1996. *1995 Statistics on HNTIDZs.* Beijing: State Science and Technology Commission.

Peterson, T. 1990. Suddenly, High Tech Is a Three-Way Race. *Business Week* (June 16):118–123.

Progress of Economic and Technological Development Zones. 1995. Beijing: New Star Press.

Rogers, E. M., and J. Larsen. 1984. *Silicon Valley Fever: Growth of High Technology Culture.* New York: Basic Books.

Smilor, R. W., and M. D. Gill. 1986. *The New Business Incubator.* Lexington, Mass.: Lexington Books.

State Administration of Taxation. 1991. Regulations on the Tax Policy for the National Development Zones for New and High Technology Industries. In *Introduction to China's High and New Technology Industrial Development Zones.* Beijing: State Science and Technology Commission.

State Council. 1984. *Provisional Regulations on Corporate Income Tax and Industrial and Commercial Consolidated Tax for Joint Ventures, Cooperative Enterprises, and Enterprises with Sole Overseas Investment in Special Economic Zones and the Economic Technological Zones in 14 Coastal Cities.* Beijing.

State Science and Technology Commission. 1991. Interim Regulations on Relevant Policies for the National Development Zones for New and High Technology Industries. In *Introduction to China's High and New Technology Industrial Development Zones.* Beijing: State Science and Technology Commission.

_____. 1992. *The Main Programs of Science and Technology.* Beijing.

Tasuno, S. 1986. *The Technopolis Strategy: Japan, High Technology, and the Control of the Twenty-First Century.* New York: Prentice Hall.

Tyebjee, T. T., and A. V. Bruno. 1984a. Venture Capital: Investor and Investee Perspectives. *Technovation* 2:185–208.

_____. 1984b. A Model of Venture Capitalist Investment Activity. *Management Science* 30:1051–1066.

Wang, J., and J. H. Bradbury. 1986. The Changing Industrial Geography of the Chinese Special Economic Zones. *Economic Geography* 62:307–320.

Worthington, J. 1982. Industrial and Science Parks: Accommodating Knowledge-Based Industries. *Planning for Enterprise, Proceedings of an International Seminar.* Swansea.

Zhang, C. Y. 1993. *Cradles of China's High and New Technology Industries.* Beijing: State Science and Technology Commission.

10

Foreign Direct Investment in the North China Coastal Region

Chi Kin Leung

China has become a major destination of international investment in recent years. During 1979–1995, China attained a total of US$229 billion in realized foreign investment, which refers to investments from both foreign countries and Chinese territories with different economic systems, including Hong Kong. Approximately three-fourths of the investment was rendered between 1990 and 1995 as a result of the further opening of the Chinese economy and the growing sophistication of the investment environment (SSB 1996). The primary form of foreign investment in China is direct investment, which is of three basic types: sole proprietorship, equity joint venture, and contractual joint venture. Together these three types accounted for 58.1 percent of the total amount of realized foreign investment (hereafter total realized foreign investment) during 1979–1995 (SSB 1996). Foreign investment in China concentrates geographically in the southern and eastern coastal regions, with Guangdong, Fujian, Jiangsu, and Shanghai sharing 61.1 percent of the total amount of realized foreign direct investment (hereafter referred to as total realized direct investment) during 1979–1995. Nevertheless, foreign investment has gradually diffused to other coastal provinces and municipalities in China. The proportions of total realized direct investment during 1979–1989 and 1990–1995 shared by the North China coastal region, which comprises the municipalities of Beijing and Tianjin and the province of Hebei, were 7.0 and 8.4 percent, respectively (SSB 1986–1996, 1990a).

Despite these developments, there remains a lack of detailed understanding of the locational and sectoral composition of foreign investment in China at the subregional or subprovincial levels; the factors underlying the foreign investment patterns and the corresponding local economic impacts within regions or provinces cannot be discerned. Based on an analysis of provincial foreign investment data and the information of 2,417 direct ventures in the North China coastal region, this chapter has three primary objectives. The first is to detail the locational and sectoral composition of foreign direct investment (FDI) in the North China coastal region. The spatial distribution of FDI with respect to three types of areas—city proper and inner suburbs, outer suburbs, and county (or prefecture-level cities, county-level cities, and counties for Hebei)—will be specifically elaborated. The second is to examine the locational and sectoral determinants of FDI in the region. The subtle factors underlying the patterns of FDI in the region will be highlighted. The third is to assess the local economic impacts of FDI in the region. The development potential of the localities in the region will be briefly evaluated.

This chapter has five major sections. The first provides a brief conceptual overview of the locational and sectoral patterns of foreign investment at the subregional levels and the corresponding determinants. The second elaborates the environment for foreign investment in the North China coastal region and the data and methodology. The third section details the locational and sectoral composition of FDI in the region. The fourth section then examines the corresponding locational and sectoral determinants, followed by the last section on the local economic impacts of foreign investment in the region.

Conceptual Background

The locational distribution of foreign investment at the subregional levels generally exhibits two major tendencies—concentration in major economic centers and in special investment zones (Hall 1988; Hill and Munday 1992). The concentration in major economic centers is attributable primarily to the better infrastructures and local market potential of these areas, whereas that in special investment zones is attributable to the lower tax and tariff schedules of these localities (Hill and Munday 1994; Lim and Fong 1991). These locations offer foreign firms not only lower investment overhead but also reduced local market entry and foreign trade barriers. Foreign firms, however, may incur higher wage expenditures in major economic centers or developed special zones, but these cost disadvantages can be partly compensated for by better labor quality and productivity in these locations (Coughlin, Terza, and Arromdee 1991; Glickman and Woodward 1987). The tendency to concentrate in major economic centers and special zones is especially evident in developing regions or economies, as the provision of infrastructure and incentives is usually confined to these two types of locations (Abumere 1982; Herrin and Pernia 1987). The tendency to concentrate in major economic centers is also especially evident among nonmanufacturing activities. This is so because of the greater local market orientation and information requirements of these industries (Bagchi-Sen 1991; Leung 1990).

With respect to sectoral composition, foreign investment in major economic centers or special zones usually congregates around certain industries. Such congregation is attributable mainly to the specific structure and locational specialization of indus-tries in major economic centers and to the peculiar development orientation and incentive structures of special zones (Leung 1996). The sectoral composition of for-eign investment in these locations is further determined by the constitution of major source economies there and the industrial specialization of the major source economies (Dunning 1988, 1993). Locations with inflows dominated by a few major source economies tend to have a sectoral composition of foreign investment mirror-ing the industrial structures of those economies. Nevertheless, the sectoral composi-tion of foreign investment in major economic centers is usually more diversified. This is so because of the greater economic complexities of these central locations.

The Environment for Foreign Investment in the North China Coastal Region

The North China coast, defined in this study as the area comprising the municipal-ities of Beijing and Tianjin and the province of Hebei, is a major economic region in China. In 1995 this region shared 8.9 and 8.7 percent of the nation's gross domes-tic product and total gross value of industrial output (GVIO), respectively (SSB 1996, 1997). The cities of Beijing, Tianjin, and to a lesser extent Shijiazhuang, Tangshan, and Handan are the major economic or industrial centers in the region. At the beginning of major foreign investment inflows in 1986, these five major cities combined shared 74.2 percent of the total GVIO of the region (SSB 1987a). The major industries in the cities of Beijing and Tianjin included chemicals, machinery, ferrous metal processing, textiles, transport equipment, and electronics, whereas those in Shijiazhuang, Tangshan, and Handan included textiles, ferrous metal pro-cessing, and machinery. In aggregate, of the six major industries indicated above, chemicals, ferrous metal processing, electronics, and to a lesser extent transport equipment had higher degrees of locational specialization.

With the designation of Tianjin as an open coastal city in 1984, an increasing number of special investment zones were established to promote foreign investment in the region (Figure 10.1). These special zones are of two major types: Economic and Technological Development Zones (ETDZs) and High and New Technology Industry Development Zones (HNTIDZs). They can be designated either by the central, provincial, or local governments. Both types of special zones generally offer a low corporate income tax rate of 15 percent and a tax-free period of at least two years, with HNTIDZs emphasizing foreign investment in advanced technology industries such as electronics and telecommunications. The largest special invest-ment zones in the region include the Tianjin ETDZ, the Beijing HNTIDZ, and the Beijing ETDZ. By mid-1993, the region had a total of twenty-six national- and provincial-level ETDZs and HNTIDZs (Editorial Board of DFDVC 1994). In addition, the region has a Free Trade Zone (FTZ) located at the port of Tianjin, which was established in 1991 for promoting international trading ventures.

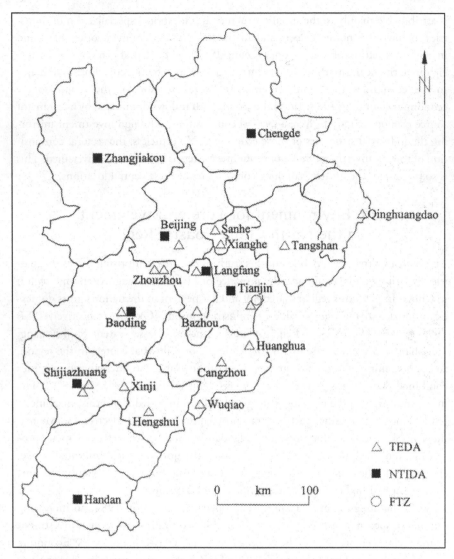

FIGURE 10.1   Special Investment Zones in the North China Coastal Region

Data and Methodology

This study employs two sets of data to investigate the detailed locational and sectoral composition of foreign investment in the North China coastal region. The first set of data was compiled from the provincial statistical yearbooks of Beijing, Tianjin, and Hebei; it contains the total realized direct investment for all the city districts and counties of Beijing and the total realized foreign investment for all the cities and counties of

Hebei for the period 1993 to 1995 (BJMSB 1994–1996; HPSB 1994–1996). It also contains the total amount of intended direct investment for all the city districts and counties of Tianjin for the period 1994 to 1995 (TJMSB 1995–1996). This set of data provides a relatively comprehensive picture of the locational patterns of foreign investment in the region, as the period accounted for approximately 70 percent of the inflows during 1979–1995 (SSB 1986–1996, 1990a). The second set of data was compiled from the *Directory of Foreign Direct Ventures in China* (Editorial Board of DFDVC 1994). This set of data contains the city or county location, product or service characteristics, source economy composition, form of investment, amount of registered capital, and year of establishment of a total of 2,417 direct ventures in the region during 1979 to mid-1993. This set of data accounted for approximately 12.9 and 15.1 percent of the total number and the total amount of registered capital (hereafter total registered capital) of enterprises with foreign investment in the region in that period, respectively; it is the single most comprehensive source of information concerning the detailed locational and sectoral characteristics of FDI in the region (SSB 1994, 1995).

In the following sections, the data is tabulated based primarily on a three-tier locational framework. For Beijing and Tianjin, this includes 1) the city proper and inner suburbs (inner suburbs are the districts that border the respective city proper area), 2) outer suburbs, and 3) county areas; for Hebei, this includes 1) prefecture-level cities, 2) county-level cities, and 3) counties. The data is also tabulated based on the sectoral framework established by the China Standard Industrial Classification at the two-digit level (NTSB 1984), with multiple regression analysis constituting the basis of investigation on the locational and sectoral determinants of FDI in the region. Given its greater significance on long-term economic growth, the following analysis focuses on foreign manufacturing investment in the region.

The Development of FDI in the North China Coastal Region

During 1979–1995, the North China coastal region attained a total of US$18.8 billion in realized foreign investment, accounting for 8.2 percent of the national total (SSB 1996). Approximately 59 percent of the realized investment was direct investment, with equity joint venture the most preferred mode of entry by foreign firms. A significant portion of the inflows occurred between 1990 and 1995. This five-year period accounted for 77.5 percent of the total realized foreign investment during 1979–1995. There were 21,107 registered enterprises receiving foreign capital in the region by 1995, of which 9,115 were direct ventures with US$11.1 billion of total realized direct investment (BJMSB 1996; HPSB 1996; SSB 1996; TJMSB 1996).

The majority of direct ventures in the region were manufacturing investments. However, manufacturing activities constituted only 54.1 percent of total realized direct investment because of the large capital outlay of several dozen major hotel and real estate ventures (Editorial Board of DFDVC 1994). The three largest source economies were Hong Kong, Japan, and the United States, accounting for approximately 40, 11, and 10 percent of the total realized direct investment in the region during 1979–1995, respectively (BJMSB 1989–1996; HPSB 1989–1996; TJMSB 1996).

Locational Composition

Foreign direct investment in the North China coastal region concentrated in the municipalities of Beijing and to a lesser extent Tianjin. These two municipalities shared 82.0 percent of the total realized direct investment and 73.5 percent of the total number of direct ventures in the region (Table 10.1). Foreign direct ventures in Beijing clustered in the city proper and inner suburbs, whereas those in Tianjin clustered in both the city proper and inner suburban area and in the Tianjin TEDA. The counties of these two municipalities that had high levels of foreign investment include Shunyi, Pinggu, Tongxian, Wuqing, and Jinghai, most of which are situated along the major transportation arteries between the two cities. Of the two activity categories, manufacturing investment was slightly more county-oriented than non-manufacturing investments in the two municipalities (Table 10.2).

In Hebei, foreign direct investment concentrated in the provincial capital of Shijiazhuang and in the major prefecture-level cities of Tangshan and Qinghuangdao (Figure 10.2). Together these three cities shared 38.3 percent of the total realized foreign investment in the province during 1993–1995. The other areas that had moderate levels of foreign investment included the cities of Baoding, Langfang, Zhuozhou, and Fengnan. Given the larger geographical extent, foreign investment in Hebei was spatially more dispersed, with approximately 36 percent of the registered capital of the analyzed ventures located in the county-level cities and counties in the province.

As with the overall pattern, direct investments from each of the three major source economies—Hong Kong, Japan, and the United States—concentrated in Beijing and to a lesser extent Tianjin. The three economies, however, differed slightly in the provincial orientation of their investments, with Hong Kong and Japan focusing more in Beijing than the United States. The proportions of total real-ized direct investment from Hong Kong, Japan, and the United States in the Beijing region during 1990–1995 were 53.4, 57.7, and 42.8 percent, respectively (BJMSB 1991–1996; HPSB 1991–1996; TJMSB 1991–1996). The three economies also differed slightly in the subprovincial distribution of their investments, as Hong Kong's investments in Beijing, especially manufacturing investments, were more dis-persed in the county area of the municipality than Japanese or American invest-ments (Table 10.3). By contrast, Hong Kong's investments in Hebei were spatially more concentrated in the prefecture-level cities than Japanese or American invest-ments. The differences in the subprovincial distribution of the investments in Tianjin between the three source economies cannot be detailed because of data

TABLE 10.1   Provincial Distribution of Foreign Investment in the North China Coastal Region, 1995

| | Realized Investment[a] | | | Number of Enterprises[b] | | |
| | | Direct Investment | | | Direct Investment | |
	Total	Total	Manufacturing	Total	Total	Manufacturing
Beijing	8.0	5.3	2.2	8,855	4,002	3,088
Tianjin	7.9	3.8	2.4	7,599	2,701	2,035
Hebei	2.9	2.0	1.4	4,653	2,412	2,083
Total	18.8	11.1	6.0	21,107	9,115	7,206

[a]Amount of realized investment during 1979–1995, in US$ billions. Total (columns 1 and 4) includes foreign loans and direct and indirect ventures.
[b]Number of enterprises receiving foreign capital.

SOURCES: BJMSB 1989–1996; HPSB 1996; SSB 1986–1996, 1990a; TJMSB 1996.

TABLE 10.2   Subregional Distribution of FDI in the North China Coastal Region

| | | Percentage of Registered Capital | | | | |
| | | | Municipality/Prefecture-Level City | | | |
	Registered Capital (US$ billion)	Total	City Proper/ Inner Suburbs	Outer Suburbs	County-Level City	County
Beijing						
Total	1.67	78.1	76.6	1.5	n.a.	21.9
Manufacturing	1.00	72.2	69.8	2.4	n.a.	27.8
Tianjin[a]						
Total	1.34	88.2	42.9	45.3 (31.3)	n.a.	11.8
Manufacturing	0.94	84.6	44.6	40.0 (33.9)	n.a.	15.4
Hebei						
Total	0.47	64.0	n.a.	n.a.	12.2	23.8
Manufacturing	0.45	64.5	n.a.	n.a.	12.6	22.9

[a]Figures within parentheses are the corresponding data for Tianjin TEDA.
SOURCE: Compiled from the 2,417 direct ventures listed in the Directory of Foreign Direct Ventures in China (Editorial Board of DFDVC 1994).

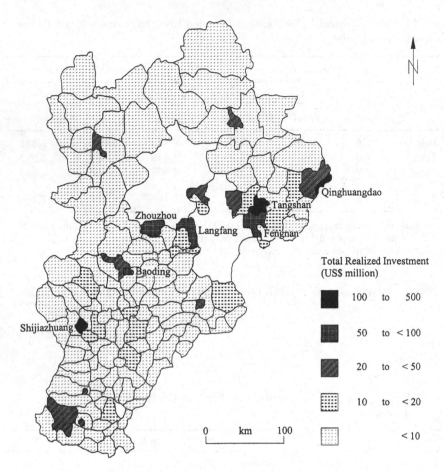

FIGURE 10.2   Foreign Direct Investment in the North China Coastal Region

constraints, but American investments in the municipality were more concentrated in the Tianjin TEDA than Hong Kong or Japanese investments (SPB 1995).

Sectoral Composition

With respect to sectoral composition, foreign direct investment in manufacturing in the North China coastal region concentrated in electronics and to a lesser extent transport equipment, textiles, metal products, apparel, nonmetallic mineral processing, and plastic products. Together these seven industries shared 38.9 percent of the total registered capital of the 2,417 ventures compiled from the *Directory of Foreign Direct Ventures in China.* In nonmanufacturing, foreign direct investment focused on real estate and hotels, which accounted for another 22.8 percent of the total registered capital of the ventures.

TABLE 10.3   Subregional Distribution of FDI in Beijing and Hebei by Major Source Economies ($billion)

Municipality/Province	Hong Kong	Japan	United States	Other Asian	Other Western
Beijing					
	0.77	0.27	0.26	0.08	0.15
	(0.30)	(0.23)	(0.18)	(0.04)	(0.14)
City proper/	71.8	82.4	80.4	90.2	89.1
inner suburbs (%)	(54.6)	(80.4)	(72.5)	(79.6)	(90.5)
Outer suburbs (%)	2.5	0.9	0.6	0.3	0.8
	(6.4)	(1.0)	(0.3)	(0.7)	(0.8)
County (%)	25.7	16.7	19.0	9.5	10.1
	(39.0)	(18.6)	(27.2)	(19.7)	(8.7)
Hebei					
	0.27		0.20		
	(0.26)		(0.18)		
Prefecture-level cities	68.4		58.1		
(%)	(67.3)		(60.2)		
County-level cities	12.5		12.0		
(%)	(12.9)		(12.5)		
Counties (%)	19.1		29.9		
	(19.8)		(27.3)		

NOTES: Figures outside parentheses are the corresponding data for all ventures; those within are for manufacturing ventures. Tianjin is not included in this table because of a lack of source economy information in the municipality data set.

SOURCE: Compiled from the 2,417 direct ventures listed in the *Directory of Foreign Direct Ventures in China* (Editorial Board of DFDVC 1994).

Given the dominance of the two municipalities, a significant portion of the investment in the electronics, transport equipment, and real estate and hotel sectors was located in Beijing and Tianjin, with most of the activities clustering in the city proper and inner suburbs. The only exception to this pattern was investment in the electronics sector in Tianjin, which was located primarily in the outer suburbs where the TEDA is located. The other industries that had high levels of direct investment in Beijing included apparel and ferrous metal processing, whereas those in Tianjin included textiles, metal products, and import-export trade. Unlike their heavy industrial or real estate counterparts, investments in the apparel industry in Beijing were located mainly in the county area, especially Pinggu, while foreign trade ventures in Tianjin clustered primarily in the FTZ.

Foreign direct investment in Hebei concentrated in textiles, nonferrous metal processing, metal products, and nonmetallic mineral processing. The majority of investments in these sectors were located in prefecture-level cities, with Handan,

Qinghuangdao, and Tangshan accounting for a high proportion of investments in textiles, nonferrous metal processing and metal products, and nonmetallic mineral processing in the province, respectively. The sectoral composition of FDI was more diversified in Shijiazhuang than in other prefecture-level cities, although the provincial capital had a larger share of the electronics investments in the province.

As regards the sectoral composition of the major source economies, Hong Kong's investments in the region focused on real estate and hotels and to a lesser extent transport equipment and the apparel industry, whereas Japanese and American investments both concentrated in the transport equipment and electronics industries. The major transport equipment ventures of Hong Kong and the United States are located in Beijing. They include Beijing Light Automotive, Beijing Jeep, and Beijing Warner Gear. Japan, on the other hand, has its major transport equipment venture—Tianjin Toyota Motor—in Tianjin. In the electronics industry, the most important American venture—Motorola (China) Electronics—is in Tianjin's TEDA, while the largest Japanese venture—Beijing Matsushita Color Lanescope—is in Beijing's Chaoyang district. The other prominent ventures in Beijing include Beijing Shougang Baosheng Strip Steel, Beijing International Switching, and Babcock & Wilcox Beijing; those in Tianjin include Tianjin Sanfeng Minibus and China Tianjin Otis Lifts. As regards nonmanufacturing activities, most of Hong Kong's real estate and hotel investments are in the city proper and inner suburbs of the Beijing municipality. Although the sectoral composition of the major source economies in Hebei cannot be detailed because of data constraints, limited information indicates Hong Kong's investments in the province concentrate in nonferrous metal processing and the textile industry (Editorial Board of ACFERT 1990–1995).

It is important to note that since mid-1993 the region has received a significant number of large manufacturing ventures. For example, Novo Nordisk (a biotechnology company) and Nestlé (a food processor) have established facilities in Tianjin, while a Hong Kong firm has built an engine parts venture and a British pipe-casting business has invested in Hebei. Some of the other recent large ventures include Beijing Ericsson Communication Systems, Tianjin Yingchong Musical Instruments, and Baoding Jinfeng Battery.

In addition to these large investments, the majority of global companies have also established their ventures or offices in the region. Of the Global Fortune 500 manufacturing companies that have not previously been mentioned, Akzo Nobel, AT&T, Hewlett Packard, IBM, and Siemens each have at least three operations in the Beijing-Tianjin area. Other Global 500 manufacturing companies that have a strong presence in the region include Daewoo, Mitsubishi, NEC, and Philips.

Locational and Sectoral Determinants

The locational composition of FDI in the North China coastal region generally corresponds to the patterns of foreign investment in other regions, with major economic centers as the primary areas of concentration. To examine the relative importance of major locational determinants of FDI in the region, the total realized foreign investment during 1993–1995 as a dependent variable was regressed on a number of population and economic parameters, which are listed in Table 10.4, for all the cities and counties in Hebei. Data constraints precluded the city and county units of Beijing and Tianjin from being analyzed. The analysis generated the following regression model, with the figures in parentheses representing the standardized regression coefficients for the corresponding variables:

$$FI = 0.00062 + 0.1873 \text{ NAPOPN} + 0.0137 \text{ SINV}$$
$$(0.7537) \qquad\qquad (0.1160)$$
$$(R^2 = 0.6883, \text{ F} = 161.189, \text{ P} < 0.001)$$

The model shows that strong infrastructures and local market potential (indicated by NAPOPN) were the most important determinant of the subprovincial distribution of foreign investment in Hebei. The inclusion of NAPOPN rather than TGVIO in the model suggests that human infrastructural factors (such as structural diversity of the nonagricultural workforce) were more important than their physical counterparts (such as scale of industrial production) in affecting the locational considerations of especially foreign manufacturing ventures in the province. The productivity and average annual wage of workers, however, did not appear to be significant location decision factors. This situation is reflected by the weak to moderate correlation coefficients between FI and PROD ($r = 0.3233$), and FI and WAGE ($r = 0.4715$). Structural diversity of the nonagricultural workforce, or human resources endowment, is similarly evident as a primary locational determinant of foreign investment in the Beijing-Tianjin area. This is not surprising, for this area is one of the most important educational and research centers in China. In 1995 the area had 190,690 scientists and engineers, accounting for 14.1 percent of the national total (SSB 1997a). This concentration of technology personnel is attractive to especially foreign complex technology investments, as reflected by the clustering of electronics ventures, including research-oriented ventures, in the area.

The model also shows that the presence of special investment zones (indicated by SINV) was another important determinant of the subprovincial distribution of foreign investment in Hebei. Aside from lower tax and tariff schedules, these zones further provide foreign firms necessary infrastructure and simpler investment environments, as they are administered by a limited number of designated agencies for the provision of a wide range of services and support activities (Editorial Board of DFDVC 1994; SPB 1995). The significance of special zones in promoting foreign investment is also evident in the Beijing-Tianjin area, especially Tianjin. Since all the major special zones in the North China coastal region are situated within the respective city limits, the growth of these zones reinforces the concentration of FDI

TABLE 10.4   Variables for Modeling the Locational Determinants of Foreign
Investment in the North China Coastal Region

Variable	Unit	Description
Dependent		
FI	US$ billion	Total amount of realized foreign investment during 1993–1995
Independent		
TPOP	million persons	Total population, 1990
NAPOPN	million persons	Nonagricultural population, 1990
TGVIO	billion yuan	Total gross value of industrial output, including the output of village and individual enterprises, 1990
PSGVIO	percent	Gross value of industrial output of the state sector as a percentage of total gross value of industrial output, 1990
SINV	0 or 1	Special investment incentives (1 if a city or county has a TEDA, NTIDA, or FTZ. 0 otherwise)
PROD	yuan/person-year	Labor productivity, 1990
WAGE	yuan	Average annual wage of workers, 1990

in the city proper and suburban areas of the two municipalities and in the key cities of the province.

The sectoral composition of FDI in the North China coastal region also generally corresponds to the patterns of foreign investment in other regions, with the sectors of higher degrees of locational specialization as the primary areas of congregation. Although attempts to model the sectoral determinants of foreign investment did not materialize because of data constraints, the importance of locational specialization as a determinant was highlighted by its moderate to strong association with the level of foreign investment in the region. Of the five manufacturing industries with the largest shares of the total registered capital of the analyzed ventures, only textiles lacked adequate levels of locational specialization.

It should be noted that the locational and sectoral composition of foreign investment in the region was also influenced in part by the peculiar organizational environment of the two municipalities, especially Beijing. As the nation's capital, Beijing is the location of all state organs and a number of key state enterprises. These state organs and enterprises usually have extensive resource bases and domestic networks; they are capable of collaborating with large foreign companies, especially global companies, in major ventures. These ventures in turn engender further foreign investment inflows because of their demands on semi-manufactures. This is especially true for ventures that have local contents requirements because of their domestic market orientation. Beijing Jeep, for example, is an equity joint venture between Chrysler and Beijing Automotive (a key state enterprise); it led to a number of automobile parts supplier ventures including Beijing Warner Gear (a Sino-American equity joint venture), Beijing Aohua Automotive Parts (a Sino-Australian equity joint venture), and Beijing Changsheng Muffler (an equity joint venture by a Hong

Kong firm) (Editorial Board of DFDVC 1994). In 1986 before the inflows of major ventures, Beijing had a total of 416 large and medium-sized, mostly state-owned manufacturing enterprises. Together with Tianjin, the area shared 8.4 percent of the total number of large and medium-sized manufacturing enterprises in the nation at that time (SSB 1987). Nevertheless, it is important to emphasize that large state enterprises or state-dominated local economies are not necessarily conducive to foreign investment inflows because of their higher degree of structural rigidity, as reflected by the weak relationship ($r = 0.3933$) between the total realized foreign investment during 1993–1995 and the percentage of total GVIO accounted for by the state sector in 1990 for all the cities and counties in Hebei.

Local Economic Impacts

The inflows of FDI have led to an expansion and change of the local industrial systems in the region, especially in the Beijing-Tianjin area where direct ventures are concentrated. During 1990–1995, the GVIO of foreign ventures in the two municipalities combined increased more than tenfold from 8.1 to 86.3 billion yuan, contributing significantly to the average annual GVIO growth rate of 13.0 percent in Beijing and of 17.9 percent in Tianjin in that period. The two industries where foreign ventures congregated—electronics and transport equipment—were the centers of growth. The GVIO of the two industries combined in the area increased from 14.4 billion yuan in 1990 to 61.3 billion yuan in 1995. As a result, the levels of foreign ownership and locational specialization of the industrial systems in the Beijing-Tianjin area have increased. This is especially true in Tianjin, where the GVIO of foreign ventures surpassed that of state enterprises in 1995.

The inflows of FDI have also led to a technologically more sophisticated workforce in the Beijing-Tianjin area, as a significant number of ventures there are complex technology producers. In addition to production activities, most complex technology ventures maintain certain in-house research and development functions for the Chinese and Asian markets. Several large global ventures such as Motorola have further established research centers or ties with major universities in the area. In 1995, foreign direct manufacturing ventures employed approximately 550,000 persons in the area, accounting for 19.1 percent of the combined manufacturing workforce in the two municipalities (BJMSB 1996; TJMSB 1996).

The impacts of foreign investment on the local industrial systems in Hebei, on the other hand, were limited. During 1990–1994, the GVIO of foreign ventures in the province increased by approximately 13.1 billion yuan (at current prices; 9.9 billion yuan at 1990 constant prices), accounting for 11.3 percent of the GVIO growth at and above the township level in the period. Most of the impacts were further confined to the key cities of Shijiazhuang, Tangshan, and Qinghuangdao. Together these three city areas shared approximately 57 percent of the GVIO growth of foreign ventures during 1990–1994 (Table 10.5). Nevertheless, the province had an average

TABLE 10.5   Impact of Foreign Investment on Local Industrial Growth in Hebei, 1990–1994

City Area[a]	GVIO[b]		Absolute[c] GVIO Growth of Foreign Ventures 1990–1994 (billion yuan)	Contribution of Foreign Ventures to GVIO Growth 1990–1994 (%)	GVIO by Foreign Ventures(%)	
	1990	1994			1990	1994
Shijiazhuang	16.1	39.3	2.1	9.1	*	5.9
Tangshan	13.5	39.3	2.5	9.9	*	6.5
Qinghuangdao	3.2	10.6	2.9	39.2	*	28.3
Handan	10.4	24.6	1.2	8.5	*	4.9
Xingtai	6.0	12.9	0.2	2.9	*	1.6
Baoding	7.2	16.6	1.4	14.9	*	8.6
Zhangjiakou	6.4	12.9	0.4	6.2	*	3.0
Chengde	3.3	7.2	0.4	10.3	*	6.0
Cangzhou	4.5	12.7	1.3	15.9	*	10.5
Langfang	2.2	7.3	0.3	5.9	*	4.4
Hangshui Prefecture	3.2	8.2	0.4	8.0	*	4.9
Province total	76.0	191.6	13.1	11.3	*	6.8

[a] Including the counties that a city administers.
[b] GVIO data are at or above the township level, in billion yuan at current prices.
[c] At current prices.
* Insignificant.

SOURCES: HPSB 1991, 1995; SSB 1995.

annual GVIO growth rate in the period comparable to that of the Beijing-Tianjin area because of significant manufacturing investments from domestic sources. In aggregate, foreign direct investment has accentuated the economic disparities within the Beijing-Tianjin area and within Hebei, but it has not widened the gap in industrial production between the two municipalities and the province.

Conclusion

Foreign direct manufacturing investment in the North China coastal region concentrates in major economic centers and in complex technology sectors, with the city proper and inner suburbs of Beijing and Tianjin, the Tianjin TEDA, and the key cities of Hebei as the locational foci, and with electronics and transport equipment as the sectoral foci. The concentration in major cities is attributable primarily to strong infrastructure and local market potential and to the presence of special investment zones, whereas the congregation around electronics and transport equipment is ascribable mainly to the high degrees of locational specialization of the region in these two industries. In particular, human infrastructural factors such as the structural diversity of the nonagricultural workforce are important determinants of the subprovincial distribution of foreign investment in the region. This is especially true in the Beijing-Tianjin area, where a significant portion of China's technology personnel is concentrated. In nonmanufacturing activities, FDI concentrates in the city proper and inner suburbs of the two municipalities; it focuses on the real estate and hotel sectors.

The concentration of FDI in especially the Beijing-Tianjin area has led to a significant structural change of the industrial systems there, with foreign ventures and the electronics and transport equipment industries accounting for an increasing portion of local industrial outputs. Foreign investment has also led to a technologically more sophisticated workforce in the area, enhancing the long-term development capabilities of the two municipalities. The impacts of foreign investment on the local industrial systems in Hebei are limited and are confined to Shijiazhuang, Tangshan, and Qinghuangdao. Although foreign investments have accentuated the economic disparities within the Beijing-Tianjin area and within Hebei, they have not widened the gap in industrial production between the two municipalities and the province as a result of extensive domestic investments in manufacturing activities in the latter.

Given the determinants, the concentration of FDI in the major economic centers and special investment zones in the region is likely to continue. The diffusion of FDI to other parts of the region will be limited and will be confined primarily to the vicinities of the Beijing-Tianjin corridor and of the key cities in Hebei. Significant improvements in local infrastructure and the provision of extensive incentives and supports will be necessary in order to increase the inflows of foreign investment to other parts of the region, especially the less developed areas of Zhangjiakou and Chengde.

References

Abumere, S. I. 1982. Multinationals and industrialization in a developing economy: The case of Nigeria. In *The Geography of Multinationals,* ed. M. Taylor and N. Thrift, pp. 158–178. London: Croom Helm.

Bagchi-Sen, S. 1991. The location of foreign direct investment in finance, insurance, and real estate in the United States. *Geografiska Annaler* 73B(3):187–197.

BJMSB (Beijing Municipality Statistical Bureau). 1989–1996. *Beijing Statistical Yearbook.* Beijing: China Statistical Publications.

Coughlin, C. C., J. V. Terza, and V. Arromdee. 1991. State characteristics and the location of foreign direct investment within the United States. *Review of Economics and Statistics* 73(4):675–683.

Dunning, J. H. 1988. *Explaining International Production.* London: Unwin Hyman.
_____. 1993. *The Globalization of Business.* London: Routledge.
Editorial Board of ACFERT. 1990–1995. *Almanac of China's Foreign Economic Relations and Trade.* Beijing: China Society Publications.
Editorial Board of DFDVC. 1994. *Directory of Foreign Direct Ventures in China.* Volume 1. Beijing: Jinhua Publications.
Glickman, N. J., and D. P. Woodward. 1987. *Regional Patterns of Manufacturing Foreign Direct Investment in the United States.* Special Report for U.S. Department of Commerce, Economic Development Administration, Washington, D.C..
Hall, H. 1988. *Foreign Investment and Industrialization in Indonesia.* Singapore: Oxford University Press.
HPSB (Hebei Province Statistical Bureau). 1989–1996. *Hebei Statistical Yearbook.* Beijing: China Statistical Publications.
Herrin, A. N., and E. M. Pernia. 1987. Factors influencing the choice of location: Local and foreign firms in the Philippines. *Regional Studies* 21:531–541.
Hill, S., and M. Munday. 1992. The UK regional distribution of foreign direct investment: Analysis and determinants. *Regional Studies* 26(6):535–544.
Hill, S., and M. Munday. 1994. Foreign manufacturing investment in France and the UK: A regional analysis of locational determinants. *Tijdschrift voor Economische en Sociale Geografie* 86(4):311–327.
Leung, C. K. 1990. Locational characteristics of foreign equity joint venture investment in China, 1979–1985. *Professional Geographer* 42(4):403–421.
_____. 1996. Foreign manufacturing investment and regional industrial growth in Guangdong province, China. *Environment and Planning A* 28(3):513–536.
Lim, L.Y.C., and P. E. Fong. 1991. *Foreign Direct Investment and Industrialization in Malaysia, Singapore, Taiwan, and Thailand.* Paris: OECD.
NTSB (National Technology Supervisory Bureau). 1984. *Classification and Codes for National Economic Activities.* Beijing: China Standards Publications.
SPB (Statistics and Planning Bureau). 1995. *Annual Statistical Report of Tianjin Technology Economic Development Area.* Tianjin: Statistics and Planning Bureau.
SSB (State Statistical Bureau). 1986–1997. *China Statistical Yearbook.* Beijing: China Statistical Publications.
_____. 1987a. *China Urban Statistical Yearbook.* Beijing: China Statistical Publications.
_____. 1990a. *China Commerce and Foreign Economic Statistics 1952–1988.* Beijing: China Statistical Publications.
_____. 1997a. *China Statistical Yearbook on Science and Technology.* Beijing: China Statistical Publications.
TJMSB (Tianjin Municipality Statistical Bureau). 1991–1996. *Tianjin Statistical Yearbook.* Beijing: China Statistical Publications.

11

China in the Pacific Rim: Trade and Investment Links

Gang Xu

China is playing an increasingly important role in the world economy. This growing role stems from two related developments in China's economy since 1978: its rapid growth and its increasing integration with the world economy. Between 1978 and 1996 China's GDP grew at nearly 10 percent a year. China has thus quadrupled its real output within less than two decades. Even more impressive is the recent expansion of China's external sector. Between 1978 and 1996, China's merchandise trade—exports plus imports—expanded at a rate of 17 percent a year. China has also achieved remarkable success in attracting foreign direct investment (FDI). In the 1985–1996 period, realized FDI in China increased at 36 percent annually, with net FDI inflows in 1996 reaching $42 billion.

China's continued growth and opening up have established its importance in the world economy. Using purchasing power parity (PPP) exchange rates, China is now the world's third-largest economy, after the United States and Japan. It has become the world's tenth-largest trading nation and second-largest recipient of FDI, after the United States.

The recent rapid expansion of China's external sector has benefited from the industrial restructuring of major East Asian economies. China's continued growth and opening up have in turn contributed to the economic dynamism of the Pacific Rim region. The purpose of this chapter is to review recent developments in China's integration with the world economy, to explore China's role in the development of trade and investment links in the Pacific Rim region, and to speculate on its future development. Pacific Rim here refers to the region that encompasses 1) East Asia—Japan, China, the Asian NIEs (Newly Industrializing Economies), and the ASEAN-4; 2)

173

North America—the United States and Canada; and 3) Australasia—Australia and New Zealand. Unless otherwise stated, the Asian NIEs refer to South Korea, Hong Kong, Taiwan, and Singapore; the ASEAN-4 (Association of Southeast Asian Nations) refers to Malaysia, Indonesia, Thailand, and the Philippines.

China as a Trading Nation in the Pacific Rim

Foreign Trade and Exchange Regimes

Prior to the 1978 reform, China was largely isolated from the world economy. For nearly three decades after the founding of the People's Republic, the country pursued an inward-oriented industrialization strategy characterized by "self-reliance" and an import-substitution trade regime. The central government maintained a monopoly on foreign trade and an overvalued exchange rate. Foreign trade was simply a balancing item in China's national plans. The volume of imports was set by the state on the basis of domestic supply gaps in key industrial sectors. More than 90 percent of all imports were producer goods. Exports were seen as a means to generate foreign exchange to finance imports. More than half of exports were primary products. Twelve state-owned foreign trade corporations (FTCs) controlled all trading activities. Moreover, China's domestic market was kept insulated from the international market through an "airlock" system; administratively set domestic prices of traded goods bore no relation to world market prices (World Bank 1988).

The Chinese currency (renminbi) was greatly overvalued at the inception of China's trade reforms—a result of low domestic prices of consumer goods because of state subsidies. The overvalued exchange rate, combined with price distortions in the domestic market, built a strong bias toward imports and against exports. Exporting manufactured goods was rarely a profitable activity. The domestic currency losses incurred on exports were covered either by the profits from reselling imported manufactured goods in the domestic market or by state subsidies.

As a result of this rigid trade regime, foreign trade played a limited role in China's economy before 1978. In 1977 China's trade turnover was $14.8 billion, and its trade-to-GDP ratio was only 8.9 percent—one of the lowest in the world (State Statistical Bureau of China 1994; Lardy 1994).

A key element of China's economic reform drive has been its strategy of seeking greater integration with the world economy. Since the mid-1980s China has aggressively pursued an outward-oriented development strategy and has gradually reformed its foreign trade and exchange regimes.

Decentralization of Foreign Trade Authority.   China's trade reform began with a decentralization of foreign trade management. Two reform measures were crucial. The first was the sharp reduction in the share of trade controlled by state mandatory plans. The second was the decentralization of trading activities. By the mid-1990s the share of exports and imports covered by state plans had fallen to below 15 percent. The number of FTCs grew dramatically, from 12 in 1977 to 1,500 in 1986, and reached more than 5,000 in 1995. In addition to the FTCs, more than 140,000 foreign-funded enterprises (FFEs) and around 7,000 large state-owned enterprises (SOEs) have been granted the right to export and import on their own account (State Committee for Economic Restructuring 1995).

Provision of Export Incentives.   To counteract the export-discouraging effects of the overvalued exchange rate, China adopted a foreign exchange retention system in 1979 to allow local enterprises and governments to retain a certain portion of the foreign exchange they earned from exports. This provided an incentive for enterprises to increase their export sales because foreign exchange entitlements were essentially import entitlements that entailed a premium. Foreign exchange retention ratios were initially negotiated on a case-by-case basis and were raised in the course of China's trade reforms. By the mid-1980s the portion of the foreign exchange retained by local enterprises and governments had increased to over 40 percent. To meet the need for foreign exchange transactions, local swap markets were opened, first in Shenzhen in 1985 and later in all major commercial centers in China.

Foreign Trade Contract System.   The removal of administrative barriers to trade, combined with the domestic economic boom, stimulated a rapid growth of China's merchandise trade in the mid-1980s. However, China's partially reformed trade regime continued to favor imports over exports because of the disparities between domestic and international prices and the continued overvaluation of the renminbi. Moreover, under the retention system the amount of foreign exchange local enterprises could earn depended on their volume of exports, not the profitability of their operations. As a result, local enterprises sought to increase their export sales regardless of the size of their losses. This caused growing domestic currency losses. The *agency system,* which was introduced in 1984 and aimed to bring the domestic prices of traded goods closer to world market prices, proved to be of little relevance to most exporters (Fukasaku and Wall 1994). To limit its financial liability for trade losses, the central government applied a contract responsibility system to foreign trade in 1988. Since the amount of domestic currency the central government would provide to subsidize trade losses was fixed in central-local contracts, the contract system imposed a relatively hard budget constraint on export producers and helped limit trade losses.

Devaluation.   The root cause of the weak incentive for exports in the early years of China's trade reform was the overvaluation of the renminbi. Although administrative decentralization encouraged an increasing number of local enterprises to engage in foreign trade, it did little to change the incentive structure for trading activities; the system's traditional bias against exports continued. To get its exchange rate right, China has substantially devalued the renminbi in the course of its trade reforms. Devaluation of the renminbi began in 1981 with the adoption of an "internal settlement rate," which sets the yuan at 2.8 to the dollar (as opposed to the official exchange rate of 1.5) for foreign trade transactions. In January 1985, partly in response to the pressures from the International Monetary Fund (IMF) and from China's trading partners who considered the internal rate a form of export subsidy, China eliminated the dual exchange rates by devaluing the official rate to the internal rate (Shirk 1994; Fukasaku and Wall 1994). The need for further devaluation of the renminbi became apparent by late 1985 when China began to run a huge trade deficit and saw a dangerous drawdown of its foreign exchange reserves. To improve its current account balance and replenish its depleted foreign exchange reserves, China devalued the renminbi five times between 1985 and 1993 (Figure 11.1). Finally, effective on January 1, 1994, China replaced the dual-track system of exchange rates with a unified managed floating exchange rate system. Thus, by 1994 the value of the renminbi had been brought down to below 17 percent of what it had been in 1980. Since then, the renminbi has come under pressure to appreciate.

In summary, two decades of foreign trade reforms have freed up an increasing number of Chinese enterprises from central control and encouraged them to engage in foreign trade. More importantly, China's exchange rate reforms, combined with domestic price reforms, have helped reduce the distortions in domestic resource allocation. The devaluation of the renminbi has enhanced the competitiveness of Chinese exports in the international market, as is discussed in the next section.

Growth and Trade Balance

The recent growth in China's merchandise trade is impressive. Between 1978 and 1996 China's trade turnover increased from $20 billion to $290 billion (Figure 11.2). China has thus moved up from the thirtieth-largest to the tenth-largest trading nation in the world. China now accounts for about 3 percent of world trade. According to the IMF, China has become the largest exporter to the world market among developing economies.

Three broad trends in the recent development of China's merchandise trade are noteworthy. First, China's trade growth gained momentum in the mid-1980s, and the pace of this growth has accelerated since then. As shown in Figure 11.2, China's trade turnover hovered around $40 billion before 1984. This figure jumped to $70 billion in 1985 and to $115 billion in 1990. In the 1991–1996 period, China's trade turnover more than doubled.

Three factors are crucial in explaining China's trade growth since the mid-1980s. First, China accelerated the pace of its foreign trade and exchange reforms after 1984. Trade decentralization and currency devaluations boosted trading activities. Second, the export-oriented development strategy adopted since the mid-1980s has

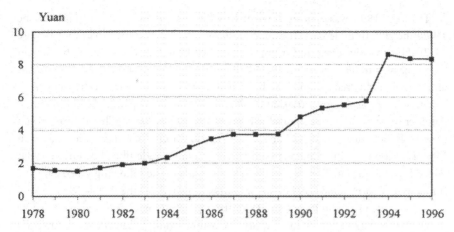

FIGURE 11.1   Renminbi-Dollar Exchange Rate, 1978–1996 (yuan per dollar)
SOURCE: *China Statistical Yearbook* (various issues).

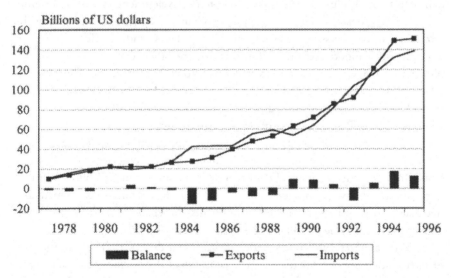

FIGURE 11.2   China's Merchandise Trade, 1978–1996
SOURCE: *China Statistical Yearbook* (various issues).

resulted in a rapid increase in the inflow of FDI, particularly to the Chinese coastal region. This in turn generated large trade volumes—both imports and exports. Third, China shifted the focus of its domestic reforms from agriculture to the industrial sector in the mid-1980s. Not only has rapid industrial growth brought China's economy into a high-growth track but it has also spurred a significant expansion of Chinese imports and exports.

The year 1990 was a turning point in China's trade balance. Throughout the 1980s the growth in demand for imports far exceeded the export expansions (Figure 11.2). The excessive demand for imports resulted from China's domestic boom and led to massive trade deficits. China is a resource-poor country on a per capita basis, and its new industrialization drive has relied heavily on imported technology. As a result, China's rapid industrialization has increased its dependence on imports. Another important factor responsible for the rapid expansion of China's imports has been its import liberalization. The sudden deterioration of China's trade balance in 1985 is a case in point. China started a new round of trade liberalization in 1984, which led to a substantial reduction in import tariffs in the following years. The lifting of tariffs, combined with strong demand triggered by the 1984–1985 boom, caused a sudden surge in imports in 1985. In that year, China's imports increased by 54 percent to $42 billion, while exports went up by only 5 percent to $27 billion, resulting in the record trade deficit of $15 billion.

The year 1990 saw a turnaround in China's trade balance. Except for 1993, China has enjoyed sizable trade surpluses since 1990. The transition from trade deficit to trade surplus is a significant long-term development in China's external economic relationships. The trade surpluses China enjoyed in the 1990s can be attributed to its success at exporting, on the one hand, and its effective demand management—import licensing and quotas—on the other. The trade deficit of $12 billion in 1993 was caused by a slowdown in demand for certain categories of Chinese exports in overseas markets and, to a lesser extent, by unfavorable terms-of-trade changes in primary commodity markets. Moreover, the 1993 economic boom at home—a GDP growth rate of 13.4 percent—not only stimulated a strong demand for imports but also diverted a considerable portion of exportable goods to the Chinese domestic market. In addition, there was a surge in FDI inflows to China in 1993, with realized FDI in that year amounting to $27 billion. In that same year, the imports associated with FDI inflows—in the form of equipment and materials as part of foreign partners' investments—reached $17 billion, which was an increase of more than 100 percent from the previous year.[1]

A third basic trend in China's trade growth is that during 1978–1996, the level of merchandise imports fluctuated more widely than that of merchandise exports. This is clearly a reflection of China's reform-driven, "stop-go" cycles in the external sector. Since the inception of its economic reforms, China has seen five major "boom-bust" cycles. These "boom-bust" cycles are characterized by periods of rapid growth, with growing inflationary pressures and external imbalances, and sharp slowdowns in periods of retrenchment (Bell et al. 1993; Yusuf 1994). In each of these cycles, reform initiatives first stimulated excessive demand for investment and growth; at later stages, rising inflation and the fear of overheating often called for the need to contain reform initiatives and curb growth by deflating demand. As part of these cycles, China's imports have also experienced periods of rapid growth (1979/1980, 1984/1985, 1987/1988, and 1993) and retrenchment.

Commodity Composition

China's trade growth since 1978 has been accompanied by remarkable changes in the composition of traded commodities. Most remarkable is the rapidly rising share of manufactured products in total exports, which climbed from 46 percent to 86 percent (Table 11.1).

Much of the increase in the relative share of manufactured exports occurred after the mid-1980s. Although exports of manufactured goods, in particular basic manufactures (Class 6 in the Standard International Trade Classification or SITC), increased considerably in the early 1980s, their share in total exports changed little. This is because the increase in manufactured exports in this period was largely offset by China's increased exports of primary goods. It should be noted that despite domestic energy shortages, China increased its oil exports after the mid-1970s to take advantage of rising oil prices in the world market, and it continued to be a major oil exporter through 1985 (Lardy 1994; Naughton 1997). Primary products (SITC 0 through 4, plus 68) continued to account for about one-half of China's exports throughout the first half of the 1980s. The surge in domestic demand triggered by the 1984–1985 domestic boom even led to a drop in the relative share of manufactured exports, as large amounts of exportables were diverted to domestic consumption.

After the mid-1980s, China shifted away from exporting petroleum to exporting manufactured products. A key factor in this shift was the adoption of the so-called coastal development strategy, which aimed to develop an export-led economy in the Chinese coastal region by using foreign manufacturing investment. As FFEs began to make significant contributions to China's exports in the late 1980s, the relative share of manufactured exports expanded rapidly. Price reforms in the domestic market and the substantial devaluation of the renminbi after the mid-1980s also allowed market forces to play an increasingly important role in determining China's export structure.

The bulk of the increase in China's manufactured exports after 1985 consisted of labor-intensive manufactures. By 1995, basic manufactures (SITC 6) and miscellaneous manufactured goods (SITC 8) accounted for about 60 percent of China's exports. Labor-intensive goods, such as textiles, apparel, footwear, travel goods, and toys, are China's major export products. These labor-intensive products started the export drives of Japan, South Korea, and Taiwan. They have also characterized China's success so far. Significant changes have also been recorded in the category of machinery and transport equipment (SITC 7), whose share in China's total exports expanded from 3 percent in 1985 to 21 percent in 1995. Much of this increase was accounted for by increased exports of low-end telecommunications equipment and domestic electrical products. Since the production of these goods involves primarily processing and assembly activities in China, they too were labor-intensive.

The commodity composition of China's imports has also changed. Although the share of primary products in total imports increased between 1979 and 1982, from 27 percent to nearly 40 percent, this share declined substantially to about 18 percent

TABLE 11.1   Commodity Composition of China's Merchandise Trade, 1985–1995 (percent)

	SITC[a]	1978	1980	1985	1987	1989	1990	1991	1992	199?
Total exports	0–9	100.0	100.0	100.0	100.0	100.0	100.0	100.0	100.0	100.(
Primary commodities	0 to 4 plus 68	53.5	50.2	50.6	33.6	28.7	25.6	22.5	20.0	18.?
All manufactured goods	5 to 8 less 68	46.5	48.7	36.9	47.7	54.7	55.7	58.5	80.0	81.?
Chemicals	5	2.4	6.2	5.0	5.7	6.1	6.0	5.3	5.1	5.(
Basic manufactures	6 (less 68)	4.6	22.1	16.4	21.7	20.7	20.3	20.1	19.0	17.?
Machinery & transport equipment	7	3.4	4.7	2.8	4.4	7.4	9.0	10.0	15.6	16.?
Misc. manufactured goods	8	36.1	15.7	12.7	15.9	20.5	20.4	23.1	40.3	42.?
Unclassified goods	9	0.0	1.1	12.5	18.7	16.6	18.7	19.0	0.0	0.(
Total imports	0–9	100.0	100.0	100.0	100.0	100.0	100.0	100.0	100.0	100.(
Primary commodities	0 to 4 plus 68	27.4	34.2	12.3	15.4	19.6	16.4	16.8	16.4	13.?
All manufactured goods	5 to 8 less 68	70.8	63.5	81.8	72.1	67.7	64.5	65.6	83.4	86.?
Chemicals	5	10.2	14.5	10.7	11.6	12.7	12.4	14.6	13.8	9.?
Basic manufactures	6 (less 68)	30.6	21.0	28.2	22.5	20.8	16.7	16.5	23.9	27.?
Machinery & transport equipment	7	25.5	25.5	38.4	33.8	30.8	31.5	30.7	38.8	43.?
Misc. manufactured goods	8	4.5	2.5	4.5	4.2	3.4	3.9	3.8	6.9	6.?
Unclassified goods	9	1.8	2.3	5.9	12.5	12.7	19.1	17.6	0.2	0.?

SOURCE: Asian Development Bank, *Key Indicators of Developing Asian and Pacific Countries* (various issues); Sta
Statistical Bureau of China, *China Foreign Economic Statistical Yearbook* (various issues).
[a]SITC = Standard International Trade Classification

in 1985 and has remained at that level thereafter. The share of manufactured goods in China's imports, on the other hand, increased from 70 percent in 1979 to 80 percent in 1995. Among manufactured products, imports of machinery and transport equipment increased most significantly. By 1995 machinery and transport equipment accounted for 40 percent of China's total imports.

In summary, within less than two decades China has successfully converted its trade pattern to one that is increasingly determined by the country's comparative advantages. And China has become a major exporter of labor-intensive manufactures.

China's Trade Links in the Pacific Rim

The bulk of China's trade is with other economies in the Pacific Rim. Data on the direction of trade indicate that China's dependence on the Pacific Rim markets is high and growing (Table 11.2). Over the 1978–1996 period, China increased its share of trade with the region, for both exports and imports. In 1996 nearly three-quarters of China's merchandise trade was accounted for by the Pacific Rim nations. In absolute terms, China's trade with its partners in the Pacific Rim reached $212 billion by 1996. This figure represents a fourfold increase from 1985. Table 11.3 indicates that China has become a major trading partner for a number of economies in the Pacific Rim.

As a destination for exports, China absorbed nearly $100 billion worth of goods from the region in 1996, up from $28 billion in 1985. Over this period the share of

TABLE 11.2   Direction of China's Merchandise Trade, 1985–1996 (percent)

	1985	1987	1989	1990	1991	1992	1993	1994	1995
Exports	100.0	100.0	100.0	100.0	100.0	100.0	100.0	100.0	100.0
Pacific Rim	68.9	66.5	73.0	75.1	78.8	77.9	72.3	75.6	74.7
East Asia	56.0	54.4	60.5	62.2	65.6	63.6	48.5	52.2	52.2
NIEs	33.7	38.2	44.6	47.6	51.3	49.9	31.3	34.4	33.1
Japan	22.3	16.2	15.9	14.6	14.3	13.7	17.2	17.8	19.1
ASEAN-4	2.7	2.5	2.5	2.9	2.9	2.6	2.7	3.1	3.7
North America	9.4	8.7	9.1	9.2	9.4	10.8	19.8	18.9	17.6
Australia & New Zealand	0.8	0.9	0.9	0.8	0.9	0.9	1.3	1.4	1.2
Rest of the world	31.1	33.5	27.0	24.9	21.2	22.1	27.7	24.4	25.3
Imports	100.0	100.0	100.0	100.0	100.0	100.0	100.0	100.0	100.0
Pacific Rim	67.5	65.5	63.2	69.2	74.3	72.9	69.8	71.6	71.0
East Asia	47.5	44.2	41.5	47.4	52.2	53.7	52.9	51.7	50.1
NIEs	11.8	20.9	23.7	33.2	36.5	37.0	30.4	28.9	28.1
Japan	35.7	23.3	17.8	14.2	15.7	16.7	22.5	22.8	22.0
ASEAN-4	2.1	3.3	3.6	4.0	4.3	3.6	3.2	3.8	4.5
North America	14.9	14.4	15.1	15.0	15.1	13.2	11.6	13.7	14.2
Australia & New Zealand	3.0	3.6	3.0	2.8	2.7	2.4	2.1	2.4	2.2
Rest of the world	32.5	34.5	36.8	30.8	25.7	27.1	30.2	28.4	29.0

SOURCE: IMF, *Direction of Trade Statistics Yearbook* (various issues).

China's imports from the Pacific Rim increased from 68 to 71 percent. The period after 1985 also saw remarkable changes in the pattern of China's import markets. First, China has diversified the source of its imports. In 1985 the top three import markets (Japan, the United States, and Hong Kong) accounted for 62 percent of China's total imports; this share declined to 44 percent by 1996. Second, the Asian NIEs have made rapid inroads into the Chinese market. In this respect, the rise of Taiwan and Korea as China's major trading partners is most noticeable. Between 1985 and 1996 China's imports from the Pacific Rim increased by $70 billion; more than 40 percent of this increase was accounted for by Taiwan (23 percent) and Korea (18 percent). Taiwan's share in China's imports expanded to about 12 percent by 1996, and Korea's to 9 percent. Taiwan has thus become the mainland's second-largest import market, next to Japan (21 percent), while Korea is the fourth-largest import market, after the United States (12 percent).

China's export dependence on the Pacific Rim's market is higher than its import dependence. By 1996 the share of Chinese exports destined to the Pacific Rim reached 75 percent. This figure represents an increase of 6 percentage points since 1985. In absolute terms, China's exports to the Pacific Rim increased from $19 billion in 1985 to $113 billion in 1996. The bulk of this increase was to the United States, Japan, and Hong Kong.

The success of China's foreign trade owes much to Hong Kong's expertise in marketing Chinese products in the international market. As China's exports shifted away from primary products to manufactured goods after 1985, Hong Kong's role as an

TABLE 11.3   China as a Trading Partner in the Pacific Rim

| | Exports to China | | | | Imports from China | | | |
| | 1985 | | 1996 | | 1985 | | 1996 | |
	Percent	Rank	Percent	Rank	Percent	Rank	Percent	Rank
Japan	7.1	2	5.3	5	5.0	5	11.6	2
Korea	–	–	8.8	3	–	–	5.7	3
Hong Kong	26.2	2	34.4	1	25.3	1	37.1	1
Singapore	1.5	13	2.7	10	8.8	4	3.3	9
Thailand	3.8	7	3.5	6	2.2	12	2.7	8
Malaysia	1.0	16	2.4	11	2.4	11	2.4	10
Indonesia	0.5	14	4.4	4	2.9	8	3.7	7
Philippines	1.8	11	1.6	11	5.6	4	2.1	12
United States	1.8	14	1.9	15	1.2	17	6.7	4
Canada	1.0	4	1.1	5	0.4	16	2.1	5
Australia	3.8	4	4.9	5	1.3	15	5.2	5
New Zealand	2.4	6	2.6	8	0.8	15	3.4	6

SOURCE: IMF, *Direction of Trade Statistics Yearbook* (various issues).

entrepôt became crucial to China's export expansion. Between 1985 and 1991 the share of exports to Hong Kong increased from 26 percent to 45 percent. The bulk of the Chinese products that went to Hong Kong were re-exported to other destinations, especially to the United States and European countries. In recent years, however, China has increased the sale of its products through domestic firms in the international market, which has led to a sharp decline in the share of its exports being marketed through Hong Kong.

The role of the United States as a major market for Chinese exports has become increasingly pronounced. On a country-of-destination basis, China sent 18 percent of its total exports to the United States in 1996—a 10-percentage point gain since 1985. This increase is remarkable because it occurred when the share of China's exports to Japan and Hong Kong declined by 2 and 4 percent respectively over the same period. Because a considerable portion of the Chinese products that go to Hong Kong are re-exported to the United States, the actual dependence of Chinese exports on the U.S. market tends to be much higher than conventional statistics might suggest.[2]

In absolute terms, Japan's imports from China have also increased substantially since 1985. In relative terms, however, Japan's share in China's exports declined between 1985 and 1992, from 22 percent to 14 percent. Although this share has begun to rebound recently, it has remained below the 1985 level.

China's trade with the ASEAN countries has remained very low, as is its trade with Australia and New Zealand. In 1996, the ASEAN–4 accounted for only about 3 percent of China's exports. Australia and New Zealand combined had a share of merely 1.3 percent of China's exports.

The nature of bilateral trade relationships between China and the major subgroups of economies in the Pacific Rim is diverse. And it also has changed remarkably over the past decade or so (Figure 11.3). Japan seems to have benefited most

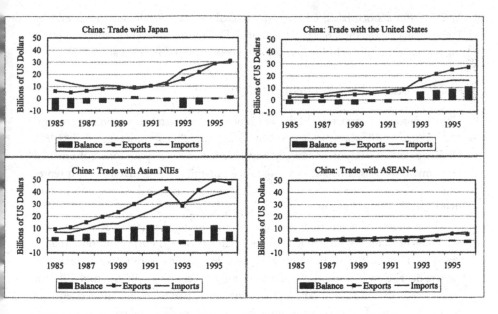

FIGURE 11.3   Bilateral Trade Relationships Between China and Major
Economies in the Pacific Rim
SOURCE: IMF, *Direction of Trade Statistics Yearbook* (various issues).

from the recent expansion of the Chinese market. With minor exceptions, Japan has
enjoyed persistent, large trade surpluses with China in the past decade. These surpluses were especially large in the 1985–1987 and 1993–1994 periods.

China's trade surplus with the Asian NIEs as a group is huge. Except for 1993,
China enjoyed persistent trade surpluses with the Asian NIEs in the past decade.
Especially in the 1985–1991 period, China's trade surplus with the Asian NIEs was
huge and growing. After 1991, however, this surplus stagnated.

Especially noteworthy is the recent change in the Sino-U.S. trade relationship.
Due to the different accounting approaches the United States and China use in
reporting the part of Chinese exports that flow through Hong Kong, data on Sino-
U.S. bilateral trade have caused much confusion. Nevertheless, as Figure 11.3 shows,
China's own trade data indicate that since 1993 China has enjoyed huge and grow-
ing trade surpluses with the United States.

China incurred persistent deficits in its trade with the ASEAN–4. However, the size
of these deficits remained modest due obviously to the low level of their bilateral trade.

China's Coastal Development
Strategy and FDI

China's rapid growth in trade has been accompanied by big increases in FDI inflows.
On a flow basis, realized FDI in China grew from $1.2 billion in 1984 to $43 billion

in 1996. On a stock basis, China had attracted over $175 billion in total FDI by 1996. China is now the largest recipient of all FDI in the developing world, accounting for one-third of total FDI to developing countries (Tables 11.4 and 11.5).

The Coastal Development Strategy

China's success at attracting FDI owes much to its coastal development strategy. This strategy took shape in 1986–1987 and was officially announced in 1988. China's coastal development strategy was designed to develop an outward-oriented economy in its coastal region by using foreign manufacturing investment. It represents a policy response to the new domestic and international economic situation that China faced at that time.

China adopted foreign trade reforms to promote exports immediately after the launching of its reform program. The goal of China's early export promotion strategy was quite limited. Export promotion was seen as a means to ease foreign exchange constraints and to increase the capacity to import technology. By the mid-1980s it became clear that the initial export promotion policy had been only moderately successful. The rate of China's export growth remained modest; the composition of its exports had changed little; and the burden of export subsidies was huge and growing. Moreover, China's traditional export-producing bases, such as Shanghai, lacked the strength to expand further, as domestic supplies of raw materials were increasingly constrained by rising regionalism as well as infrastructure bottlenecks.

China's initial opening to FDI was quite limited. Although the ban on foreign investment was lifted in 1979, the Chinese authorities remained cautious toward FDI. The Special Economic Zones (SEZs), which were formally established in 1980, were initially modeled after the export-processing zones established in other East Asia developing countries. While the Shenzhen SEZ provided a "laboratory" for China's market reform experiments, it was less successful in bringing high-technology investment into the Chinese economy. By the mid-1980s the Shenzhen SEZ was importing far more than it exported, and it was absorbing more domestic resources than foreign investments (Fukasaku and Wall 1994).

The second half of the 1980s saw a dramatic change in East Asia's economic environment. Japan's huge trade surpluses with the United States led to a sharp appreciation of the Japanese yen after the Plaza Accord in 1985. To counteract the effects of a strong yen, Japanese corporations accelerated the pace of their overseas investment. At the same time, the strengthening of the yen substantially improved the international competitiveness of Asian NIEs' exports, which had already achieved remarkable success under their export-oriented strategy since the 1970s (World Bank 1993; Kwan 1994). As a result, the Asian NIEs' exports surged, and they began to build up huge surpluses in their trade with the United States after 1985. Partly in response to mounting pressures from the United States, the Asian NIEs began to revalue their currencies in 1986. Currency realignments, coupled with rising wages, also forced the

TABLE 11.4 Realized FDI in China, 1979–1996

	1979–1983	1984	1985	1986	1987	1988	1989	1990	1991	1992	1993	1994	1995	1996
In billions of US dollars														
Flows	1.8	1.3	1.7	1.9	2.3	3.2	3.4	3.5	4.4	11.0	27.5	33.8	37.5	42.4
Stock	1.8	3.1	4.7	6.6	8.9	12.1	15.5	19.0	23.3	34.4	61.9	95.6	133.2	175.6
FDI flows by origin (percent)														
Total	100.0	100.0	100.0	100.0	100.0	100.0	100.0	100.0	100.0	100.0	100.0	100.0	100.0	
Hong Kong	58.0	53.0	48.8	60.4	69.1	65.6	61.2	54.8	57.0	70.0	65.0	59.9	54.5	
Taiwan	–	–	–	–	–	–	–	0.6	1.1	9.3	11.3	10.0	8.5	
Japan	12.8	15.8	16.1	10.7	9.5	16.1	10.5	14.4	12.2	6.6	4.9	6.2	8.5	
United States	11.5	18.1	18.2	16.8	11.4	7.4	8.4	13.1	7.4	4.6	7.4	7.4	8.2	
Rest of the world	17.7	13.1	16.9	12.1	10.0	10.9	19.9	17.1	22.3	9.5	11.4	16.5	20.3	

SOURCE: State Statistical Bureau of China, *China Statistical Yearbook* (various issues).

TABLE 11.5 China's Share of FDI Inflows to Developing Countries, 1984–1995 (percent)

	1984	1986	1988	1990	1992	1994	1995
All developing countries	100.0	100.0	100.0	100.0	100.0	100.0	100.0
China	7.9	15.1	14.6	12.2	23.2	36.9	33.3
Asian developing countries	100.0	100.0	100.0	100.0	100.0	100.0	100.0
China	26.1	31.6	25.2	18.9	43.4	64.4	60.1

SOURCE: IMF, *Balance of Payments Statistics Yearbook* (various issues).

Asian NIEs to move out of unskilled labor-intensive manufacturing industries, in which they were losing comparative advantage. Thus, the industrial restructuring of Japan and the Asian NIEs after 1985 created plenty of opportunities for other developing economies in the East Asian region.

The Chinese government responded to these opportunities by launching the so-called coastal development strategy. China's coastal development strategy had three broad objectives. First, by providing a platform for the relocation of labor-intensive manufacturing from more advanced East Asian economies, especially Hong Kong and Taiwan, China was hoping to develop an export-led economy in its coastal region. Second, by having "both ends outside"—using foreign resources to manufacture products for foreign markets, the Chinese coastal region was expected to cede much of the domestic market to inland producers. Third, entering highly competitive world markets would also require Chinese enterprises to improve their efficiency and modernize their production technologies.

To ensure the success of the coastal development strategy, China further opened its economy to foreign trade and investment in 1987–1988. Key policy measures adopted under the coastal development strategy included 1) a substantial decentralization of decisionmaking power to local authorities, which allowed them to set conditions for foreign investment and trade, 2) deregulation of FDI regimes, and 3) provision of tariff exemptions and rebates for imports of raw materials, intermediate inputs, and capital goods, so long as they were used to produce exports.

The Surge in FDI and Its Effects

China's coastal development strategy has proven hugely successful in a number of aspects.

FDI and Industrial Growth. The growth in FDI inflows in China since the 1990s has been remarkable (Table 11.4). Although the initial effects of China's coastal development strategy were dampened by the political incident of 1989, net inflows of FDI to China jumped from $4.4 billion in 1991 to $11 billion in 1992. Between 1991 and 1996 realized FDI in China grew 60 percent a year. The continued inflow of FDI has contributed to China's growth. By 1995 FDI accounted for about 15 percent of China's total capital formation, and FFEs contributed about 14 percent of Chinese industrial output. The contribution of FDI is also substantial in other ways. Industrialists from Hong Kong and Taiwan have infused production technologies and managerial expertise into the Chinese economy. They have also helped forge the links between Chinese and overseas markets. Multinational enterprises from Japan and the United States have introduced China to advanced transport and electronics technology. Thus the inflow of FDI has played an important role in stimulating China's productivity growth over the past decade.

FDI and Exports.   The continued inflow of FDI to China has been especially important to its export growth. Much of China's success at exporting is due to the rapid expansion of manufactured exports. The contribution of FFEs to this expansion has been significant. While in 1986 FFEs accounted for less than 2 percent of China's exports, this share expanded to 26 percent in 1993 and reached one-third by 1995. The vast majority of exports by FFEs were products produced by processing duty-free imports.

The Chinese Economic Area.   An even more significant outcome of China's coastal development strategy is the emergence of a Chinese Economic Area that comprises coastal China, Hong Kong, and Taiwan. Hong Kong's role as an entrepôt for channeling goods and capital in and out of the Chinese mainland has been well documented (Sung 1997). The relocation of labor-intensive light manufacturing activities from Hong Kong to the Pearl Delta region in South China played a crucial role in the recent expansion of China's manufactured exports. Although conventional statistics on the FDI from Hong Kong included "round-tripping" domestic capital as well as foreign capital that went to China via Hong Kong, FDI by Hong Kong firms in the mainland has been substantial. Since the appreciation of the Taiwanese currency in 1987, there has been a surge in Taiwanese investment in the mainland. On a flow basis, Taiwan has become the second-largest source of realized FDI in the mainland, accounting for about 10 percent of the total in recent years. As a large part of labor-intensive manufacturing industries were relocated from Hong Kong and Taiwan to coastal China, the trade and investment links between them have become intensified. In fact these three areas have become so integrated that they are close to being a single economic unit. The strengthening of the trade and investment ties between coastal China and Hong Kong and Taiwan has been a key factor in China's export growth since the mid-1980s.

Challenges and Prospects

China and the Pacific Dynamism

Two decades of foreign trade and investment reforms have transformed China from a virtually closed economy to a major participant in international commodity and capital markets. Ratios of intraregional trade and investment indicate that China has become increasingly integrated into the Pacific economy. The strengthening of trade and investment ties between China and other Pacific Rim economies promoted economic growth in the region.

Long-term projections suggest that if recent trends continue, China's GDP could become the world's second-largest on a PPP basis by 2020. And China's merchandise exports could rise to 6–8 percent of the world's total, ahead of Japan (World Bank

1997a,b). Given the attractiveness of China's domestic market, it can also be expected that FDI inflows to China will continue to grow for some time. China's continued growth, combined with greater integration in the Chinese Economic Area, could lead to China's playing the role of a new growth pole in the world economy.

The rise of China as a major player in the world economy is bound to have a tremendous effect on the rest of the world. The economies in the Pacific Rim will feel the greatest impact because they are China's major trading and investment partners. In the long run, China's emergence as a leading trading nation will offer a huge and growing market to its trading partners. In the medium or short term, however, China's further integration with the world economy will generate friction and painful adjustments, especially for those developing countries that are in direct competition with China for export markets.

China as an Importer.   China has a vast potential for imports. The country needs large improvements in its infrastructure as well as upgrading of its industrial equipment. Furthermore, a more affluent population will certainly generate an increasing demand for consumer goods. While consumer products are likely to be increasingly produced at home, China's imports of capital- and knowledge-intensive products will continue to grow. Industrial countries stand to gain from exporting to China's vast and largely untapped markets. They will also benefit from favorable terms-of-trade changes because China's huge import demand for capital- and knowledge-intensive products is likely to drive up the prices of these products relative to prices of labor-intensive products. Both geography and comparative advantage lead one to expect that the United States and Japan, and increasingly the Asian NIEs, will be the main beneficiaries of China's future growth.

China as an Exporter.   China has emerged as a major supplier of relatively labor-intensive manufactured goods in the world market, and the country has a vast potential to export. The vast army of unemployed and underemployed labor and the growing integration of the national labor market will have a dampening effect on wage demands, which will help China to sustain international competitiveness in the near future. China also possesses considerable capability to upgrade its industrial structure and, therefore, export structure because of accumulated industrial learning. The business expertise and thick market networks developed by overseas Chinese are yet another asset into which China could tap.

If China continues its efforts to exploit the international comparative advantage offered by its abundant labor, it can be expected that Chinese exports will continue to be concentrated in labor-intensive manufactures for some time to come. Further expansions of China's exports will generate more pressure on its competitors in the international market. Although labor-intensive manufacturing in developed countries will also come under increasing competitive pressure from China, the real challenge will be largely for the developing countries. Specific to the Pacific Rim, the competitive pressure from Chinese exports will probably be felt most strongly in Southeast Asian developing economies (except for Singapore), which are in direct competition with China for export markets. As China's coastal region eventually moves up the ladder of comparative advantage to exports of intermediate technology manufactures, many of those labor-intensive manufacturing industries abandoned by coastal China are likely to migrate to inland China, rather than low-cost locations in other developing countries.

China's strong export performance since the mid-1980s owes much to the devaluation of the renminbi. Price competitiveness has worked for China so far. In the coming years, however, China has to deal with the indirect effect on its exports of the Asian financial crisis—the devaluation of the currencies in several ASEAN countries and Korea.

China as a Destination for FDI.   China has become an increasingly attractive destination for FDI. As Table 11.5 shows, China has increased its share in total FDI inflows to the Asian developing countries. In the future, while other Asian developing economies may remain attractive to FDI due to their favorable cost structures, China is likely to continue to attract a large portion of FDI inflows to the East Asian region. This is because, in addition to low labor costs, China possesses certain "locational advantages"—its vast domestic market, considerable technological capability, and a broad resource base—which are increasingly important to FDI (UNCTD 1993).

Studies of multinational enterprise (MNE) activities suggest that as FDI regimes become broadly similar around the globe, the ability of a host country to create fundamentally sound macroeconomic conditions is key to attracting the inflow of FDI (UNCTD 1993). Therefore, maintaining a stable macroeconomic climate and developing efficient economic institutions and infrastructure will be crucial to sustaining FDI in China. Moreover, the degree of openness of the Chinese domestic market will become an increasingly important determinant of FDI inflows to the country.

Overseas investment by mainland Chinese enterprises has also expanded considerably in recent years. International experience suggests that the emergence of a major economic power is accompanied by major involvement in world finance at a later stage. Therefore, China's position in international finance can be expected to change. From a long-term perspective, China's role as an investing nation should not be underestimated.

Challenges of Deeper Integration

How Open Is the Chinese Economy?   Measured by conventional statistics like trade ratios, China seems to have made impressive progress in integrating itself with the world economy (Figure 11.4). Such statistics should be treated with some caution,

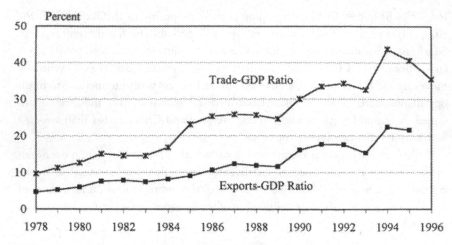

FIGURE 11.4   China's Openness to Foreign Trade
SOURCE: *China Statistical Yearbook* (various issues).

however. The reason is twofold. First, China's GDP is widely believed to be underestimated. Second, China's trade volume is overestimated by including the full value of exports produced from imported inputs. If the figures on GDP and trade are adjusted, then China is much less open than conventional statistics might suggest (World Bank 1997a,b).

Moreover, China's economy has so far been only shallowly integrated into the world economy. China's domestic market remains protected. Not all enterprises are allowed to trade directly with foreign partners. The renminbi is only partially convertible. Portfolio investment, which was not allowed until 1991, is subject to heavy regulation. China will be facing big challenges in seeking further integration with the world economy in the years to come.

Deepening Trade Integration.   The opening of the U.S. market to Asian exports proved crucial to the "East Asian Miracle." Japan and the Asian NIEs relied on external demand as a major engine of their economic growth for an extended period of time (World Bank 1993). It appears that China will not experience this development pattern for two reasons. First, very large economies like China's cannot rely exclusively on external demand to sustain a rapid growth rate. Second, trade issues have become increasingly complicated and sensitive. Further expansions of China's exports will inevitably meet increasing protectionism in industrialized countries. China may have to turn to its domestic market at a much earlier stage.

Even in the medium or short term, China's export growth will not be sustainable without large open markets. It is very likely that China will need to further liberalize its import regimes and absorb more exports from its trading partners in exchange for market access in major industrial countries. Opening the domestic market, however, will create difficulties for a large number of domestic enterprises. Since the mid-1980s China has adopted a full set of commercial policies to shield its key domestic industries from international competition. Because these industries are dominated by state-run enterprises, further import liberalization will require the central government to deal with the consequences of painful industrial adjustments. Rising unemployment and increasing insolvency of SOEs could easily become politically sensitive issues. Entering international markets will also require China to increase the transparency and predictability of its trade regime, accept labor and environmental standards, and protect intellectual property rights.

Improving FDI Efficiency.   Up to now, China's success at attracting FDI has been achieved without creating significant linkages between its domestic and external sectors. Overdependence on imported inputs for export production has led to a very low domestic content of China's manufactured exports, in particular those by FFEs. As a result, the spillovers from China's exporting sector to the rest of the economy seem very limited. To capture more benefits from FDI and increase the gains from exporting, China needs to develop an integrated domestic market and enhance domestic supply capability. Through investing in a foreign country, MNEs deliver a package of tangible and intangible assets. In this package, the capital component— traditionally seen as the principal benefit of FDI—is of diminishing importance. In the future, China's FDI policy should be geared more to capturing the dynamic spillovers of technology transfer so as to accelerate the virtuous cycle among FDI, technology diffusion, and long-term development.

International Integration and External Shocks.   Over the past two decades, national economies have become increasingly integrated at regional and global levels. Before the onset of the Asian financial crisis in mid-1997, the case for greater international integration seemed quite straightforward. Today, the magnitude of the damaging effects of the Asian financial turmoil leads governments of developing countries to think twice before moving toward deeper integration with the international economy.

China is the one East Asian economy that has not been directly affected by the Asian financial turmoil. A number of factors help explain why China remained relatively resilient to the Asian financial crisis. First, a strong export performance made it possible for China to maintain a favorable trade position in recent years. Second, the period after 1990 saw strong growth in FDI inflows to China. Third, China's favorable current account balance and continued FDI inflows have in turn led to a huge build-up of its foreign exchange reserves (Figure 11.5). Fourth, China's external finances were very different from those of the countries hit by the crisis. Portfolio investment—a factor that

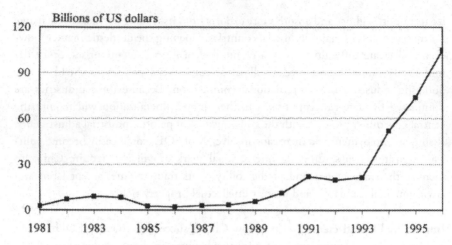

FIGURE 11.5   China's Foreign Exchange Reserves, 1978–1996
SOURCE: IMF, *Balance of Payments Statistics Yearbook* (various issues).

is widely believed to be largely responsible for the Asian financial crisis—accounted for only a very small part of foreign capital inflows to China.

Recent evidence suggests that the Asian financial turmoil seems to have reinforced China's cautious approach to liberalizing its capital account. In order to prevent a similar economic meltdown from happening in China, the Chinese government will likely continue to pursue trade surpluses, prefer FDI rather than portfolio investment, and maintain a sufficiently high level of foreign exchange reserves.

As China's exposure to foreign trade and investment increases, external developments will inevitably have an increasingly significant impact on China's future growth. How to strengthen economic fundamentals and enhance institutional capacity so as to maintain macroeconomic stability and manage possible external shocks will be a lasting issue in China's long-term development.

Conclusion

China is emerging as a leading Pacific Rim economy. More than two decades of rapid growth and opening up has brought China to the center stage of East Asian economic dynamism. China's further growth could result in the country's playing the role of a new growth pole in the Pacific Rim. China has emerged as a major supplier of labor-intensive manufactures. Its export dependence on the markets of the Pacific Rim is high and growing. China's rising import demand will create a huge market for its trading partners. On the other hand, further expansion of China's exports will generate increasing competitive pressures on some developing countries.

China has become an attractive destination for FDI. The bulk of FDI in China has also come from the Pacific Rim countries. As China possesses a range of locational advantages that are important to FDI, it can be expected that FDI inflows to China will continue to grow in the foreseeable future.

China has made impressive progress in integrating itself with the world economy. China has so far integrated most closely with Hong Kong and Taiwan. The emergence of the Chinese Economic Area that encompasses coastal China, Hong Kong, and Taiwan will have significant implications for the Pacific region. On the other hand, China's economy has so far been only shallowly integrated into the world economy. Deepening trade integration will require China to deal with the domestic consequences of further import liberalization. Moreover, as its exposure to foreign trade and investment increases, China needs to strengthen its economic fundamentals and enhance its institutional capacity to maintain macroeconomic stability and manage possible external shocks. It seems certain that further strengthening of the trade and investment links between China and its partners will promote economic growth in the Pacific Rim. To the extent that economic integration can help reduce potential conflicts, China's further growth and integration with the world economy will become an important factor in the region's prosperity and stability.

Notes

1. In 1993, the imports by newly established FFEs accounted for 16 percent of China's total imports. State Statistical Bureau of China (1994, p. 220).

2. Lardy (1994) argues that China's export dependence on the U.S. market probably reached 30 percent in the early 1990s.

References

Bell, W. M., et al. 1993. *China at the Threshold of a Market Economy.* Washington, D.C: IMF Occasional Paper 107.

Fukasaku, K., and D. Wall. 1994. *China's Long March to an Open Economy.* Paris: OECD.

Kwan, C. H. 1994. *Economic Interdependence in the Asia-Pacific Region.* London: Routledge.

Lardy, N. R. 1994. *China in the World Economy.* Washington, D.C.: Institute for International Economics.

Naughton, B., ed. 1997. *The China Circle.* Washington, D.C.: Brookings Institution Press.

Shirk, S. L. 1994. *How China Opened Its Door.* Washington, D.C.: Brookings Institution.

State Committee for Economic Restructuring. 1995. *Reforms, 1996.* Beijing: Reform Press.

State Statistical Bureau of China. 1994. *China Development Report, 1994.* Beijing: China Statistics Press.

Sung, Y. 1997. Hong Kong and the Economic Integration of the China Circle. In *The China Circle,* ed. B. Naughton, 41–80. Washington, D.C.: Brookings Institution Press.

UNCTD (UN Conference on Trade and Development). 1993. *World Investment Report, 1993.* New York: UN.

World Bank. 1988. *China: External Trade and Capital.* Washington, D.C.: World Bank.

_____. 1993. *The East Asian Miracle.* Oxford: Oxford University Press.

_____. 1997a. *China Engaged: Integration with the Global Economy.* Washington, D.C.: World Bank.

_____. 1997b. *China 2020: Development Challenges in the New Century.* Washington, D.C.: World Bank.

Yusuf, S. 1994. China's Macroeconomic Performance and Management during Transition. *Journal of Economic Perspective* 8(2): 71–92.

PART TWO

Social Changes

12

Changes in the Chinese Population: Demography, Distribution, and Policy

Chiao-min Hsieh

The population of China is notable for its sheer size. Its fifth and most recent census shows that by November 1, 2000, China's total population was 1,265.83 million, which does not include the 6.78 million people of Hong Kong, the 440,000 residents of Macao, or Taiwan's 22.28 million people. China is home to about one-fourth of mankind and about 64 percent of the population in Asia, which comprises more than forty countries.

Counting the Chinese people is not an easy task. However, demographic surveys are nothing new in China. As early as the year A.D. 2, the Western Han dynasty counted 59 million people. By A.D. 1100, the total may have reached 100 million, and by A.D. 1800, perhaps 300 million inhabited China. In 1949, China's population was about 475 million. No definite figures have existed, though both Chinese and foreign scholars have come up with about fifty different estimates based on land use, salt consumption, and even postal service.

The People's Republic conducted its first modern census on June 30, 1953. It reported a total population of 582,602,417 (excluding Taiwan). In modern statistical terms, the 1953 census was fairly simple, but it represented the first modern attempt to solve the riddle of the Chinese population and thus caught the attention of the whole world. Since this first census, there have been others in 1964, 1982, 1990, and 2000.

This chapter explores the shifts in population composition during the past forty-seven years, the pattern of population redistribution, and the impact these changes have wrought on government policy. It is divided into three parts. First, I examine and compare the five censuses to illustrate Chinese demographic changes. Next, I explore the patterns of population redistribution and analyze the population density revealed in the 1990 census. Finally, I review China's population policies, which wavered between encouraging a large population and advocating family planning until enforcement of the one-child family was implemented following the 1982 census.

The Five Modern Censuses

The 1953 census was conducted on the *de jure* population (those habitually and legally in residence); it was not a *de facto* counting (the number of persons actually present at a particular place on the day of the census). Also, the census workers asked only five simple questions. Demographers may consider this census too simple or incomplete, but it represents the first complete census attempted by China and the most accurate source of population data at that time. Moreover, the census proved very helpful in comprehensive economic planning efforts for the country. It is interesting to note that the age composition section of the census indicated that China was a young country, with 45 percent of the total population under 20 years of age, 35 percent between 21 and 44, 13 percent between 45 and 59, and only 7 percent over 60 years of age. A young population means more people of working age and of greater productive capacity, a desirable result for a country seeking to undergo nationwide industrialization (Figure 12.1).

In 1964 China carried out its second census, which was reportedly more detailed than the first. Unfortunately, the results have not been published; information from various articles and citations are the only sources available. The evidence suggests that this 1964 census reported a population of 694 million, an increase of 112 million or 22.3 percent over the 1953 figures.

On October 27, 1982, China announced the preliminary results of the third modern census: It reported a total population of over 1 billion and included data on urban-rural residents, minorities, education levels, birth and death rates, and population size of various political units. This census was also primarily designed as input for comprehensive state planning.

In July of 1990, the People's Republic announced the fourth census, which reported the total population of China as 1,133.6 million (excluding Taiwan). This census included figures for population according to sex (584,967,421 males, or 51.6 percent of the total; and 548,632,579 females, or 48.4 percent); racial group (Han population = 1,042,482,187 or 96.96 percent of the total, and minority groups = 91,200,031 or 8.04 percent of the total); and birth and death rates (birth rate of 20.9 per thousand; death rate of 6.3 per thousand; and natural increase rate of 14.6 per thousand). Education levels and urban population were also reported (Table 12.1).

The 1982 census can be considered of high quality by international standards. It was reported that after the general census registration, a sample quality check showed

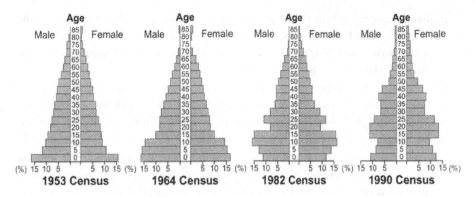

FIGURE 12.1   Population Pyramids of China

TABLE 12.1   Five Modern Censuses in China

	Total Population (in millions)	Density (Persons/ sq. mi.)	Sex Ratio (Female =100)	Birth Rate (per 1,000)	Death Rate (per 1,000)	Natural Rate (per 1,000)	Urbani- zation[a] (%)	Family Size	Illiteracy (%)	College[b] (per 100,000)	Minorities[c] (%)
1953	582.6	161.8	107.56	37.0	17.0	20.0	13.26	4.33			6.06
1964	694.5	192.7	105.46	39.3	11.5	27.8	18.30	4.43	33.56	416	5.76
1982	1,008.1	280.0	106.30	20.9	6.4	14.5	20.60	4.41	22.81	615	6.68
1990	1,133.6	314.7	106.60	20.9	6.3	14.6	26.23	3.96	15.88	1,422	8.04
2000	1,265.8	351.3	106.74	17.0	7.0	10.0	36.09	3.44	6.72	3,611	8.41

[a]Proportion of urban total population
[b]Persons in junior college or higher
[c]Minority nationalities to total population

repeated registration at only 0.71 percent and omitted registration at 0.56 percent. Thus the difference was only 0.15 percent, which certainly is quite low when compared with the population censuses of many, and perhaps most, other countries. It is also worth mentioning that this modern census confirmed that the country's annual population statistics were rather accurate, for the numbers for total population from the census were very close to those of the annual statistical report. On April 20, 1982, the State Statistical Bureau reported that by the end of 1981 the total population in China was 996,220,000. If we subtract from the total census figure of 1,008,175,288 the 4,240,000 military personnel and the natural increase in population for the last half of 1982, or 6,580,000, which was calculated from the 14.55 percent natural increase rate, we arrive at 997,366,288 persons. Therefore the discrepancy between the totals in these two reports is only 1,100,288 or 1.10 percent of the total population as shown by the census.

Preparation for the census began at the end of 1979, and the period from July 1 to July 10, 1982, was chosen as the standard census time. About 4 million survey takers and 1 million supervisors, as well as more than 100,000 computer analysts, participated in the undertaking. China thus mobilized for the census more than twelve times the manpower used to build the Great Wall about two thousand years ago.

A comparison of China's population between 1964 and 1982 reveals an increase of 313,593,529 persons, or 45.1 percent, an annual rate of 2.1 percent. However, there is a difference in growth rates if we divide the eighteen years into two periods. During the first nine years, from 1964 to 1973, the total increase was 186,660,000, which averages 20,740,000 persons annually to give a growth rate of 2.68 percent. The subsequent nine years, from 1973 to 1982, showed an increase of only 126,930,000, giving an average of 14,100,000 persons annually, with an increase rate of only 1.51 percent. This clearly reflects that China's program of population control, initiated in the early 1970s, was having the desired effect.

From this new census it was learned that total births in 1982 equaled 20,689,704 or a rate of 20.91 per thousand, while deaths for the same year totaled 6,290,103 or a death rate of 6.26 per thousand. Thus, the natural increase in population was 14,399,601, or a rate of 14.65 per thousand. In 1964 the birth rate was 39.34 per thousand, the death rate was 11.56 per thousand, and the natural increase rate was 27.78 per thousand. Such a drastic reduction of the natural increase rate from 27.78 per thousand to 14.65 per thousand within eighteen years appears to be quite rare in the world's population evolution.

As for urbanization, the number of urban residents in 1982 had increased by 79,485,541 persons since 1964 or 62.5 percent, which is higher than the total population increase of 45.1 percent. During these eighteen years, the percentage of urban inhabitants within the total population increased from 18.3 percent in 1964 to 20.6 percent in 1982, a growth of less than 2.2 percent. Thus, the rate of urbanization in China has been rather low compared to that of other countries. This may relate to regulations that made rural-urban migration difficult and to the fact that many urbanites, especially young people, were "resettled" to rural areas during the upheavals of the Cultural Revolution.

Racially, from 1964 to 1982, the Han population increased only 285,407,456 persons, or 43.8 percent, while all minority groups increased 27,309,518 persons or 68.4 percent. The percentage of minorities to the total population of China has risen from 5.8 percent in 1964 to 6.7 percent in 1982. The main reason for this is due to the government policy of supporting the economic development of minority groups, allowing them more autonomy, and loosening the enforcement of birth control measures.

For every 100,000 persons, the education level between 1964 and 1982 changed as follows: University graduates increased from 416 to 615 persons; high school graduates from 1,319 to 6,622 persons, junior high school graduates from 4,680 to 17,758 persons, and elementary school graduates from 28,330 to 35,377 persons. At the same time, the percentage of illiteracy decreased from 33.56 percent to 22.81 percent. All these data reflect the fact that, in spite of a ten-year interruption by the Cultural Revolution, education still advanced to a certain degree. In most industrial countries, there are several thousands or even tens of thousands of college graduates for every 100,000 people, whereas in China the total still remains quite low. Thus, China still has to expend much effort in education in order to implement its modernization and industrial programs.

In comparing the census of 1990 to that of 1982, the total population of 1990 was 1,133.6 million compared to 1,008.1 million in 1982, an increase of 12.45 percent in that period. The population density in 1982 was 280 per square mile; by 1990 it had increased to 314.7 per square mile. The male to female ratio (female = 100) changed from 106.3 in 1982 to 106.6 in 1990. The birth rate increased slightly from 20.91 per thousand in 1982 to 20.98 in 1990, and the death rate declined from 6.36 in 1982 per thousand to 6.28 in 1990. The natural increase thus rose from 14.5 per thousand in 1982 to 14.6 in 1990. During this period from 1982 to 1990, the Han Chinese increased from 940 million to 1 billion or about 0.94 percent. Minority groups increased from 67.2 million to 91.2 million, or 35.52 percent. The urban population increased from 206 million to 296 million or from 20.6 percent to 26.2 percent of the whole of China.

The natural increase of population, according to the 1990 census, was lowest in Shanghai, only 4.96 per thousand, and in Beijing, Zhejiang, Tianjin, and Liaoning it was lower than 10 per thousand. There were eight provinces in which the increase rates were between 10 and 15 per thousand. They were Sichuan, Heilongjiang, Jilin, Shandong, Hebei, Jiangsu, Inner Mongolia, and Guangxi. The increase rate was more than 15 per thousand in seventeen provinces and autonomous regions including Tibet, Xinjiang, Ningxia, and Gansu, among others.

Among the 56 minority groups, those numbering over a million increased from 16 groups in 1982 to 19 in 1990. Those numbering between 100,000 and 1,000,000 increased from 13 in 1982 to 15 in 1990. During this period, the minority groups that at least doubled their populations included Russians, Manchus, Tujia, Gelo, Xibe, Lhopa, Monba, and Hezhens.

During the forty-one years from 1949 to 1990, the Chinese population changed in the following ways:

First, during this period, China experienced two peak periods of births. The first was from 1953 to 1957, when the population increased by 71,710,000 persons, with an average increase of 14,340,000 persons a year. The second stage was from 1962 to 1971, when the increase was 193,700,000, with an average increase of 19,370,000 persons a year. During these two periods the birth rates were over 30 per thousand.

Second, the death rate has decreased from 17 per thousand in 1953 to 11.56 in 1964, to 6.36 in 1982, and to 6.28 in 1990.

Third, the young age composition of the Chinese population has increased, thus increasing the labor force.

Fourth, Chinese traditional culture may have an effect on the birth rate. There have been few single females, a low divorce rate, and early marriage. The Chinese consider males to be more important than females.

Fifth, average life expectancy has increased from 35 years in 1936 to 67.88 years in 1982 and to 71 in 1990.

Sixth, the educational status of the population has been raised. Before 1949, educated people made up only one-fourth of the whole population but now two-thirds are educated. The percentage of high school graduates has increased from 2 percent to 29 percent.

Seventh, due to the process of industrialization, urbanization has accelerated. In 1949, the urban population was only 10.61 percent of the total population, but by 1989 it had increased to 51.7 percent. This has generated many problems related to housing, transportation, public utilities, and pollution.

In 2000, China conducted its fifth national population census. This census was different from the four previous ones in that it was the first census of the new millennium and the first census conducted since China had initiated its market economic system. Also, the census was by far the largest population census ever conducted in human history. About 10 million people participated directly in the census work. In addition, new technology, such as optical character reading (OCR), was adopted in the design of the census questionnaire.

The major findings of the 2000 census are as follows.

1. In 2000, China had a population of 1,265.83 million (excluding Taiwan, Hong Kong, and Macao).

2. The population increased by 132.15 million persons, or 11.66 percent, over the 1990 census population of 1,133.68 million during the previous ten years and four months. The average annual growth was 12.79 million persons, or a growth rate of 1.07 percent.

3. As for sex composition, 653.55 million persons, or 51.63 percent, were males, while 612.28 million persons, or 48.37 percent, were females. The sex ratio (female = 100) was 106.74.

4. In age composition, 289.79 million persons were in the age group of 0–14, accounting for 22.89 percent of the total population, 887.93 million people were in the age group of 15–64, accounting for 70.15 percent, and 88.11 million persons were in the age group of 65 and over, accounting for 6.96 percent. As compared with the 1990 population census, the age group of 0–14 was down by 4.82 percent, and the age group of 65 and over was up by 1.39 percent.

5. There were 455.94 million urban residents, accounting for 36.09 percent of the total population; the number of rural residents stood at 807.39 mil-

lion, accounting for 63.91 percent. Compared with the 1990 census, the number of urban residents rose by 9.86 percent.

6. The average size of family household was 3.44 persons, or 0.52 persons less than the family size of 3.96 persons of the 1990 population census.

7. In regard to literacy, 85.07 million persons, or 6.72 percent, remained illiterate, compared with 15.88 percent in the 1990 census, a drop of 9.16 percent.

8. About 45.71 million persons held a college degree, or 3,611 per 100,000. That compares with 1,422 university graduates per 100,000 in the 1990 census.

9. As to the racial composition, 1,159.40 million persons, or 91.59 percent, were of Han nationality, and 106.43 million persons, or 8.41 percent, were of various national minorities. Compared with the 1990 census, the Han population increased by 116.92 million persons, or 11.22 percent, while minorities increased by 15.23 million persons, or 16.7 percent (Table 12.1).

Patterns of Population Distribution

A distinctive feature of the Chinese population is its uneven distribution. The deserts of Western China are among the emptiest regions of the globe, while few places can match the crowded conditions in the delta area of the Yangtze River. China's geographers have long realized this uneven spatial pattern of the Chinese population.

Following the eastern provincial boundaries of Inner Mongolia, Ningxia, Gansu, Qinghai, and Xizang (Tibet), China can be divided into two parts: southeastern and northwestern. The territories of both sections are not too different: roughly 1.83 million square miles in the northwestern section and 1.44 million square miles in the southeastern section. Yet the population size for each section varies greatly. The southeastern part has 1.063 billion inhabitants, or 93.08 percent of the total population of China, whereas only 70 million people make their home in the northwestern part, or only 6.2 percent of the country's total population.

If the 1990 population data is examined in more detail, it can be seen that the eastern coastal provinces (Liaoning, Hebei, Shandong, Jiangsu, Zhejiang, Fujian, and Guangdong) have a total population of 384,145,778, or 41 percent of the total population, while the sum amount of land is only 15 percent of China's total. It reveals a sharp contrast in the distribution of China's population and a high concentration of Chinese population along the coastal areas.

From statistical data of 1990, Sichuan province had the largest population in China with 107,218,173 or 9.8 percent of the total, followed by Henan and

Shandong, with 85 million and 84 million respectively, and Jiangsu, Guangdong, Hebei, and Hunan provinces each with more than 60 million. Other provinces with more than 50 million people include Anhui and Hunan. The above nine provinces combined comprise less then 20 percent of China's territory but have more than half of China's population. The rest of the twenty-one municipalities, provinces, and autonomous regions occupy 82 percent of China's total land but have only 48 percent of the population. The populations of Ningxia Hui Autonomous Region and Qinghai are 4.6 and 4.4 million respectively. The smallest population unit is the Xizang (Tibet) Autonomous Region, with only 2.1 million inhabitants.

The vertical distribution of China's population, in general, shows that the Chinese prefer to live on flat land. Exceptions are the Tibetan herdsmen, who inhabit a plateau more than 15,744 feet above sea level, and people in the Turfan Basin in Xinjiang, who live in a depression 499 feet below sea level. As a whole, the higher the land elevation, the sparser the population. In China, plateaus and mountains over 9,840 feet above sea level make up 25 percent of the territory but contain fewer than 6 million people, or 0.6 percent of the total population. Hilly land with an elevation of 1,640 feet to 9,840 feet constitutes 50 percent of the land but is occupied by less than 20 percent of the total population. Plains and hilly land less than 1,640 feet above sea level, mostly in the eastern coastal part of China, cover only 25 percent of the land but are inhabited by more than 80 percent of the population.

China's population density was only 351.3 persons per square mile in 2000. Compared with Bangladesh's 1,520 persons per square mile, the Netherlands' 860, and Japan's 769, it is not very high. However, this average figure does not portray reality adequately because the density pattern varies greatly throughout the country. For example, Jiangsu province has a population density of 1,567 persons per square mile (highest in the country), while Xinjiang or Tibet holds the other extreme, at less than 4 people per square mile. In fact, Xinjiang and Tibet form one of the least densely populated regions in the world (Table 12.1, Figure 12.2).

As a whole the total Chinese population density has increased. The population density of 351.3 people per square mile in 2000 represents an increase of 37 persons when compared to 314 persons per square miles in 1990. It increased by 71 persons since 1982, by 187 persons since 1964, and by 190 since 1963. The increase in population density also varies from region to region. Using the data from 1964 and 1982, and considering the eleven coastal political units, including Beijing, Shanghai, Tianjin, Jiangsu, Shandong, Zhejiang, Hebei, Guangdong, Liaoning, Fujian, and Guangxi, population density increased from 603 people per square mile in 1964 to 823 persons per square mile in 1982, an increase of 37.8 percent.

In the inner interior regions (including eighteen political units: Henan, Anhui, Hubei, Hunan, Jiangxi, Sichuan, Shanxi, Shaanxi, Guizhou, Jilin, Yunnan, Heilongjiang, Ningxia, Gansu, Inner Mongolia, Xinjiang, Qinghai, and Xizang), the population density increased from 123 persons per square mile in 1964 to 185 persons per square mile in 1982.

TABLE 12.2   Composition of Population, 1949–1989

	Total Population (10,000)	Male		Female		Urban		Rural	
		Population (10,000)	%	Population (10,000)	%	Population (10,000)	%	Population (10,000)	%
1949	54,167	28,145	51.96	26,022	48.04	5,765	10.6	48,402	89.4
1952	57,482	29,833	51.90	27,649	48.10	7,163	12.5	50,319	87.5
1957	64,653	33,469	51.77	31,184	48.23	9,949	15.4	54,704	84.6
1965	72,538	37,128	51.18	35,410	48.82	13,405	18.0	59,493	82.0
1978	96,259	49,567	51.49	46,692	48.51	17,245	17.9	79,014	82.1
1979	97,542	50,192	51.46	47,350	48.54	18,495	19.0	79,047	81.0
1980	98,705	50,785	51.45	47,920	48.55	19,140	19.4	79,565	80.6
1981	100,072	51,519	51.48	48,553	48.52	20,171	20.2	79,901	79.8
1982	101,590	52,319	51.50	49,271	48.50	21,131	20.8	80,459	79.2
1983	102,764	53,026	51.60	49,738	48.40	24,150	23.5	78,614	76.5
1984	103,876	53,600	51.60	50,276	48.40	33,136	31.9	70,740	68.1
1985	105,044	54,308	51.70	50,736	48.30	38,446	36.6	66,598	63.4
1986	106,529	55,075	51.70	51,454	48.30	44,103	41.4	62,426	58.6
1987	108,073	55,658	51.50	52,415	48.50	50,362	46.6	57,711	53.4
1988	109,614	56,473	51.52	53,141	48.48	54,369	49.6	55,245	50.4
1989	111,191	57,314	51.55	53,877	48.45	57,494	51.7	53,679	48.3

The change in population density between the coastal and interior regions can be explained as follows: First, it is government policy to develop the economy of the interior part of the country, and second, a large number of persons were encouraged to move from the dense coastal area to the less peopled interior. Third, the natural increase rate of the coastal population is lower than the natural increase rate of the interior population.

Table 12.2 shows the composition of the population from 1949–1989.

From Population Explosion to Birth Control

During the past fifty years, China's population has more than doubled, increasing from 582 million in 1953 to more than 1.265 billion in 2000. In this time, the rate of natural increase has averaged between 2 per thousand and 2.5 per thousand or 20 million to 25 million people per year.

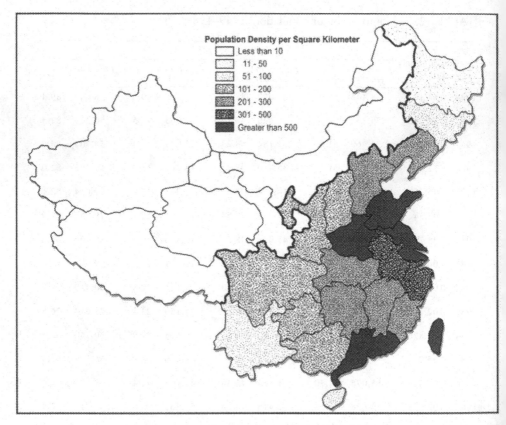

FIGURE 12.2   Population Density (per square kilometer)

This explosion of population indicates that China either wavered between conflicting policies or lacked a definitive policy altogether. In the beginning years of the People's Republic, her leaders believed the more people she had, the more powerful the country would be. Thus, a high birth rate was encouraged, and honors were bestowed on multiple-child birth mothers as heroines. As a result the population exploded.

The population problem inevitably evoked heated debate. Some economists saw the danger of uncontrolled population increase and advocated birth control. Others argued that though the rapid growth of population would lead to greater food consumption "by mouth," it should not be forgotten that an increasing population could create more productivity "by hands." From 1949 to 1968, there were twelve years in which the birth rate was far higher than before liberation. For example, in 1954 the birth rate reached 37 per thousand and in 1964 it jumped to 39.3 per thousand. During this period, demographic statistics reflected government population policy.

After 1949, many causes of death in China, such as war, disease, and famine, were reduced, and efforts were focused on improving medicine, public health services, water conservation, and so on. After 1957, the death rate in China gradually declined. With a high steady birth rate, however, the rate of natural increase still remained high. Rapid population growth soon led to serious food problems.

China feeds one-fourth of the world population with only one-seventh of the world's cultivated land. In past decades, through hard work, grain production has more than doubled—for example, from 110 million tons in 1949 to 318 million tons in 1980. However, this hard-earned output of grain satisfied only the immediate needs of the increased population rather than helping to improve the living standard.

Furthermore, China's agricultural labor force increased from 200 million in the early 1950s to over 300 million in 1980, while the cultivated land declined from 106.7 million hectares to 100 million hectares. Consequently, China's cultivated land per capita of agricultural labor force decreased from 0.41 acres in the early 1950s to 0.25 acres in early 1980. There is a limit to how much more land can be reclaimed. Excessive reclamation is already destroying the ecological balance. There is also a limit to the potential increase in grain output per capita.

Other negative consequences of rapid population growth, ranging from unemployment and lack of capital accumulation to poorer educational achievement, led China's leaders to pay attention to the overpopulation problem.

In the early 1970s, Chinese authorities advocated two children for each family, saying, "One is not too few, two is good, three is too many." Later on, this model changed to "One is best" and then to "One is enough." China now hopes that her population will be stabilized at less than 1.2 billion. In other words, the rate of natural increase would be reduced. This will be a difficult task for the country to accomplish, especially in view of the fact that China is a young country whose population has many years of fertility remaining.

In the attempt to reach zero population growth, China encourages couples to limit themselves to one child and taxes those who have more than two. In early 1979, the Chinese government initiated a campaign to promote one-child families by offering a recognition certificate, which entitles the holder to a range of economic, educational, and health benefits. The demographic pattern during these three decades changed from "high (birth), high (death), and high (increase)" in the 1950s to "high (birth), low (death), and high (increase)" in the 1960s, and to an attempt to reach "low (birth), low (death), and low (increase)" in the 1970s and beyond.

In order to implement family planning, China has mobilized demographers, economists, and sociologists to design several population models. Their best estimates suggest that one hundred years from now the most suitable population for China would be 650–700 million. To achieve that goal, the next thirty to forty years will be of crucial importance for encouraging every couple to have only one child.

What are people's attitudes toward the government birth control policy? At the moment, 65 percent of the Chinese population is under thirty years of age. The attitudes of this group of young people toward birth control are vital in determining future population growth and the success of China's family planning.

TABLE 12.3   Survey Results on the Preferred Number of Children

Preferred Number of Children	Beijing		Sichuan	
	Urban Area (%)	Rural Area (%)	Urban Area (%)	Rural Area (%)
0	5.79	1.88	0	0.75
1	57.7	23.51	61.94	22.96
2	35.76	72.73	38.06	73.33
>2	0.75	1.88	0	2.96
Total	100	100	100	100

Under the sponsorship of UNESCO, the Academy of Social Sciences in Beijing conducted a survey of China's youth regarding the birth control policy. They selected Beijing and Sichuan as the loci for their studies. The survey found that 90 percent of these Chinese youth agreed with the birth control policy and that the old concept of "the more children you have, the more happiness" apparently no longer holds in today's China. Table 12.3 shows that there is not much difference between attitudes in Beijing and Sichuan. The percentages in the urban areas of Beijing are similar to the percentages in the urban areas of Sichuan and the percentages of their rural areas are very similar as well. However, the contrast between the urban areas and rural areas is profound. For example, the percentages of individuals in urban Beijing and Sichuan who preferred one child was 57.70 percent and 61.94 percent respectively, whereas in rural areas, the corresponding statistics were only 23.51 percent and 22.96 percent. Only 35.76 percent of urban Beijing and 38.06 percent of urban Sichuan stated a preference for two children, whereas in rural areas, such preferences accounted for 72.73 percent of responses in Beijing and 73.33 percent in Sichuan.

Some fear that the one-child policy will induce social and economic problems. For example, after practicing birth control for a period of time, China's population will be dominated by the elderly. At present, those over 65 years of age constitute less than 2 percent of the total population, while 65 percent of the population is less than 30 years of age. In Western Europe, in contrast, these age groups account for about 10 percent and 20 percent respectively. In the meantime, life expectancy in China is 71, while in Europe it is 78. Therefore it will take several decades before the age composition and life expectancy of the Chinese population are similar to that of Western Europe. If China can, in fact, control its population growth during that time, the population policy can easily be adjusted to the new situation.

Moreover, an older population is not necessarily bad. With a greater life expectancy comes a larger pool of experienced and skilled labor that will increase national productivity, a favorable factor in increasing the state wealth.

Others are concerned that such a policy could result in a shortage of manpower for labor and the armed forces. At present the labor force is 600 million. If China practices a one-child-per-family policy, it is estimated that the labor force will be 760 million by 2010, 740 million by 2020, and by 2030, 550 million. All these figures are far higher than those for other countries. Moreover, the push for modernization will naturally increase technology and capital and decrease the demand for labor. It is safe to state that China will not have a manpower shortage problem in the next century. If she wants to select soldiers from young people between the ages of eighteen and twenty, there will be a pool of 181 million.

No doubt some problems will emerge as China institutes stricter birth control practices under its family planning program. However, in spite of the many difficulties that lie ahead, China has no choice but to limit its rapid population growth.

References

Aird, J. S. 1982. Population Studies and Population Policy in China. *Population and Development Review* 8:267–297.

Chan, A. 1974. Rural Chinese Women and the Socialist Revolution. *Journal of Contemporary Asia* 4(2):197–203.

Chen, C., and C. W. Tyler. 1982. Computer Simulation of Migration and Small-Area Population. *Modeling and Simulation* 13:1199–1203.

_____. 1983. Demographic Implications of Family Size Alternatives in the People's Republic of China. *China Quarterly.*

Chen, M. 1979. Birth Planning in China. *International Family Planning Perspectives and Digest,* September, 92–101.

Chen, P. 1980. Three in Ten Chinese Couples with One Child Apply for Certificates Pledging They Will Have No More. *International Family Planning Perspectives and Digest* 70.

Chinese Sociological Association. 1980. A Survey Report on Child-Bearing Desire and Family Pattern Preference of the Chinese Youth.

Ching, C. C. 1982. The One-Child Family in China: The Need for Psychosocial Research. *Studies in Family Planning* 13:208–212.

Coale, A. J. 1981. Population Trends, Population Policy, and Population Studies in China. *Population and Development Review* 7:84–109.

Goodstadt, L. F. 1981. China's One-Child Family: Policy and Public Response. *Population and Development Review* 8:37–56.

Hai, F. 1973. Women's Movement in Communist China. Hong Kong: Union Research Institute.

Keyfitz, N. 1984. The Population of China. *Scientific America* 250:38–47.

Letour-Leau, C. 1926. *The Evolution of Marriage.*

Li, D. 1957. Birth Control Is a Complex Task. *People's Daily,* March 8.

Li, S. 1982. Developmental Trends in Chinese Population Growth. *Beijing Review* 2, January 11.

Li, W. L. 1970. A Matrix Model of Population Redistribution. *Proceedings of the American Statistical Association,* Social Statistics Section, 189–193.

Liu, W. F. 1974. Family Changes and Family Planning in the People's Republic of China. Paper given at the Annual Meeting of the Population Association of America. New York City, April 18.

Liu, Z. 1982. Population Planning in China. *China Reconstructs,* February.

Ma, Y. C. 1981. *New Theory on Population.* Beijing: Publication Society (in Chinese).

Petersen, W. 1971. The Malthus-Godwin Debate. *Demography* 8:13–26.

Population Information Program, John Hopkins University. 1982. Population and Birth Planning in the People's Republic of China. *Population Reports* 25: J577-J618.

Rogers, E. M. 1973. *Communication Strategies for Family Planning.* New York: Free Press.

Saith, A. 1981. Economic Incentives for the One-Child Family in Rural China. *China Quarterly* 89:493–500.

Song, J. 1980. On the Target of China's Population Development. *People's Daily,* March 7.

Wang, D. 1980. Prospects of Employment. *Beijing Review* 2, January 11.

Wu, J. 1957. A New Viewpoint on the Chinese Population Problem. *China Reconstructs* 102.

Xu, X. 1982. Families Carry Out the Population Control Policy in Rural Areas. *People's Daily,* February 5.

Yang, D. Q., et al. 1979. *Population.* People's Publishing House of Hebei (in Chinese).

_____. 1981. *Population Theory.* People's Publishing House of Henan (in Chinese).

Zhang, Z. Y., et al. 1981. *China's Youth Intentions Toward Birth Control: A Survey Report from Cities and Rural Areas between Beijing and Sichuan.* Beijing: Academy of Social Sciences (in Chinese).

13

Population Characteristics and Ethnic Diversity

David W. S. Wong and Kevin Matthews

Although the most obvious changes in China since the reform have been in the economic realm, other aspects of the Chinese society have responded directly or indirectly
to the changing economic conditions. Population is one such dimension that has been
influenced significantly by the economic reform. During the era of reform and change,
China also implemented population policies to parallel developments taking place in
economic policies. Because of its significant impacts and implications, the controversial one-child policy has attracted a great deal of attention (for example, Smith 1991).
However, the discussion of racial and ethnic characteristics, a very important dimension of population, has been rather sparse and superficial. Particularly needed are
examinations of changes in regional ethnic composition and in the geographical distributions of ethnic groups within China because some minority issues have political
and economic importance. Although a long-term trend in the Chinese population is
increasing racial integration, the isolation of some minority groups persists. After two
decades of economic reform and rapid change, it is time to evaluate the level of racial
integration and the magnitude of spatial separation among China's ethnic groups.

Since the official release of the 1990 census data in 1993, numerous studies have
been completed on the current status of population and minority nationals in China.
Many studies also have compared the changes in population between the 1982 census and the 1990 census, with special attention to minorities. This chapter focuses on
changes in several spatial dimensions of population and minority nationals between
1982 and 1990, including segregation levels and ethnic diversity. Historically,
minorities have been concentrated in peripheral regions of the country. In different
eras, Chinese governments adopted various policies to contain, assimilate, or subdue

minority groups. Recent policies have sought to integrate them with other groups, especially the Han. The research results reported in this chapter may shed light on these developments as well as help to evaluate the current population policies toward minority groups.

The chapter begins by assessing previous studies of the Chinese minority population based primarily on census data. It then provides a brief discussion of the characteristics of the minority population between the two census periods. The third section focuses on the spatial distributions of the minority nationals between 1982 and 1990. Section four introduces concepts of segregation measurement, which are used in the next section to examine the level of ethnic mixing. Both of these sections examine the issue at the national and provincial levels. The last section summarizes the major findings and discusses briefly some implications related to these results.

Relevant Chinese Population Studies

Since the release of the Chinese census data for 1982 and 1990, numerous studies have analyzed demographic changes and the spatial distribution of minority groups in China. After the release of the 1982 census data, Banister (1987) compared changes in the size of different ethnic groups since the 1950s. Because the Chinese government encouraged minorities to reinstate their origins, the minority population counts reversed the previous declining trends. However, some minority groups such as the Miao, Manchu, Bouyei, and Dong had been significantly assimilated by Hans. Similarly, the China Financial and Economic Publishing House (1988) compared changes in ethnic composition within China over three decades and examined changes in other demographic characteristics. This study reached conclusions similar to Banister's (1987); that is, the size of minority nationals had increased rapidly, and their shares of the Chinese population had been rising. The increase in their percentages was partly related to the census reclassification process, but it was also due to the fact that most minority groups have higher birth and hence natural growth rates than do the Hans.

Focusing on the growth of minority populations, Hsu (1993) reported that the number of recognized races increased from 41 in 1953 to 55 in 1979, and those groups with over 1 million members increased from 10 in 1953 to 18 in 1990. Hsu (1993) argued that the increased minority population size was due to the relaxed birth control policies for minority groups and the reinstating of racial identity. Che-Alford (1986) examined several demographic statistics relating to major Chinese ethnic groups and identified the major geographical clusters of each group. As expected, the minority groups were concentrated in peripheral and autonomous regions. This study was similar to the one conducted by Pien (1990), who also included data on education. Pien concluded that minority populations, especially women, had a significantly higher illiteracy rate than the national average, partly because of their peripheral locations. Poston and Shu (1987) also used the 1982 census data to provide a very detailed geographical description of the distribution of different ethnic groups in China. They generated a series of maps showing the provincial dominance of major ethnic groups.

More recently, Gladney (1995) examined the degree of ethnic diversity within the Chinese population based partially on the 1990 census data. He argued that the Han are rather culturally and linguistically diverse. Also using the 1990 census, Zhang (1992) examined the demographic characteristics of minority populations. Since the 1980s, fertility rates of minority women have been declining (except for the Manchu), while the mortality rate among minority groups remains higher than that of the Hans. Yuan et al. (1997) provided a brief description of geographical distribution of minority groups and discussed their levels of educational achievement using the 1990 census data. A comprehensive review of the Chinese minority population issues can also be found in Smith (1991). Most of these studies included some discussion concerning the location of minority groups. Few, however, have provided any measures of ethnic diversity or segregation levels among these groups or changes in the population geography. This study analyzes the changes in segregation and ethnic diversity in the Chinese population between the two censuses.

The Population Geography of China

At the time of the 1982 census, there were 29 provinces or provincial-level units, including 21 provinces, 4 autonomous regions, and 3 municipalities. In the late 1980s, Hainan Island, a large island off the southern coast of China and formerly part of Guangdong province, was officially recognized as a province. For some comparisons between the 1982 and 1990 censuses, it is convenient to aggregate the data for Hainan with Guangdong. In terms of population, provinces are usually more populous than autonomous regions. Between the two censuses, most provinces did not change their rankings according to population size. While most autonomous regions are small, Guangxi ranked eleventh in 1982 and tenth in 1990. Because Hans account for more than 90 percent of the entire Chinese population, most provinces are dominated by them, with the exceptions of Xizang and Xinjiang, in which the majority are Tibetans and Uyghurs, respectively.

Between the two censuses, provincial populations grew at different rates (Table 13.1). Henan had the largest absolute increase, followed by Guangdong. The smallest absolute growth was found among the smaller autonomous regions and the municipalities. It is not clear what factors contributed to large increases in some provinces but not others. However, most high-growth provinces have large cities. It is possible that due to the recent economic reform, cities have been thriving and thus attracted more people as the central government relaxed internal migration control. Jiangxi ranked first in terms of rate of increase in minorities, although its rank according to absolute increase is rather low. It is very likely that the high rate of increase in minorities was not attributable to demographic factors such as birth, but rather due to the migration of minority populations into the area. Fujian and Jiangsu are other provinces experiencing a similar growth.

The rankings by absolute growth (Table 13.1) are little different from those by population size. However, with respect to growth rates, and using the 1982 population size as the base, the five autonomous regions all had relatively high growth rates (with the exception of Inner Mongolia). Ningxia and Xizang ranked first and second, while Guangxi and Xinjiang ranked sixth and seventh, respectively. Among the municipalities, Beijing experienced the fastest growth (ranked third), while the other two (Shanghai and Tianjin) did not grow as fast as most of the provinces.

As noted above, the growth of minority populations outpaced that of the Han between the two censuses, and in general, minority groups are concentrated geographically. What is unique about China's minorities compared with those of Western countries is that they are generally not thought of as belonging to different races but are distinguishable from one another on the basis of cultural and linguistic differences. As Poston and Shu (1987) summarized, ethnic conflicts have characterized Chinese history for more than two thousand years as the Han, originally one of the many tribes located in central China, started expanding and conquering non-Han people through military power and cultural assimilation. Many non-Han tribes fled to the peripheral regions, while the expansion of the Han to the south was easier than to the west and north. Still, the non-Han people tended to choose inaccessible locations or areas with a difficult physical environment to distance themselves from the Han majority. Some of these groups have consequently preserved their cultural characteristics to a significant degree, including relatively high growth rates as compared to the growth rate of the Hans.

Many studies have described the general distribution of minority populations (e.g., Poston and Shu 1987; Pien 1990). The overall distribution of non-Han people did not change significantly between 1982 and 1990. East and Central China have always had the lowest percentages of minority people, and minority nationals cluster in the peripheral regions. But there are several major differences between the two census years.

Several provinces experienced an increase in the percentage of minorities between 1982 and 1990. This group includes Inner Mongolia, Hebei, and Beijing in the north, Liaoning in the northeast, and Sichuan in the southwest. Anhui and Shandong also experienced small increases. Only two provinces, Xizang and Guangdong, experienced a decline in the percentage of minorities. The reason for the decline in Xizang is not clear, but the change in Guangdong is due to the separation of Hainan Island from the province.

Characteristics of the Minority Population: 1982–1990

The provincial-level minority population data used in this study are derived from the 1982 and the 1990 censuses. Though county-level analysis would have been preferable, digital county-level data for 1982 are not yet available. The provincial boundary file for cartographic representation used in this study is available from the

TABLE 13.1   Population Rankings of Provinces

Rank in Absolute Growth	Rank in % Growth	Rate of Growth (total)	Rank in Increase in Minorities	Rate of Increase	% Growth in Minorities
Henan	Ningxia	0.1951	Guizhou	Jiangxi	3.5761
Guangdong	Xizang	0.1784	Liaoning	Hebei	1.8228
Shandong	Beijing	0.1721	Guangxi	Hunan	1.1914
Hebei	Guangdong	0.1701	Hunan	Liaoning	1.1187
Sichuan	Fujian	0.1614	Yunnan	Fujian	0.8560
Hunan	Guangxi	0.1599	Xinjiang	Guizhou	0.5144
Jiangsu	Xinjiang	0.1586	Hebei	Guangdong	0.3912
Anhui	Hebei	0.1524	Sichuan	Inner Mongolia	0.3904
Hubei	Henan	0.1493	Inner	Jiangsu	0.3863
Guangxi	Qinghai	0.1441	Jilin	Jilin	0.3802
Jiangxi	Gansu	0.1432	Guangdong	Sichuan	0.3356
Yunnan	Shaanxi	0.1376	Heilongjiang	Zhejiang	0.3165
Fujian	Shanxi	0.1371	Hubei	Shanxi	0.2898
Shaanxi	Jiangxi	0.1363	Xizang	Beijing	0.2842
Guizhou	Yunnan	0.1357	Qinghai	Henan	0.2623
Liaoning	Guizhou	0.1344	Ningxia	Shanghai	0.2495
Shanxi	Shandong	0.1340	Gansu	Ningxia	0.2450
Gansu	Tianjin	0.1315	Fujian	Shandong	0.2398
Zhejiang	Anhui	0.1312	Henan	Anhui	0.2389
Heilongjiang	Hubei	0.1289	Shandong	Heilongjiang	0.2383
Inner Mongolia	Shanghai	0.1250	Anhui	Tianjin	0.2337
Jilin	Hunan	0.1231	Jiangxi	Qinghai	0.2229
Xinjiang	Inner Mongolia	0.1132	Anhui	Xinjiang	0.2134
Beijing	Jiangsu	0.1080	Zhejiang	Hubei	0.2034
Shanghai	Liaoning	0.1046	Jiangsu	Yunnan	0.1972
Tianjin	Jilin	0.0931	Tianjin	Gansu	0.1944
Ningxia	Heilongjiang	0.0781	Shaanxi	Xizang	0.1935
Qinghai	Sichuan	0.0753	Shanxi	Guangxi	0.1896
Xizang	Zhejiang	0.0659	Shanghai	Shaanxi	0.1752

Consortium for International Earth Science Information Network (CIESIN). The data cover territories directly under the control of the People's Republic of China. Taiwan, Hong Kong, and Macau are regarded as parts of China but are listed in the database with missing values. Hainan Island, a part of Guongdong Province in 1982, became a separate province in the 1990 census.

In both the 1982 and 1990 censuses, the Chinese government recognized 58 minority groups, including an "unknown" category of probably 350 other ethnic groups (Gladney 1995). The "naturalized" group is also one of the 58 groups. From 1982 to 1990, the total Chinese population increased from slightly over 1 billion to 1.13 billion, an increase of approximately 12.6 percent over the eight years. Among the 1 billion or so people, the majority are Hans, an ethnic group which itself is relatively diverse (Gladney 1995). From 1982 to 1990, the Han population increased from 931 million to 1,039 million, an 11.6 percent increase, which is below the national rate. Thus the share of the Hans in the entire population declined from

93.55 percent in 1982 to 91.92 percent in 1990, although the Han population registered the greatest absolute increase. The second-largest ethnic group is the Zhuang, at approximately 15 million people, about twice as many as the third-largest group, the Hui in 1982 and the Manchu in 1990.

Figure 13.1 shows the population size of each minority group in the two censuses. Because the Han population is disproportionally large, it is not included in this figure so that the differences in population size among other groups are more visible. The ethnic groups are arranged in ascending order according to their 1990 population sizes. With only two exceptions (naturalized and unknown), all ethnic groups experienced an absolute increase between 1982 and 1990. The increases are much smaller than that experienced by the Hans (108 million) and range from several hundred to several million people. The increase of each group as a percentage of its 1982 population is reported in Figure 13.2. It is clear that some groups grew faster than others. Eight groups (Tujia, Xibe, Lhopa, Manchu, Hezhen, Russian, Monba, and Gelo) grew more than 100 percent between 1982 and 1990—certainly more than they could have grown by natural increase alone. In general, the population growth rates of the minority groups are higher than that of the Han (Pien 1990), though they have recently slowed (Zhang 1992). But the relatively high rate of natural increase cannot be responsible for such dramatic growth of the minority populations between the two censuses. Several scholars have discussed the phenomenon of "coming out," whereby minority nationals claim back their ethnic identities due to the new population policies. This phenomenon will be discussed later.

Ethnic groups with a growth rate lower than the national rate of 12.6 percent include the Han, Naxi, Korean, unknown, and naturalized. Zhuang, a group exhibiting a very small percentage increase, had an absolute increase of approximately 2 million—a number much larger than the total population of many small ethnic groups.

Because of the different growth rates, rankings of minority groups according to size also changed from 1982 to 1990. Han and Zhuang were the largest and the second largest in both years. Lhopa was the smallest in both years. Fourteen other groups retained their ranks in the eight-year period. Major shifts in rank were experienced by Uygur (from 4th to 6th), Manchu (from 7th to 3rd), Korean (from 12th to 14th), and Gelo (from 37th to 23rd). Among these, the upward shift of Gelo from 37th to 23rd is the most dramatic. Most other groups shifted up or down by only two to three positions.

Dominance of the Hans

As described above, the Han people dominate eastern and central China, whereas the minority groups are mostly found on the periphery. Intuitively, the Han and minorities do not mix well geographically. Assessing the level of segregation among them can reflect whether the Han and minorities are spatially separated. In this section, we explore changes in the dominance of the Han and changes in segregation levels between the 1982 and 1990 censuses.

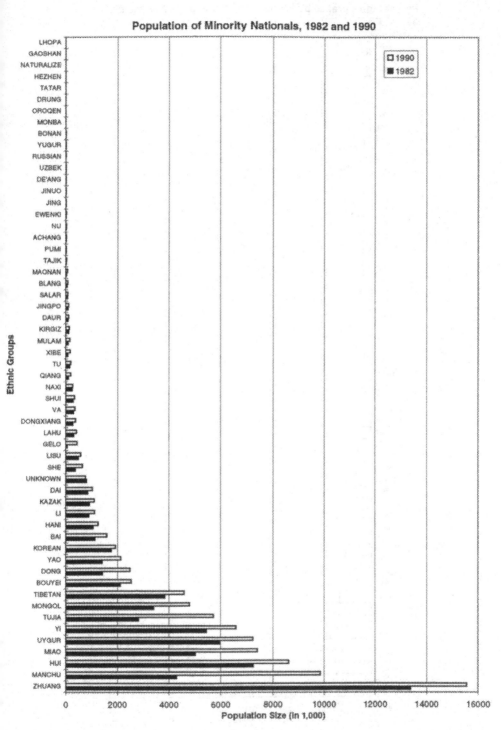

FIGURE 13.1   Population of Minority Nationals, 1982 and 1990

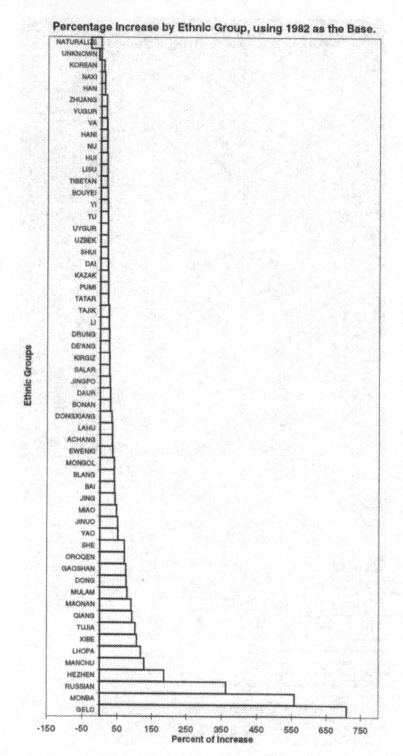

FIGURE 13.2   Percentage Increase in Ethnic Groups, Using 1982 as the Base

A commonly used measure of segregation is the segregation index or the index of dissimilarity (D), introduced by Duncan and Duncan (1955).[1] The index is frequently applied to compare white (W) and black (B) populations in the context of North America. It reflects how the two groups are distributed across areal units within a study region. If the two groups distribute across all areal units in the same fashion, then the index will be zero, indicating no segregation. Perfect segregation (D = 1) usually means that each areal unit is occupied entirely by one group. The index is a summary measure of segregation over the entire region with no recognition of the possibly various levels of segregation within the entire region.

The segregation index can handle only two groups. In order to apply D to China, we grouped all minorities into one group and calculated D based upon Han and non-Han for both censuses. Because historically, minority populations were pushed to the periphery of the country as the Han became dominant in most parts of China, a high overall segregation level is expected. In 1982, D was 0.6179, but it declined slightly to 0.6029 in 1990. This small decline in segregation levels is not surprising, even as minorities experienced an enormous increase during the eight-year period. Because Han people are so numerous, even the high growth rates of minorities did not increase the relative size of minority populations significantly, and thus the increases in their proportions cannot alter the overall pattern of ethnic mixing and the dominance of the Han.

Though widely used, the two-group D index is not very effective for a multiethnic society like China. Even in North America, the index has its limitations. Morgan (1975) introduced a multigroup version of D, which is labeled as D(m) in order to differentiate it from the traditional two-group index. The index is very similar to the chi-squared statistic that compares the observed with expected frequencies.[2]

The interpretation of D(m) is identical to D. It is also a summary measure of the level of segregation for an entire area. Including all 58 ethnic groups in the Chinese population censuses, the D(m) for 1982 was 0.7349, while the D(m) for 1990 was 0.7231. The difference between the two census years is very small, though there is some evidence that segregation declined slightly.

As we can see from the two segregation indices, the level of segregation is primarily determined by the difference in the distribution pattern of different groups over all provinces. Given the dominance of the Han (93.55 percent in 1982 and 91.92 percent in 1990), the small decline in the share of Han between the two censuses did not alter the segregation level substantially. Extraordinary increases in several minority groups (such as the Gelo, Monba, Russian, Hezhen, and Manchu), which amounted to several millions of people in these groups, were relatively small in comparison to the approximately 1 billion Han. This small change in the segregation level is partly because the measures used so far are summary statistics describing segregation levels for the entire country. It is unrealistic to have expected any significant overall change for the country between the two censuses. Still, the slight decline in segregation levels signified that Han and non-Han mingled more intensely. There appeared to have been more opportunities for mixing among groups and greater potential for enhanced levels of integration. Changes in segregation or diversity at the local scale, which are reflected in the two segregation measures discussed above, are more likely.

Due to the limitations of D and D(m) to reveal local-scale segregation, Wong (1996) suggested decomposing D by using the difference between two ratios to produce D_i. Thus, each areal unit would have a measure showing the difference in the proportion of the two groups in that areal unit. D_i for areal unit i can be defined as

$$D_i = \frac{b_i}{B} - \frac{w_i}{W},$$

where b_i and w_i in the context of North America are commonly referred to as black and white population counts in areal unit i, and B and W are total black and white populations of the entire study region. This measure can indicate which group has a larger than proportional share of the population mix in the given areal unit and can also highlight the extent of the dominance. Figure 13.3 includes two maps showing D_i for all areal units in the two census years and another map comparing the D_i in the two years. D_i is calculated by comparing the proportions of Han and non-Han in provincial units. A positive value means that the Han are more concentrated than non-Han in that areal unit, and vice versa.

In general, regions with negative values (i.e., where non-Han are more dominant) are found mostly on the periphery of the country in both years, except in the northeast (i.e., Heilongjiang and Jilin, where the dominant groups are Manchu and Korean). The strong dominance of the Han is apparent in the central part of eastern China, as well as in Sichuan and Guangdong (only for 1990, after Hainan Island was separated from it). Comparing the maps for 1982 and 1990 and also referring to the map comparing the D_i for the two years (1990–1982), there is clear evidence of Han dominance in Guangdong—an artifact created by the separation of Hainan Island from the province. Otherwise, several southwestern provinces, the west, and a strip cutting across the country from the west to the eastern coast display a moderate increase in the dominance of the Han (i.e., the category of 0 to 0.012 increase in D_i). These results indicate that the dominance of some minority groups concentrated in the south and west (such as Bai, Hani, and Yi in Yunnan, Tibetan in Xizang, Uygur in Xinjiang, and Zhuang in Guangxi) has diminished in those regions.

In contrast, Liaoning, Hebei, Hunan, and Guizhou experienced the largest decrease in the dominance of the Han (–0.03 to –0.014). Most of these provinces or regions have high to moderate levels of non-Han population, and the growth of minority populations in these areas was relatively high (Table 13.1). Major minority groups that concentrated in those provinces include Manchu, Mongolian, and Korean in the northeast, and Bouyei, Dong, Miao, Tujia, Yao, and Yi in the southwest. The results of this analysis seem to suggest that these groups are returning to prominence in those regions. Inner Mongolia, two northeastern provinces (Heilongjiang and Jilin), Sichuan, and several eastern coastal provinces experienced moderate declines in Han dominance.

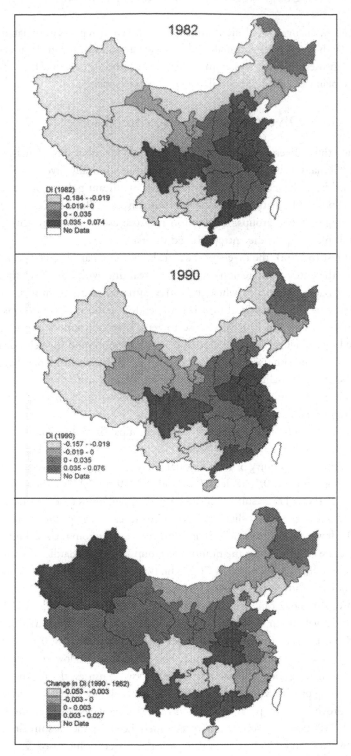

FIGURE 13.3 Modified Segregation Index, D_i, 1982 and 1990

Apparently, changes in ethnic dominance in different regions are related to the relative growth of minority nationals and reclassification of ethnic identities between the two censuses. These changes may also reflect the degree to which different groups are being absorbed or assimilated by the Han majority.

Diversity at the Provincial Level

Given the ethnic diversity and the highly localized distributions of many ethnic groups in China, it is likely that some areas are more ethnically diverse than others. As discussed later, the level of diversity may have significant policy implications. D_i can also help identify spatial variations in ethnic diversity. But like D, D_i is limited to comparing only two groups. A common measure used to describe ethnic diversity of multiple groups is the entropy-based diversity index.[3]

In contrast to $D(m)$, the diversity index is based on an areal unit. It can be calculated for the entire study region or for each areal unit within it. Thus the degree of spatial variation is revealed when the index value for each area unit is mapped. The diversity index, E_i, for areal unit i, is dependent upon the proportions of different groups in i. When E_i is very close to zero, that areal unit is almost entirely dominated by a single group—a value that indicates minimum diversity or perfect segregation. The maximum value of E_i is log(n), where n is the total number of ethnic groups. It can be achieved when each group has the same share of population within a given areal unit. In general, a diversity index with a higher value reflects a lower segregation level, and vice versa.

With the 58 officially recognized ethnic groups (including naturalized and unknown) in China, the maximum possible E_i is 1.763. We can apply the diversity measure to the entire country or to individual provincial units. The diversity index for the entire population is 0.1888 for 1982 and 0.2219 for 1990, indicating a slight increase in diversity. This result is consistent with the findings based on the segregation index, which indicates a slight decline in the level of segregation between 1982 and 1990. However, the diversity levels in both years are far from the maximum diversity that can be achieved given the number of groups. If we standardize the two diversity indices by the maximum (i.e., 1.763), the two indices become 0.107 and 0.126, which represent relatively low values compared to the theoretical maximum of 1.

The diversity index can also be calculated for each province in both 1982 and 1990, and the values can be mapped (Figure 13.4). There are wide variations in the level of diversity among provinces. In 1982, it ranged from 0.5870 in Yunnan to 0.0029 in Jiangxi. These two provinces also had the highest and lowest levels of diversity in 1990. During the eight years between the two censuses, the relative levels of diversity among provinces also produced changes in rankings. Because Hainan Island became a province, many provinces should have shifted down by one rank between 1982 and 1990. However, several provinces experienced significant shifts in ranking. These provinces can be grouped based on whether diversity increased or decreased. Guizhou, Hunan, Hebei, Fujian, and Zhejiang experienced an increase in relative

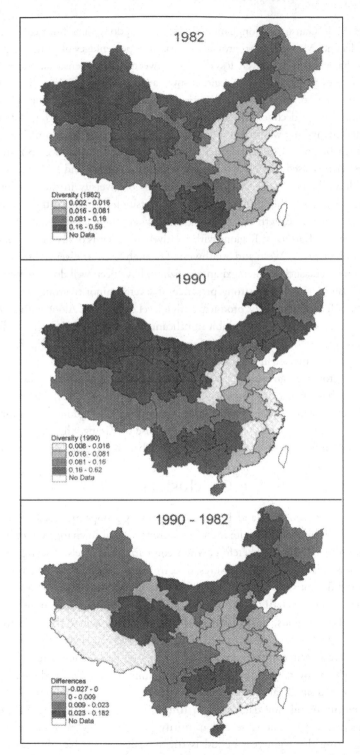

FIGURE 13.4   Level of Diversity for Each Province, 1982 and 1990

diversity, while Gansu, Heilongjiang, Xizang, Guangdong, and Shaanxi all experienced a decline. Many of these provinces have high percentages of minority populations but have become relatively less diverse between the two censuses. Nevertheless, most provinces and autonomous regions show an increase in absolute diversity during the same period. The only two exceptions are Xizang and Guangdong. The diversity level of Xizang declined slightly from 0.0998 in 1982 to 0.0961 in 1990. The decline in Guangdong was much more drastic, due to the separation of Hainan. In 1982, only seven provinces had diversity levels above the national level. But in 1990, that number increased to nine, with the addition of Liaoning and Hainan.

In addition to portraying geographical patterns of diversity in the two censuses, Figure 13.4 compares the absolute changes in diversity level between the censuses. In general, provinces in eastern and central China, where the Han dominate, have relatively low levels of diversity. Regions with relatively high levels of diversity include the Northeast, Northwest, West, and Southwest. The high levels of diversity in the southwestern provinces, such as Guangxi and Guizhou, have been well documented in the literature (Blum 1992). The strong presence of several minority groups in the same provinces with a Han majority, produces a high level of diversity. Along with provinces (Guizhou, Hunan, Hebei) that exhibit significant shifts in the relative degree of diversity, Figure 13.4 indicates that Inner Mongolia, Jilin, Liaoning, and Qinghai experienced the largest increase in absolute diversity. As the Hans are the majority group in all but two provinces, an increase in diversity implies that the share of the non-Han population has increased. Fujian, Guangxi, Yunnan, Sichuan, Xinjiang, and Heilongjiang all experienced a moderate increase. Only Guangdong and Xizang had an absolute decline in diversity level. As in the case of Xizang where the majority is Tibetan, a decline in diversity indicates that the Tibetans increased their dominance.

Conclusions

In summary, between 1982 and 1990, most minority groups increased significantly in percentage terms, but absolute increases were too small to change the ethnic mix of Han-dominated China. Therefore, when comparing the levels of segregation and ethnic diversity using different measures, data from the two censuses show very minor changes and do not show a significant decline in segregation or an increase in diversity at the national level. At the provincial scale, because minority groups tend to cluster in peripheral regions, and almost all of them grew faster than the majority Han, all but one province experienced an increase in diversity. The only exception is Xizang, where a decline in diversity reflects a reduction in the dominance of the Han.

Although the overall ethnic population in China has not changed over the two censuses, several significant issues deserve our attention. When comparing 1982 and 1990 minority population data, a major finding of this study was the dramatically high rate of increase among several minority groups. A several hundred percent increase during this eight-year period is theoretically impossible, though some of this growth is attributable to natural increase. This phenomenon has been discussed by several researchers (e.g., Banister 1987; Gladney 1995). After the 1982 census, minority groups were encouraged to "reclaim" their ethnic identity. The Chinese government's population policy gave minority groups preferential treatment with regard to conducting business and social welfare. Because of the potential economic and social benefits of being minority nationals, those who hid their ethnic identities in the 1982 census were "reclassified" in the 1990 census. This ethnic classification is an artifact that explains why some groups experienced drastic increases over the eight-year period.

Nevertheless, increases in minority groups were partly due to the higher natural increases among minority groups than among the Han. An obvious reason is that the one-child policy was not strictly enforced in regions dominated by minority groups. Minority groups were exempted from this controversial population policy. Even minority nationals living in urban areas were exempted from this policy. Thus, population growth of the minority groups has been faster than that of the Han, and their shares of the population increased together with the level of diversity.

The economic reforms that have taken place in China in the past two decades have allowed Chinese people to become more mobile. Economic prosperity in large cities and the more developed regions in East China should have encouraged minority nationals in the more remote areas to move to the east to explore new economic opportunities. However, we were unable to find strong evidence to substantiate this hypothesis. The municipalities (Beijing, Shanghai, and Tianjin) did not experience significant increases in minority population (Table 13.1). Beijing ranked third with respect to increases in total population, but ranked fourteenth with regard to increases in minority population. Tianjin experienced a similar pattern. Only Shanghai displayed a higher rank with respect to increases in minority population than its rank with regard to increases in total population. When we examine changing levels of diversity, these municipalities did not experience an increase in levels of diversity between 1982 and 1990. But there is evidence indicating that minority populations have migrated to more central (eastern and coastal) sections of China; several provinces that have not been dominated by minority populations—such as Jiangxi, Hebei, Fujian, and Jiangsu—nevertheless experienced high growth rates in minority populations. More detailed study of the migration patterns of minority populations using recent census data is needed to determine whether economic reform has brought a significant number of minority nationals to the largest cities.

Overall, the Chinese population has experienced an increase in diversity at both the national and provincial levels. Increases in diversity in some Western countries are regarded as a desirable phenomenon. Many public policies in Western countries encourage ethnic diversity. In the Chinese context, ethnic diversity has been an indicator of cultural conflict and political instability. It has been the long-term policy to assimilate the minority populations with the Han to diminish the differences among ethnic groups and to maintain political stability. As this study indicates, Xizang, an area of great concern to the Chinese government in the past because of the dominance of Tibetans, has declined in diversity (i.e., with the increasing dominance of

Tibetans). In other words, the separation between the Hans and Tibetans has not diminished significantly. Although areas of high diversity, such as Qinghai, Guizhou, Hunan, Inner Mongolia, Jilin, and Liaoning, have not received much attention recently, they experienced increases in diversity between the two censuses. The influence of the Han within these provinces has diminished, and non-Han populations have increased. The increase in diversity may reflect a greater opportunity for the Han to interact with or assimilate the non-Han. However, the increasing size of minorities could also be regarded as a threat to the dominance of the Han and thus could trigger political instability.

This study has confirmed previous findings that non-Han population groups are on the increase, especially in areas that have not recently been of great concern to the Chinese government. Depending on government policies and the attitude of the Han toward minority populations, the increases in minorities in these areas may provide an opportunity for a greater degree of assimilation and the creation of a more integrated, but perhaps diverse, Chinese population.

Notes

This research was partially funded by the International Incentive Grant Program at George Mason University. The authors would like to thank the editors for their valuable input and the editorial assistance of Joseph S. Wood of the University of Southern Maine, Portland, and Clay Mathers of the U.S. Army Corps of Engineers, Albuquerque, New Mexico.

1. The segregation index or the index of dissimilarity, D, is defined as

$$D = 0.5 * \sum_{i} \left| \frac{b_i}{B} - \frac{w_i}{W} \right|,$$

where b_i and w_i are the minority and majority population counts in areal unit i, and B and W are total populations of the two groups for the entire study area.

2. Multigroup segregation index, $D(m)$, is defined as

$$D(m) = \frac{1}{2} \frac{\sum_{i} \sum_{j} |N_{ij} - E_{ij}|}{\sum_{j} N P_j (1 - P_j)},$$

where

$$E_{ij} = \frac{N_i N_j}{N}.$$

N_{ij} is the population count of ethnic group j in areal unit i, N_i and N_j are, respectively, the total population in area i and total population in group j, N is the total population in the study area, and P_j is the proportion of population in ethnic group j. E_{ij} is the expected or average population of ethnic group j in areal unit i. Therefore, E_{ij} can be regarded as the regional-group average.

3. The entropy-based diversity index for an areal unit i is

$$E_i = -\sum_j (\frac{N_{ij}}{N} * \log (\frac{N_{ij}}{N_{i.}})),$$

or

$$E_i = -\sum_j P_{ij} * \log P_{ij},$$

where $P_{ij} = N_{ij}/N_i$, N_{ij} is the population of ethnic group j in areal unit i, and N_i is the total population in areal unit i.

References

Banister, J. 1987. *China's Changing Population*. Stanford: Stanford University Press.

Blum, S. D. 1992. Ethnic Diversity in Southwest China: Perceptions of Self and Other. *Ethnic Groups* 9:267–279.

Che-Alford, J. 1985. *Population Profile of China*. Toronto: Thompson Education.

China Financial and Economic Publishing House. 1988. *New China's Population*. New York: Macmillan.

Duncan, D., and Duncan, B. 1955. A Methodological Analysis of Segregation Indexes. *American Sociological Review* 20:210–217.

Gladney, D. C. 1995. China's Ethnic Reawakening. Analysis from the East-West Center, No. 18, AsiaPacific Issues.

Hsu, M. 1993. The Growth of Chinese Minority Populations. *GeoJournal* 30(3):279–282.

Morgan, B. S. 1975. The Segregation of Socioeconomic Groups in Urban Areas. *Urban Studies* 12:47–60.

Pien, F.-K. 1990. The Population of Chinese Minority Nationalities. *Issues and Studies* 26(4):43–62.

Poston, D. L., Jr, and J. Shu. 1987. The Demographic and Socioeconomic Composition of China's Ethnic Minorities. *Population and Development Review* 13(4):703–722.

Smith, C. 1991. *China: People and Places in the Land of One Billion*. Boulder: Westview Press.

Wong, D. W. S. 1996. Enhancing Segregation Studies Using GIS. *Computer, Environment, and Urban Systems* 20(2):99–109.

Yuan, H., Z. Zhang, and Y. Wu. 1997. Research on the Cultural Distribution of China's National Minority Population. *Social Science in China* 18:129–144.

Zhang, T. 1992. New Trends in National Minority Population Since the 1980s. *Social Science in China* 13:54–66.

14

Internal Migration

Kam Wing Chan

In the pre-reform era, China practiced a policy of rural-urban segregation and rigorously controlled rural-to-urban migration. The rapid structural change of the Chinese economy and its transition from a planned to a market economy in the recent two decades have eroded many of the previous migration barriers, resulting in a dramatic rise in population mobility. Like many other developing countries, China now faces mass population exodus from the countryside. This is the joint outcome of a number of factors. Most importantly, rural decollectivization has unleashed hidden surplus labor previously locked up in the countryside. At the same time, the rapid expansion of the urban economy, especially in labor-intensive industries, has generated tens of millions of low-skilled jobs. Such a match in supply and demand was made possible by the concurrent relaxation of migratory controls and the development of urban food and labor markets. As migration started to be more prevalent, migrants have also developed extensive networks, which in turn facilitates more flows (see Mallee 1988; Nolan 1993; Chan 1994; Zhou 1996).

The mobility change has not only altered the demographics of many places but has also reshaped the configuration on which China's social and economic system used to be based and carries great importance to China's future. This chapter is an overview of internal migration in China. It first reviews the institutions controlling migration and then examines the recent migratory patterns and characteristics. The final section discusses relevant policy issues.

The *Hukou* System and Migration

Any meaningful analysis of Chinese migration must start by looking at the *hukou*, or household registration system, which affects migration in many important ways.

In China, migration has been an area of heavy state control and regulation. Those wanting to change residence are by law required to obtain permission from the public security authorities. A change in residence is deemed official and approved only when it is accompanied by a transfer of one's *hukou* to the destination. The transfer confers legal residency rights and, most importantly, eligibility for many urban jobs and accompanying subsidized welfare benefits (Cheng and Selden 1994; Mallee 1995). Such a change is granted only when there are good reasons, especially when the move serves, or at least is not at odds with, the state's interests stated in various policies, such as controlling the growth of large cities.

In essence, the *hukou* system in the pre-reform era functioned as a de facto internal passport mechanism. While approvals for migration because of marriage or for seeking support from a family member within the rural areas or within the same level of urban centers were often granted, rural to urban migration was strictly regulated and suppressed in the 1960s and 1970s. In those days, much of this type of migration was reserved for bringing in the necessary labor force in support of state-initiated programs. An approval for self-initiated relocation to a city from the countryside was only a dream for ordinary peasants. Today, peasants can move to many places, but getting a formal approval to register in a medium-sized or large city is still largely beyond their reach.

State-initiated and directed migrations, such as the cadres *xiafang* and youth rustication movements in the late 1960s, were, in large part, involuntary moves, a feature common to pre-reform China and other centrally planned economies (Chan 1994). By contrast, in the reform era, especially in recent years, almost all migratory flows, even including state-initiated migrations within the plan, are voluntary.

Based on whether or not local *hukou* is conferred in migration, three major categories of population flows can be differentiated (Chan 1996b; 2000): 1) migration with local *(hukou)* residency rights (hereafter, *hukou* migration); 2) migration without *hukou* residency rights (non-*hukou* migration); and 3) short-term movements (visiting, circulation, and commuting).

Only *hukou* migration is officially considered as *qianyi* ("migration"). The other two types of mobility are merely labeled *renkou liudong* (population movement or "floating"); the people involved in the latter are called *liudong renkou* ("floating population"). The term implies a low degree of expected permanence; the transients are not supposed to (and are legally not entitled to) stay at the destination permanently, and they are often termed, perhaps not appropriately considering the actual length of stay of many of them, "temporary" migrants. They are not the de jure residents, despite the fact that many non-*hukou* migrants may have been at the destination for years. *Hukou* migration, on the other hand, is endowed with state resources and often called "planned" migration *(jihua qianyi)*. Floaters are a "self-flowing population" *(ziliu renkou)* whose mobility takes place outside the state plans. In the eyes of many central planners, these types of movement are "anarchical" and "chaotic," which is why the officially controlled media in China often use the derogatory term *mangliu* (blind flow) in referring to non-plan mobility.

The "floating population" thus comprises those staying in places other than the place of their *hukou* registration. This is a relatively diverse bundle that includes tourists, business travelers, traders, sojourners, peasant workers contracted from other places, beggars, and other unemployed people. The *People's Daily* in 1995 reported that there was a floating population of 80 million, with about half registered with the public authorities as "temporary population" (*Renmin Ribao* 1995). Two major but different groups of floaters are most prevalent. The first group are "rural migrant workers" *(mingong)*. Most of them are unskilled laborers, and a small percentage of them are skilled craftsmen and traders, often self-employed. The other major group consists primarily of short-term visitors using urban facilities, including overnight tourists and business travelers (Li and Hu 1991). Some *mingong* are seasonal, operating in synchronization with the farm work schedule (with more outflows during the winter off-season).[1]

Social and Economic Characteristics of Migrants

As in many other developing countries, job change and family reasons are the two most important causes of migration. The socio-demographic and economic characteristics of migrants are significantly shaped by their motivations and opportunities for migration. It appears that economic factors have prevailed in most migration in China today. As stated before, *hukou* and non-*hukou* migrants face starkly different opportunities and constraints, and their contrasts are clearly shown below. The analysis that follows uses mainly data from the one-percent surveys of the 1987 and 1995 national population and from the 1990 census carried out by the State Statistical Bureau (SSB), supplemented by information from two national surveys of *mingong*, one by the Chinese Academy of Social Sciences and Agricultural Bank of China (Li 1994; Li and Han 1994) and the other by the Ministry of Agriculture (Zhang et al. 1995), and a local survey in Jinan, Shandong (Liu 1995). The SSB surveys and the census defined migrants as those crossing county-level boundaries and with a minimum stay of six months or one year in their destination. The *mingong* surveys generally used a broader definition of migrant. They often included migrants within counties and with a shorter duration of stay in their destination. Details of the data are explained in each table; interpretations of these statistics should take their definitional differences into account.

Table 14.1 indicates that non-*hukou* migration accounted for about 46 percent of all migration during the period 1985–1990 as shown in the 1990 census. The table also depicts that non-*hukou* migration is mainly employment-driven. Male non-*hukou* migrants of working age in particular were close to full participation in work, compared to only 57 percent in the same age group of *hukou* migrants (Yang 1994). While *hukou* work migrants were almost all in the "work transfer" or "assignment" categories (i.e., within-plan or approved labor transfers between enterprises), non-*hukou* migrant workers sought almost exclusively *wugong jingshang* ("employment in industry and business"). This refers to self-sought employment and self-employment totally outside

TABLE 14.1   Reasons for Migration

	1982–1987	1985–1990		
	All	All	Hukou Migrants	Non-hukou Migrants
Work Reasons				
Job transfer	20.6	14.5	18.0	4.6
Job assignment	5.1	4.7	10.4	2.7
Employment in industry and business	8.2	29.7	1.8	50.3
Family Reasons				
Migration with family	19.8	10.8	13.7	7.9
Marriage	15.8	14.2	15.6	11.3
Living with relatives and friends	13.3	10.6	6.6	13.2
Other Reasons				
Study or training	8.7	7.8	21.4	2.8
Retirement or resignation	2.6	1.5	2.1	1.0
Other	6.0	6.5	10.4	6.3
Total	100%	100%	100%	100%
Size (millions)	30.5	34.1	18.3	15.8

NOTE: For 1982–1987, migrants were defined as those crossing village, town or city boundaries staying at least for six months at the destination. The *hukou* and non-*hukou* migration figures are from the 1% migrant sample of the 1990 census tabulated by the author. Migrants were defined as those crossing county or city boundaries staying at least one year at the destination.
SOURCE: For 1982–1987, SSB 1988. For 1985–1990, the total figures are from SC and SSB 1993, volume 4.

of the state plans. Conversely, about 70 percent of all the *hukou* migrants in the 1985–1990 period moved for reasons other than starting a job. About 36 percent of *hukou* migration was due to family-related reasons (marriage, migration with family, or living with relatives or friends); marriage migration includes a significant number of rural-to-rural migrants. Another 21 percent was related to study or training. Among the non-*hukou* migrants, family-related reasons were also the second most important, accounting for about one-third of all the non-*hukou* migration.

The age and sex selectivity of migrants is clearly shown in the data. Life cycle events such as starting a job, changing jobs until one settles on a career, getting married, and going away to college are all closely associated with migration and with reaching young adulthood. The age structure of Chinese migrants is typical of a migrant population. Rural migrant labor tends to be concentrated in the most economically active age group, particularly between the ages 15 to 34. Males dominate labor migration at the national level. They are especially pronounced in the rural migrant labor population (including short-term floaters), where male migrants outnumber females by three to one. This, however, masks some notable regional exceptions such as Guangdong, where migrants from the countryside are predominantly female. Excluding the short-term floaters and including other non-work-related migrants, the 1990 census figures show that male migrants slightly outnumbered female ones. Marriage migration, however, was almost exclusively a female affair (Fan and Huang 1998).

Overall, migrants and rural migrant workers are better educated than the average population. This is partly an effect of the age structure of the migrants (young adults tend to be better educated than older adults). Despite the general similarity of the age structure between *hukou* and non-*hukou* migrants, there is a clear polarization of the two groups in educational attainment. *Hukou* migrants are disproportionately better educated (high school level and up) than non-*hukou* migrants and rural migrant labor who are heavily concentrated in the educational levels of junior high and primary schools. The most pronounced disparity is seen in the college-educated cohort. While only less than 2 percent of the nation's population aged six and above had a college education in 1990, close to one-quarter of the *hukou* migrants were college graduates! This clearly attests to the highly selective nature of the *hukou* migration.

Despite the lower educational level of rural migrant workers compared to *hukou* migrants, the former are nevertheless likely to be better educated than the average rural population. More than half of the rural migrant workers have at least junior high school education. Those who are better educated and have special vocational skills also tend to have a higher propensity to leave than people with no or little formal education (Li and Han 1994).

There are also occupational and sectoral similarities and contrasts between *hukou* and non-*hukou* migrants. The occupational structure of *hukou* migrants (who move to predominantly urban destinations except for marriage-related migrants) broadly resembles that of the urban population as a whole, but they are significantly over-represented in professional and technical positions. In contrast, 95 percent of the non-*hukou* migrants had employment at the clerical level or lower. Common jobs were manufacturing frontline workers, construction workers, nannies, and sales and service workers (Yang 1994). There are a lot of self-employed craftsmen and small vendors. In fact, self-employment has become a more favored sector for rural migrants (Liu and Chan 1998). The significant number of farmworkers among *hukou* migrants—about a quarter of total migrants—largely reflects the rural-to-rural marriage migration of women.

Among the urban migrants without *hukou,* a handful might make it and move upward through connections or entrepreneurship, but the great majority are marginalized. They are often shut out of more desirable urban positions and have to take up the dangerous and "dirty" jobs, a situation commonly faced by immigrant labor (especially illegal, undocumented workers) in many developed countries. In short, a dual urban social structure has emerged: On the one hand are people for whom jobs, housing, education, subsidized food, and medical care are an entitlement, and on the other, are those who must scramble for such goods and services or even do without them (Solinger 1995; Chan 1996a). Table 14.2 sums up the contrasts in their social and economic status due to the *hukou* divide. In many ways, this parallels the formal/informal sectoral dualism found elsewhere in the developing world and the local/foreign labor dichotomy in many developed countries.

TABLE 14.2   *Hukou* and Non-*hukou* Rural-Urban Migrants

Characteristics	Hukou Migrants	Non-hukou Migrants
Household registration status	nonagricultural and local	agricultural and nonlocal
Entitlement to state-supplied social benefits and opportunities	full	none or temporary entitlements
Legal urban resident status	yes	no or temporary
Socioeconomic sector the migrants move to	state and nonstate sectors	mostly to nonstate sector, also as temporary workers in state enterprises
Mechanism of migration	transfers determined by bureaucratic decisions within plan limits	"spontaneous," based on personal contacts and market information
Stability of moves	permanent	seasonal or semi-permanent
Labor characteristics of principal migrants		
Skill level	skilled and low-skilled workers	mostly unskilled or low-skilled laborers
Employment type	mostly permanent jobs	temporary or semi-permanent jobs in nonstate enterprises, or self-employment
Housing	same as other urban residents	low-cost shelters or homeless

SOURCE: Chan 1996a

Geography of Migration

Significant disparities in wages and living standards between the urban and rural sectors and between the coastal and inland regions underlie most of the migratory flows in the reform era. Economic factors have prevailed in most of the moves in this era, in contrast to migration in the pre-reform era where administrative factors played a decisive role. Peasant migrants in the 1990s expect to benefit from the large wage differentials, often in the range of one to three or four, between an urban unskilled job in a coastal city and a farm job in an inland province (see Liu 1994; Liu 1995). According to the 1990 census, a total of 34 million domestic migrants crossing county-level units were recorded over the 1985–1990 period, 23 million, or two-thirds, of whom moved within their respective provinces. Guangdong and Sichuan provinces had both the largest intraprovincial and interprovincial migration. But they were at the opposite ends of internal migration flows: Guangdong was the largest recipient, whereas Sichuan was the biggest sender. The same broad pattern was repeated for the 1990–1995 period (NPSSO 1997).

Available data allow us to study more specifically the flows in and out of the provinces. Based on the 1990 census data, the thirty largest interprovincial migration flows in 1985–1990 are mapped in Figure 14.1. Generally, in net migration terms, most of the coastal provinces (such as Guangdong, Beijing, Shanghai, and Jiangsu) gained from provinces in central and western regions (see Table 14.3 for definitions). The inland-coast flows are consistent with the existing large differences

TABLE 14.3   Composition of Rural Migrant Labor, 1993

| Region | Total Rural Labor (1,000) | Out-Migration Rate (%) | No. of Migrants (1,000) | % | Regional Total (%) | | |
					Within Counties	Within Provinces	Toward Urban Centers
Eastern[a]	154,505.9	8.5	13,133	25.6	28.4	66.3	82.0
Central[b]	143,295.6	15.9	22,784	44.4	40.6	70.4	83.3
Western[c]	113,755.6	13.5	15,357	30.0	37.0	76.4	66.5
All Regions	411,557.0	12.5	51,274	100	36.4	71.1	77.9

NOTES: Rural migrant labor refers to rural laborers who migrated from villages for work in 1993.
[a]Eastern = Heilongjiang, Jilin, Liaoning, Beijing, Tianjin, Hebei, Shandong, Jiangsu, Shanghai, Zhejiang, Fujian, Guangdong, Guangxi, and Hainan.
[b]Central = Neimenggu, Shanxi, Henan, Anhui, Hubei, Hunan, and Jiangxi.
[c]Western = Xinjiang, Qinghai, Gansu, Ningxia, Shaanxi, Sichuan, Guizhou, Yunnan, and Tibet.
SOURCE: Li 1994.

in the level of economic development and with the fact that a more open economy is at work in central areas. The current pattern is almost opposite to that in the pre-reform era as illustrated in Figure 14.2. In the Cultural Revolution era (1966–1976), for example, interprovincial migration was overwhelmingly toward the inland because of the government's various campaigns to send urban youth and cadres "down to the villages and up to the mountains."

The 1995 one-percent National Population Survey also provides valuable data to gauge the size of the non-*hukou* migrants who stayed more than six months in places outside of their place of original *hukou* registration. Assuming that these people had moved from their places of original *hukou* registration, one can work out the "stock" of non-*hukou* migrants as of October 1, 1995, and their presumed flows. At the inter-provincial level, the 1995 data indicate that Guangdong was the province with the largest number of non-*hukou* migrants from outside, estimated to be at 1.9 million, followed by Shanghai (666,000) and Beijing (658,400). On the other hand, Sichuan had the greatest number of out-migrants (1.5 million), followed by Henan (680,200) and Hunan (666,100) (NPSSO 1997). The pattern of interprovincial flows of non-*hukou* migration is very similar to that in Figure 14.1 because non-*hukou* migration crossing provincial boundaries tends to be more concentrated in a number of provinces and therefore dominates the largest flows. The economic forces (wage gaps) were so powerful that these migrants were willing to travel hundreds, if not thousands, of miles to very different places.[2]

The geography of non-*hukou* migrants can also be studied based on a survey of "rural migrant labor" (*mingong*) conducted by the Chinese Academy of Social Sciences in 1993 (see Table 14.3). As mentioned earlier, the concept of migrant here is different because it pertains only to labor migrants from the countryside *regardless* of their length of stay at the destination. According to this study, the stock of

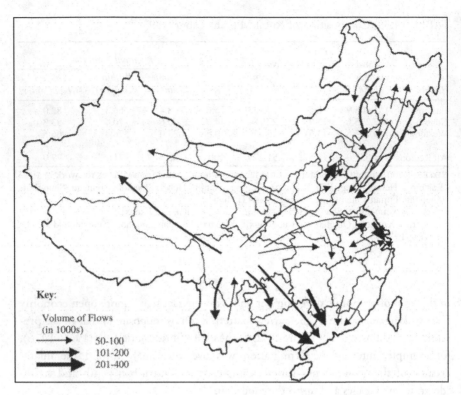

FIGURE 14.1   The 30 Largest Interprovincial Flows, 1985–1990
SOURCE: 1990 Census.

rural migrant labor, consisting of people working elsewhere *(waichu dagong)*, reached 51 million at the end of 1993, accounting for about one-eighth (12.5 percent) of the rural labor force. The central region was the largest source of rural migrant labor, with the highest labor out-migration rate (15.9 percent) and volume (22.8 million), followed by the western region (13.5 percent and 15.4 million). The eastern region had the lowest rate (only slightly more than half of that of the central region) and the smallest volume. This pattern is broadly consistent with the findings of other studies (such as Rozelle et al. 1997) of the early and mid-1990s. Because of its large labor force (population), the central region accounted for 44 percent of the estimated total outflows. The low rate of out-migration in the eastern region is attributed to the high level of development of rural enterprises in many villages and townships, which absorbed rural surplus labor. This is not the case for the central or western regions. A great portion of *mingong* movement was to urban areas (78 percent)[3] and within their own provinces (71 percent), although it appears that those crossing provincial boundaries are growing in number in the past several years. Estimates suggest that the volume of the latter rose from about 15 million in 1993 (from Table 14.3) to 25 million in 1994 and 30 million in 1995.[4]

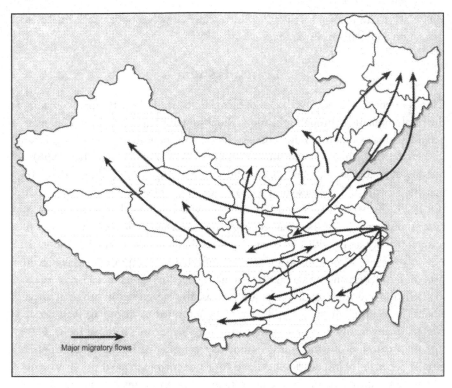

FIGURE 14.2   Estimated Major Interprovincial Migration Flows, 1966–1976
SOURCE: Chan and Yang 1996.

The pattern of *mingong* flows is reflective of the "normal" pattern of migration (predominantly short distance and toward urban centers) that China experienced in the 1950s. According to Chan (1994), there was high net rural-urban migration and rapid urbanization in the 1950s that resembled the "normal" pattern. However, this pattern was distorted in the 1960s and 1970s. There was net urban out-migration in the early 1960s and low net rural-urban migration in the Cultural Revolution period (1966–1976) because of the heavy government intervention in the urbanization process in pursuit of a socialist development strategy. Rural-urban migration accelerated in the late 1970s and gained momentum in the mid-1980s after more open policies were implemented.

Based primarily on the one-year residence criterion, the 1990 census shows that rural-urban migration made up about half of all migration crossing county-level units in 1985–1990 with urban-to-urban migration accounting for another one-third. For 1990–1995, based primarily on the six-month residence criterion, rural-urban migration and urban-urban migration each accounted for about 36 percent of all migration crossing county boundaries. It appears that rural-to-rural migration rose substantially in the 1990s, from about 13 percent of the total migration in 1985–1990 to 24 percent in

1990–1995. This may be a sign of increasing regional specialization of the rural sector (both farm and nonfarm work) (*Singtao Daily* 1997).

Policy Issues

China's recent economic growth cannot be separated from the active participation of the rural population through migration. Rural migrant labor is really the "muscle " behind China's economic might (Gilley 1997). Migrants have exerted significant impacts on both the rural and urban economies (see Smith 1996; Chan 1998). A number of major development policy issues relate to migration. First, China's development strategy and policies continue to be generally tilted against the rural sector (Oi 1993; Chan 1994; West and Wong 1997), which has not helped bring about a faster transformation of the rural sector. A reconsideration and reorientation of the sectoral priority, along with a set of rural policies that favor agricultural development, rural investment, income growth, and rural infrastructure improvement, is urgently needed. Further expanding agriculture, including opening up marginal land, expanding cash crop and non-crop farming, will increase the capacity to absorb rural surplus labor. The land tenure system will have to be reformed to discourage half-hearted farmers (many of whom now work outside their villages on a seasonal basis). Given that the expansion of agriculture is ultimately constrained by the finite supply of farmland, rural off-farm development and rural out-migration are the only ways out. The push to move out, or to vote by feet, to escape from unfavorable policies will remain enormous in the coming several years. This requires designing a set of appropriate migration policies that take a broad view. Current migration is by and large beneficial to both the rural and urban sectors. It is hard to imagine the recent rapid economic growth in the coastal provinces without those labor transfers. The issue then is how to continue facilitating this absorbable intersectoral flow and at the same time mitigate both the short-term and emerging long-term problems.

In spite of their indisputable contributions to the urban economy, non-*hukou* transients are often treated as outcasts by local governments under the *hukou* system, and they face intense discrimination by many urban residents. This expanding group is often denied access to many basic urban services or forced to pay high fees for such services, such as education and medical care. In fact, the needs of these "have-nots" are often totally left out of service provision and infrastructural planning in many cities (Chan 1997). Such a division of population within cities into those with urban entitlements and those without is bound to nurture social conflicts in the long term and must be changed.

Many of these migrant workers are now an integral part of the urban labor force, but the uncertainties surrounding their residency status give rise to opportunities for employers to exploit migrants' powerless "temporary" position. In many ways, the nature of such exploitation is similar to that of using "temps" and immigrant workers elsewhere; it is probably more blatant and "effective" in China, for many political and social institutions discriminate against the transients. The unclear residency status has also fostered some negative behavior among migrants; a sense of alienation and of not being accepted as an urban citizen has arguably contributed to some migrants' less-than-constructive conduct, including vandalism and more disruptive crimes.

An overnight elimination of this discrimination is unrealistic as the increased costs of providing even the basic services to the "have-nots" are beyond the capacity of most urban governments. Such action may run the great risk of triggering a huge exodus from the countryside in the short term, thus crippling the system. But discrimination against this segment of the population needs to be mitigated for equity and efficiency as well as political reasons. The solution lies in doing away with policies that favor certain segments of the population (which are also a major contributor of rural-urban disparities). The government should continue to reduce urban subsidies (e.g., by charging the full costs of urban services) and free up resources that could otherwise be made available for equalization transfers. At the same time, urban governments should develop affordable services for recent migrants and, ultimately, incorporate the entire urban population, regardless of residency status, in their development plans and budgets. China can look to other Asian developing countries for useful lessons and experiences in providing low-cost urban services (Yeung 1991).

Helping non-*hukou* migrants become permanent residents and encouraging their assimilation into urban society will induce many of them to give up their half-hearted pursuit in agriculture and ease the burden on the rail systems caused by the circularity of "temporary" migrants. It is also likely that this will decrease some of the developmental impacts brought about by circular migration (the transfer of capital, know-how, and modern attitudes to rural areas). It appears, however, that circulation, even in the absence of the *hukou* barrier, will continue to prevail for quite some time at the current stage of development, as it has been in other developing countries (Skeldon 1990). The strong bonds of overseas Chinese to their hometowns, displayed in recent history, appear to support the argument that even as circular migration wanes in China, the economic and social linkages between the migrants and their home villages will not disappear altogether.

At present, there are still many obstacles, beyond the *hukou* system, to rural-urban labor mobility. These include the lack of information about urban job opportunities, restrictive administrative measures adopted by local governments to protect local workers as unemployment becomes a bigger issue, and differing social and cultural traditions. These barriers require more attention and policy action. Establishing more recruitment agencies and enacting policies that ensure the protection of rural migrants' interests can greatly help match the supply and demand of labor and reduce the costs of migration (including unnecessary trips) and job searches. The success of China's recent round of marketization reforms under Premier Zhu Rongji now critically depends on raising the productivity of the labor force. Augmenting labor mobility is crucial to achieving that goal.

There is also a great potential for using labor migration (including organized export of labor and resettlement) to combat poverty, especially in remote and resource-constrained regions. This would represent a shift from the conventional approach, which emphasizes the merits of localized economic development. As pointed out before, the rate of migration is still quite low among the poorest groups in the rural sector and in poor regions. The low rate is related to factors such as low educational achievement, lack of relevant contacts, and the expense of migration. Measures need be designed to help the rural poor overcome major short-term and long-term obstacles. These may include offering low-interest migration loans, encouraging labor recruitment agencies to go to these areas, as well as providing more long-term investment in education. China has already had some experience in this area (World Bank 1992). It should consider launching a bolder and broader migration program to take advantage of this strategy. Financial and logistical problems associated with that approach can be daunting, but what is needed most is a mind-set that sees migration as a healthy mechanism for achieving a better match of regional labor supply and demand and that recognizes migration as a means through which an individual can attain fulfillment and a higher standard of living.

Conclusion

This chapter has provided an overview of the Chinese migration system and the characteristics and spatial patterns of migration. I have demonstrated a clear, institutionally based duality in the migrant population. At the macrosocietal level, two different migration streams from different socioeconomic strata operate within fairly distinct "circuits" fixed by social and economic institutions that are based on the hukou system. Although they share some similar demographic characteristics such as age structure, they exhibit dissimilar socioeconomic characteristics and geographies because of the different opportunities and constraints they face. A number of major policy issues surrounding China's internal migration have been examined. I have argued that migration is crucial to China's current reforms and future development. At a deeper level, the "peasant invasion" of cities has challenged the established attitudes and norms of both urban and rural residents and opened up possibilities for new social change and the formation of a more diverse, tolerant, and pluralistic society. As migration relates closely to the rural and urban sectors and to regional development, it has a strategic role to play in China's development in the new millennium.

Notes

This chapter synthesizes material from a number of my previous papers and also draws on some recently released data. I would like to thank Professor Yang Yunyan for help in procuring data and Ta Liu for assisting with the production of the two maps.

1. Li and Hu (1991) estimate that about half of the floating population in large cities stayed longer than six months, and another one third stayed longer than one year.

2. For most *hukou* migrants (moving within the state sector), economic gains were generally quite small, so out-of-province *hukou* migrants were still quite "conservative," moving over shorter distances than the interprovincial non-*hukou* migrants. One explanation is that *hukou* migrants generally preferred nearby provinces in which culture, languages, and environment were similar and therefore easier to adapt to. See Chan (2000) and also Ding (1994) for detailed maps.

3. Note that this is very different from what has been argued in Qian (1996).

4. Interviews with researchers at the Chinese Academy of Social Sciences in 1994 and 1995. Rozelle et al. (1997) also report that in 1995 about one-third of rural migrant workers in their national sample moved for jobs in other provinces..

References

Chan, J.S.H. 1997. *Population Mobility and Government Policies in Post-Mao China.* Unpublished thesis, University of Hong Kong.

Chan, K. W. 1994. *Cities with Invisible Walls: Reinterpreting Urbanization in Post-1949 China.* Hong Kong: Oxford University Press.

_____. 1996a. Post-Mao China: A Two-class Urban Society in the Making. *International Journal of Urban and Regional Research* 20(1):134–150.

_____. 1996b. Internal Migration in China: An Introductory Overview. *Chinese Environment and Development* 7(1 and 2):3–13.

_____. 1997. Urbanization and Urban Infrastructure Services in the PRC. In *Financing Local Government in the People's Republic of China,* ed. Christine Wong, pp. 83–125. Hong Kong: Oxford University Press.

_____. 1998. An Analysis of Consequences and Policies Concerning Recent Population Migration in Mainland China (in Chinese). *Journal of Population Studies* (Taiwan) 19:33–52.

_____. 2000. Internal Migration in China: A Dualistic Approach. In *Internal and International Migration: Chinese Perspectives,* ed. Hein Mallee and Frank Pieke. Richmond, Surrey: Curzon Press.

Chan, K. W., and Y. Yang, 1996. Inter-provincial Migration in China in the Post-1949 Era: Types, Spatial Patterns, and Comparisons. Seattle Population Research Center Working Paper No. 96–14.

Cheng, T., and M. Selden. 1994. The Origins and Social Consequences of China's *Hukou* System. *China Quarterly* 139:644–668.

Ding, J. 1994. Zhongguo renkou shengji qingyide yuanyin bieliuchang tezheng tanxi (An analysis of inter-provincial migratory streams in China by reason). *Renkou yanjiu* 18(1):14–21.

Fan, C. C., and Y. Huang. 1998. Waves of Rural Brides: Female Migration in China. *Annals of the Association of American Geographers* 88(2):227–251.

Gilley, B. 1997. Migrant Workers Play a Key Role. In *China in Transition: Towards the New Millennium,* ed. Frank Ching, pp. 85–92. Hong Kong: Review Publishing.

Li, F. 1994. Waichu dagong renyuande guimo, liudong fanwei ji qita (The size, geographical distribution, and other characteristics of outgoing workers). *Zhongguo nongcun jingji* (Chinese Rural Economy) 9:31–35.

Li, F., and X. Han. 1994. Waichu dagong renyuande nianling jiegou he wenhua goucheng (The educational composition and age structure of outgoing workers). *Zhongguo nongcun jingji* (Chinese Rural Economy) 8:10–14.

Li, M., and Y. Hu. 1991. *Liudong renkou dui dachengshi fazhande yingxiang ji duice* (Impact of the floating population on the development of large cities and recommended policy). Beijing: Jingji Ribao Chubanshe.

Liu, Q. 1995. Rural-Urban Migration Sample Survey in Jinan Municipality of China: Sample Design and Preliminary Results. Working Paper, Australian National University.

Liu, Q., and K. W. Chan. 1998. Rural-Urban Migration Process in the Transitional Era in China. Unpublished manuscript.

Luo, Y., and J. Liu. 1994. Inter-regional Transfers of Rural Labor: Current Situation, Causes, and Measures. *Zhongguo nongcun jingji* (Chinese Rural Economy) 8:3–9.

Mallee, H. 1988. Rural-urban Migration Control in the People's Republic of China: Effects of the Recent Reform. *China Information* 2(4):12–22.

<tmbr>. 1995. China's Household Registration System under Reform. *Development and Change* 26(1):1–29.

National Population Sample Survey Office (NPSSO). 1997. *1995 Quanguo 1% renkou jiuyang diaocha ziliao* (Data on 1995 National 1% Population Sample Survey). Beijing: Tongji Chubanshe.

Nolan, P. 1993. Economic Reform, Poverty, and Migration in China. *Economic and Political Weekly* 26:1369–1377.

Oi, J. C. 1993. Reform and Urban Bias in China. *Journal of Development Studies* 29:4129–4148.

Qian, W. 1996. *Rural-Urban Migration and Its Impact on Economic Development in China*. Aldershot: Avebury.

Renmin Ribao. 1995. July 9, p. A1.

Rozelle, S., et al. 1997. Poverty, Networks, Institutions, or Education: Testing Among Competing Hypotheses on the Determinants of Migration in China. Paper presented at the Annual Meeting of the Association of Asian Studies, Chicago, March 13–16.

Singtao Daily. 1997. July 6–7, p. A5.

Skeldon, R. 1990. *Population Mobility in Developing Countries*. London: Belhaven Press.

Smith, C. 1996. Migration as an Agent of Change in Contemporary China. *Chinese Environment and Development* 7(1–2):14–55.

Solinger, D. 1995. The Floating Population in the Cities: Chances for Assimilation? In *Urban Spaces: Autonomy and Community in Contemporary China,* ed. Deborah Davis et al., pp. 113–139. Cambridge: Woodrow Wilson Center Press and Cambridge University Press.

State Statistical Bureau (SSB). 1988. *Zhongguo 1987 nian 1 percent renkou chouyan diaocha ziliao* (Tabulations of China's 1987 1 percent population sample survey). Beijing: Zhongguo Tongji Chubanshe.

_____. 1995. *Zhongguo tongji nianjian 1995* (Statistical yearbook of China, 1995). Beijing: Zhongguo Tongji Chubanshe.

State Council and State Statistical Bureau (SC and SSB). 1993. *Zhongguo 1990 nian renkou pucha ziliao* (Tabulation on the 1990 population census of the People's Republic of China). Multiple volumes. Beijing: Zhongguo Tongji Chubanshe.

West, L., and C. Wong. 1997. Equalization Issues. In *Financing Local Government in the People's Republic of China,* ed. C. Wong, pp. 283–312. Hong Kong: Oxford University Press.

Yang, Y. 1994. *Zhongguo renkou qianyi yu fazhande changqi zhanlue* (Internal migration and long-term development strategy of China). Wuhan: Wuhan Chubanshe.

Yeung, Y. 1991. *The Urban Poor and Urban Basic Infrastructure Services in Asia: Past Approaches and Emerging Challenges*. Hong Kong: Chinese University of Hong Kong.

Zhang, X., et al. 1995. 1994: Nongcun laodongli kuaquyude shezheng miaoshu (1994: An empirical description of the interregional flows of rural labor). *Zhanlüe yu guanli* 6:26–34.

Zhou, K. X. 1996. *How Farmers Changed China: Power of the People*. Boulder: Westview.

15

Gender Differences in Chinese Migration

C. Cindy Fan

Migration is one of the most studied topics in China. Yet research on gender differences in Chinese migration is meager, despite the fact that female migrants constituted 43.97 percent of all migrants during the period 1985–1990 and 52.29 percent during 1995–2000 (Population Census Office 2002) and the patterns and processes of female migration differ considerably from male migration.[1] This lack of attention is due to the widespread perception that migration is primarily led by men and that female migration is problematic. The widely held notions that men's migration propensity is higher than women's, that men travel longer distances, and that men move primarily for economic reasons and women for social reasons are mostly based on cursory observations and attributed to China's sociocultural tradition. In this chapter, I argue that as China undergoes an unprecedented transition from a centrally planned socialist economy to one in which both state plan and market operate simultaneously, gender differences can no longer be explained only as an outcome of traditional sociocultural factors. They must also be interpreted in relation to structural factors pertinent to this transition. Specifically, the unique spatial-economic changes and peculiar institutional controls characterizing China's transitional economy are critical factors of the migration process.

The empirical analysis in this chapter aims at systematically comparing the migration patterns and processes of Chinese men and women. It evaluates the three notions of gender differences outlined above, emphasizes the spatial patterns of male and female migration, and examines differentials in migrants' participation in the labor market. By doing so, I argue for greater attention to the role of gender in migration research and highlight the prominence of economic rationale and long-distance moves among female migrants in China.

Research on Gender Differences in Migration

Despite the proliferation of research on Chinese migration since the late 1980s, relatively little attention has been given to gender differences in migration. Perhaps the best evidence of this observation is Ji and Shao's (1995: 278–322) annotated bibliography, which reviews migration research published in China between 1991 and 1994. Among the ninety-four publications they included, only one article focuses specifically on gender differences, namely Li (1993). Since the mid-1990s, more researchers in and outside China have begun to examine the role of gender in migration (e.g., Cai 1997; Chiang 1999; Davin 1997, 1998, 1999; Fan 2000; Wang 2000; West and Zhao 2000; Yang and Guo 1999), but the majority of migration research lump men and women together or focus only on the migration of men. Spatial analyses of migration data, in particular, rarely differentiate between male and female migrants (e.g., Ding 1994; M. H. Li 1994; Zhu 1994), with the result that we know little about the spatial patterns of male and female migration in China. Comprehensive volumes on Chinese migration typically devote only a few pages to gender differences or to specific types of female or male migration (e.g., Shen and Tong 1992; Yang 1994). Very commonly, discussion of gender differences does not go beyond simple comparisons of the volume or rate of female and male migration (e.g., Wei 1995). Like much of the literature on migration, gender is not considered a central organizing theme but is frequently treated as one of many independent variables, such as age and education, for explaining differentials in migration volumes or rates (Pedraza 1991). These existing approaches suggest a greater emphasis on patterns rather than processes of migration and, above all, a lack of attention on the importance of gender for understanding migration and on the differential constraints and opportunities conducive to men's and women's mobility.

More frequently found are studies focused specifically on female migration without comparing it with male migration. These studies are mostly concerned with the reproduction behavior of female migrants and the difficulties of implementing birth control policies among migrants (e.g., Gu and Jian 1994: 62–79; Xu and Zhen 1992). Another popular theme is marriage. Researchers typically use an evaluative approach and discuss the pros and cons of female migration and the positive and negative impacts of female in-migrants (dubbed *wailainu* or "women from outside") (e.g., Li et al. 1991; Liu 1990; Xu and Ye 1992; Yang 1991). Underlying these approaches is the assumption that female migration is problematic and deviates from the "regular" migration led by men. On the other hand, insights about the role of gender are at best obscure when research fails to systematically compare the migration patterns and processes of men and women.

There are several widely held notions about differences between male and female migrants in China. First, men have a higher migration propensity than women. Most scholars explain this difference as an outcome of discrepancies between men's and women's social statuses and traditional roles (Li 1993; Yu and Day 1994; Zhang

1995). Second, men travel longer distances than women (e.g., Kuashiji de zhongguo . . . 1994: 253). While social status and tradition are key factors, the literature also draws heavily upon Ravenstein's "laws" of migration to explain men's greater likelihood to take risks and overcome intervening obstacles associated with long-distance migration (e.g., Yang 1994: 201). However, existing studies have provided little information on the specific spatial patterns of male and female migration. Third, men are more likely to migrate for economic reasons, and women for social reasons and as tied movers. This reflects not only the impacts of social status and tradition but also the dominance of men in the Chinese economy (e.g., Gu and Jian 1994: 24; Li 1993; Wang and Hu 1996: 91–92). The above observations are often not based on systematic empirical analyses, nor is there any discussion of how they are related to the recent changes in the Chinese spatial and political economy. As a result, there is little attention on the impact of migration on men and women or their differential experiences in the labor market,[2] although these are indeed important perspectives for a fuller understanding of the role of gender in the migration process.

In the following, I discuss how sociocultural factors, spatial-economic changes, and the *hukou* institution contribute to gender differences in migration. This discussion then forms the basis for interpreting gender differentials in migration propensity, spatial patterns of migration, reasons for migration, and migrants' occupations, which will be examined with the support of empirical data.

Factors of Gender Differences in Migration

Gender differences in migration are a function of important structural forces, which are often based on gender, that determine the status of men and women and their integration into development (Lim 1993). Although gender differences in migration are often attributed to China's sociocultural tradition, which maintains a persistent gap between the status of men and the status of women, other structural forces shaping the Chinese spatial and political economy must also be addressed for a better understanding of gender and migration. During China's transition from a centrally planned economy to one increasingly employing market mechanisms, specific spatial-economic outcomes have brought about new and varied mobility opportunities for men and women. The *hukou* institution, or household registration system, is a product of the socialist state and has unique meanings for migrants as a whole and for the differential migration processes of men and women. The following subsections describe these structural factors in China and their implications for male and female migration.

Sociocultural Factors

The traditional Chinese view of gender is one rooted in Confucianism, which prescribes individuals' roles based on their positions relative to others. The Chinese woman is defined in relation to, and is subordinate to, other males in the family—the father, the husband, and the son(s). Under the patrilocal tradition, a daughter moves out and joins her husband's family, adding to the latter's labor resources. Parents of a son, especially in rural areas, are eager to recruit the labor of a daughter-in-law, which partly accounts for the prevalence of early marriages (Croll 1987). On the other hand, the natal family has little incentive to invest in a daughter's education relative to her male siblings (L. Li 1994; Lu 1997). "Daughters married out are like water spilled out" best describes this pragmatic yet popular perception, which reinforces the persistent gender gap in education. According to the 2000 census, 13.47 percent of the female population aged fifteen or older was illiterate, compared with 4.85 percent of their male counterparts (Population Census Office 2002). Lower education translates into less access to knowledge, resources, and opportunities—all factors constraining women's mobility.

The notion of household strategy explains how collective and pragmatic decision-making contributes to gender differentials in mobility in China (Cai 1997). When migration of one or more family members is deemed favorable for bringing about increases and/or diversification of family income, men are more likely candidates because of their higher education and because of the prevailing gender discrimination in the labor market, which favors men (Knight and Song 1995). Women are more likely to remain in order to take care of household chores or work in the fields. This gendered division of labor further reinforces the practice of allocating more investment to sons than to daughters, sustaining if not widening the gender gap in education.

The Confucian view that women's place is in the family restricts their mobility. There is continued pressure for Chinese women to marry early; among the female population between the ages of 20 and 24 in 1990, more than half were married. As in many other societies, Chinese women are subject to a demanding set of gendered expectations and responsibilities associated with marriage and with their roles as wives and mothers. Men's work is traditionally considered more important, partly because of sociocultural tradition but also because their higher education correlates with higher income. Although women's mobility may increase during "marriage ages," due to the patrilocal tradition, they tend to be considerably less mobile after marriage (Li and Li 1995). When they do migrate, they are more likely than their spouses to be tied movers.

The Maoist period set itself apart from previous periods by the state's active intervention to mobilize women to engage in production as fully as men. Partly as a result of such efforts, Chinese women's level of labor-force participation is among the highest in the world (Bauer et al. 1992; Riley 1996). Nonetheless, gender continues to be a major source of inequality in China (Maurer-Fazio et al. 1997; Park 1992). The continued resistance to the one-child policy, especially among the rural population, is testimony to the persistent perception that men are more highly valued than women. The exceedingly high sex ratio at birth, reflecting aggressive efforts to ensure male offspring (e.g., gender-specific abortions, female infanticide, nonreporting of female births), illustrates a perpetual undermining of the status of women (Riley 1996). In rural areas, in particular, the gap in access to education between women and men remains large (Bauer et al. 1992; L. Li 1994). Although peasant men may improve their social and economic mobility by joining the military, going to school, and becoming cadres, many Chinese women in the countryside remain poor and uneducated.

The persistent gender gap in social status is widely considered the key factor explaining gender differences in migration. Researchers who observed that Chinese women had a lower migration propensity than men, and that women migrated shorter distances than men, generally attributed these findings to traditional gender roles and gendered division of labor (e.g., Li 1993; Zhang 1995). The prominence of sociocultural factors underlies another widely accepted notion—that men are more likely to migrate for economic reasons and women are more likely to be tied movers or move for social reasons (e.g., Kuashiji de zhongguo . . . 1994; Rowland 1994).

Nevertheless, the above findings are mostly based on cursory empirical observations, which overlook complexities in men's and women's migration processes. For example, poor and uneducated women constitute a special labor force heavily recruited to fulfill growing demands in industrial and services sectors, especially in coastal open zones. The *hukou* institution reinforces segmentation and gendered division of labor in the labor market, which affects how and where men and women migrate. Female marriage migrants move long distances from the poor western part of China to the rich eastern region, evidence of the importance of spatial-economic factors and of the *hukou* institution. None of these can be fully understood simply on the basis of sociocultural factors; they must be interpreted by integrating socio-cultural factors with the spatial-economic changes taking place in the Chinese economy and the *hukou* institution.

Spatial-Economic Changes

The success of China's rural reform, namely, the household responsibility system, has unleashed a large and growing agricultural labor surplus, to which the state responded in 1984 by relaxing its control over temporary migration, allowing peasants to obtain "temporary residence permits" *(zhanzhuzheng)* ("Guowuyuan dui nongmin . . ." 1984). Subsequently, waves of migrants flocked to towns and cities to look for jobs in the industrial and service sectors. The model of "leaving the land but not the village" *(litu bu lixiang)* has been adopted by many peasants who work in towns adjacent to farm-land and continue to engage in farmwork (e.g., during harvests). "Leaving the land and the village" *(litu you lixiang)* has become increasingly popular among peasant migrants who have moved to towns and cities farther away. As discussed earlier, because of their greater opportunities in the labor market, men are usually considered more favorable candidates for migration that is aimed at augmenting household income. In contrast, many women are left behind to take care of the farmland, resulting in the feminization of agriculture already observed by many researchers (Bossen 1994; Gao 1994a). Some households contract their land to other workers, or hire workers from other areas to farm their land, triggering further waves of migration among rural areas.

At the same time, China's reform and open-door policies have produced a spatial economy characterized by a rapidly growing coastal region and by lagging inland areas, providing a strong incentive to both men and women in poorer areas to migrate to coastal areas (Fan 1995; 1996). In many coastal open zones, foreign investment, in conjunction with local cadres, has brought about a proliferation of industrial enterprises, many of which are found in rural areas adjacent to cities and towns. Rural areas in coastal provinces are especially attractive to foreign investors because of preferential policies, availability of space, more relaxed environmental regulations, and wage rates that are lower than those of urban areas. A peculiar mixed industrial-agricultural landscape (dubbed *desakota* by McGee [1991]) characterizes these rural enterprises. The Pearl River Delta is a prominent example (Lin 1997: 115).

Many rural enterprises recruit migrant workers to take up low-skilled, labor-intensive, low-paying jobs. The availability of large numbers of poor uneducated peasant women, and the widely accepted notion that they are docile and tolerant of long hours of work and poor working conditions, has made them perfect candidates for these rural enterprises. In areas with extensive rural industrialization, opportunities and wages have improved for men, whereas new employment opportunities in export-processing industries have channeled women into the lowest-paid and least secure new jobs, such as in the apparel, footwear, and electronic industries (Gao 1994b; L. Li 1994). Nevertheless, the economic reforms have indeed introduced new, perhaps unprecedented, incentives and opportunities for Chinese women to migrate on their own rather than as tied movers (Kuashiji de zhongguo . . . 1994: 253). Their sheer volume prompted the invention of the term *dagongmei*,[3] which challenges the notion that women migrate primarily for social motives. Many *dagongmei* migrated from poor inland provinces to coastal open zones, contradicting another widely held notion that women move primarily short distances.

Another type of migration heavily influenced by changes in the spatial economy is marriage migration. Much research on marriage migration assumes that marriage is the motive and migration is its by-product and that women are more likely than men to migrate in response to marriage. Although this is largely true in China, where the patrilocal tradition prescribes that women move to their husbands' households, recent studies by Fan and Huang (1998) and Fan and Li (2002) questioned the assumption that marriage migration is primarily social rather than economic in nature. Due to a lack of other opportunities, women in poor areas tend to resort to marriage as a strategy for achieving social mobility (Bossen 1984; Honig and Hershatter 1988; Wang and Hu 1996: 287). In their study, Fan and Huang (1998) documented well-defined streams of female marriage migration from the poorer southwestern provinces to the more developed eastern region. This suggests a strong economic rationale and presence of women's agency behind such migration and questions the notions that women move primarily for social reasons and over short distances. Both *dagongmei* and marriage migrants are prominent among female migrants, whose large volumes and unique migration processes are a strong testament to the critical role of spatial-economic changes in Chinese migration.

The Hukou Institution

Hukou is a form of population registration formally introduced in China in 1958 (Chan and Zhang 1999; Cheng and Selden 1994; China reforms 2001; Yu 2002: 18). Every Chinese household has a *hukou* registration "book" that records every household member and their rural (agricultural) or urban (nonagricultural) *hukou* status. The *hukou* institution is a key factor for defining an individual's opportunities and socioeconomic positions in China (Cheng and Seldon 1994; Christiansen 1990). It has for decades tied Chinese peasants to the countryside, and it underlies the segmentation and slow emergence of China's labor market (Knight and Song 1995). Without the proper *hukou* in the destination, a migrant is excluded from the benefits that local people are entitled to (such as housing and education) and is ineligible for many desirable jobs (especially in the formal sector) offering better pay and greater security.

The 1984 directive, outlined earlier, allowed peasants to move as temporary migrants to cities and towns but did not permit transfer of their *hukou* to their destination. These "non-*hukou* migrants," or "temporary migrants," constitute a less formal type of migration, largely based on migrants' initiatives rather than state planning. It contrasts with the more formal, government-regulated "*hukou* migration," or "permanent migration," such as a job transfer from one work unit to another, or a job assignment from the state after college graduation. The temporary migrants, mostly peasants, are generally considered outsiders and are not welcome to enter the local labor market unless they take on the particular kinds of jobs that natives are unwilling to assume (Chan 1996; Mallee 1996; Smith 1996; Solinger 1995; Yang and Guo 1996). Despite their active participation in the destination's labor force, migrants without local *hukou* are shut out from many desirable jobs and benefits local people are entitled to. Many are confined to the bottom rungs and most marginal sectors of the economic hierarchy. The distinctions among urban natives, permanent migrants, and temporary migrants further reinforce the segmentation of the labor market (Fan 2002).

Expansion of the urban economy and increasing consumerism have created new demands for temporary migrants, who by their work could ease the lives of urban residents and allow the latter to participate more fully in their preferred work (Roberts 1997). Gendered markets have come into being targeting male or female temporary migrants for specific kinds of jobs. Physically demanding jobs, such as those involved in construction and manual loading and transporting, for example, are virtually exclusively taken by male migrants; many service jobs requiring low skills and affording low pay, such as those in restaurants and hotels, are primarily assumed by female migrants (Yang and Guo 1996). Markets for nannies and maids now exist in many Chinese cities, drawing migrant women to work in urban homes, enabling female spouses to work outside the home. These examples suggest that the *hukou* institution and labor market segmentation have created new and separate occupational niches for male and female migrants (Fan 2003).

Under the current system, it is very difficult to shift one's *hukou* unless one joins the army or goes away to college.[4] However, marriage migration is considered permanent, and one is permitted to transfer his or her *hukou* to a new destination in that circumstance, especially if the destination is a rural area.[5] This opportunity, in conjunction with the economic rationale of marriage migration discussed earlier, is particularly attractive to peasant women from poor areas who wish to migrate to more desirable, albeit rural, areas. Not only can a migrant woman benefit from the resources in her husband's family, she can also achieve permanent residence and be in a better position to gain access to resources in the destination, including employment opportunities and social benefits provided by employers and local governments. In other words, both economic and institutional benefits are relevant to explaining the prevalence of female marriage migration. This, in conjunction with the gendered labor market processes for male and female migrants, underscores the importance of *hukou* for understanding Chinese migration.

Empirical Analysis

The above discussion has sought to establish that gender differences in migration are more complicated than what is revealed by a simple comparison of mobility rates. Various gendered constraints and opportunities shape the migration processes of men and women. The empirical analysis below documents the differences between male and female migration in China and interprets these differences by drawing upon the sociocultural, spatial-economic, and institutional factors outlined above. Specifically, it examines gender differences in migration propensity, spatial patterns of migration, reasons for migration, and the occupation of migrants.

Data for the empirical analysis are mainly drawn from a one-percent village-level sample of the 1990 census,[6] and accordingly the numerical data presented below have been multiplied by 100. The 1990 census defines a migrant as an individual five years or older whose usual place of residence on July 1, 1985, was in a different urban district, county-level city, or county than that on July 1, 1990. Although this definition underestimates the actual volume of migration by excluding multiple moves, return migrants, migrants younger than five years old, and moves within the same county-level unit, the 1990 census remains by far the most comprehensive source of national migration data in China.

Migration Propensity

Migration is a selective process. Who migrates and who does not reflect differentials in the population in terms of motivation to migrate, access to resources, knowledge and opportunities about migration, and ability to overcome the obstacles to migration. Table 15.1 illustrates some aspects of gender differences in migration selectivity. Among the 35.3 million migrants between 1985 and 1990, 56.03 percent were male and 43.97 percent were female. Male and female migrants accounted for 3.74 percent

TABLE 15.1   Gender Differentials in Socioeconomic Characteristics

	Nonmigrants		Migrants		Male-Female Ratio	
	Male (1)	Female (2)	Male (3)	Female (4)	Nonmigrants (5)[a]	Migrants (6)[b]
Number						
millions	509.65	484.86	19.81	15.54		
%	96.26	96.89	3.74	3.11	99	120
Age						
Mean (years)	31.64	32.22	27.34	26.67		
15–29 (%)	33.93	34.11	64.00	66.35	99	96
Hukou(%)						
Agricultural	78.88	81.56	44.19	57.49	97	77
Education Level (%)						
6+						
Junior high or above	42.70	28.56	71.98	58.29	150	123
15–29						
Junior high or above	67.47	52.32	85.23	70.61	129	121

NOTE: This table refers to population aged 5 and above.
[a](5) = (1) / (2) * 100
[b](6) = (3) / (4) * 100
SOURCE: 1990 Census One-Percent Sample.

and 3.11 percent of their respective populations aged five or above in 1990. This table employs the male-female ratio (columns 5 and 6), a technique based on the logic of the sex ratio, in order to evaluate differentials in men's and women's relative proportions in various categories. A male-female ratio larger than 100 indicates a larger proportion of men than women, and a ratio smaller than 100 indicates a larger proportion of women than men. This technique controls for the distribution of men and women in the general population and permits an unbiased comparison of male and female migrants. For example, in 1990, the sex ratio of Chinese migrants was 127, which reflects gender differentials in migration propensity as well as a high sex ratio in the general population (106 for the population aged 5+). By employing the male-female ratio, which compares the relative proportions of male (3.74 percent) and female (3.11 percent) migrants rather than their actual numbers, we control for the sex-ratio effect of the general population. After correcting for the effect of gender distribution in the general population, for every 100 female migrants there were 120 male migrants.

Mean ages of migrants (27.34 for men and 26.67 for women) and nonmigrants (31.64 for men and 32.22 for women) show that the former were significantly younger than the latter. Migration was especially selective of the 15–29 age group, which accounted for 64 percent of male migrants and 66.35 percent of female migrants. The selectivity of young adults will be discussed more fully below.

Agricultural *hukou* represents primarily the rural population in China and accounted for the majority of male (78.88 percent) and female (81.56 percent) nonmigrants. The proportions of migrants with agricultural *hukou* were substantially lower. However, the male-female ratio of 77 indicates that a significantly higher proportion of female migrants than male migrants had agricultural *hukou* and suggests that female migrants were more likely than their male counterparts to come from rural backgrounds.

The education statistics in Table 15.1 show that migration is indeed selective of the skilled and able. For both the 6+ and 15–29 age groups, the proportions of migrants with junior-high-or-above education were significantly higher than their nonmigrant counterparts. The male-female ratios for the nonmigrant population indicate a large gender gap in education—a ratio of 150 for nonmigrants aged 6 or older means that men were one and a half times more likely than women to have received education at or above the junior-high level. The imbalance was smaller among migrants, however, as illustrated by smaller ratios for both the 6+ (123) and 15–29 (121) age groups. Even though the ratios were still larger than 100, their smaller size indicates that migration was especially selective of women substantially more educated than their nonmigrant counterparts. In other words, female migrants had gained proportionally more education relative to male migrants, which suggests that the former's improvement in human capital was a means to make up for the sociocultural constraints on women's mobility.

Figure 15.1 shows age-specific migration rates, defined as the number of migrants per 100 people. Regardless of gender, and for both intraprovincial and interprovincial moves, age-specific migration rates reached their peaks at the 20–24 year age group. Not surprisingly, migration rates for intraprovincial moves were higher than for interprovincial moves, since the latter likely involved longer distances and more severe intervening obstacles. In terms of gender differentials, men's migration rates were higher than women's across almost all age groups, except for ages 70 and above where the sex ratio of the Chinese population significantly favored women (see also Figure 15.2 and below).

The differential effects of the life cycle on men and women are depicted in Figure 15.2, which plots the sex ratio of the Chinese population and migrants with age. A higher life expectancy among Chinese women than men contributed to sex ratios lower than 100 among the population aged 65 and above. For all other age groups, the sex ratio remained higher than 100, making China one of the few countries in the world where men outnumbered women and reflecting a persistent gap in social status between men and women. The sex ratio of migrants was markedly higher than that of the general population, except for the 70+ ages, again confirming that men had a higher migration propensity than women.

Two age periods of particularly high migrant sex ratios stand out. First, the sex ratio increased through the young-adult ages and encountered its first peak at 25–29 for interprovincial migrants and 30–34 for intraprovincial migrants, which coincided with the "just-married" ages of the majority of Chinese women and reflected marriage's constraints on women's mobility. The majority of Chinese women between the ages of 20 and 24 (50.94 percent), and 94.10 percent of those aged 25–29, were married. Men's age of marriage tended to be somewhat older, hence the relatively lower proportions of married men for age groups 20–24 (30.53 percent) and 25–29 (79.97 percent). By ages 30–34, 98.33 percent of women and 91.13 percent of men were married. Migration did not seem to have delayed women's marriage, as respectively

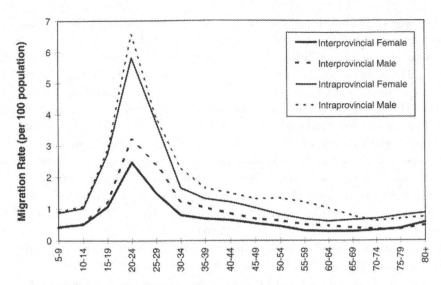

FIGURE 15.1   Gender Differentials in Age-Specific Migration Rate
SOURCE: 1990 Census One-Percent Sample.

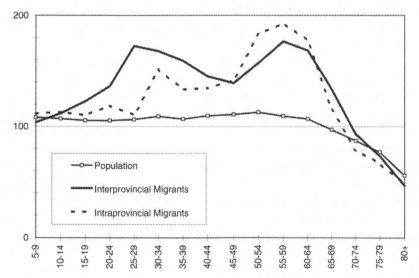

FIGURE 15.2   Sex Ratio of Chinese Population and Migrants
SOURCE: 1990 Census One-Percent Sample.

52.96 percent and 92.16 percent of female migrants aged 20–24 and 25–29 were married, similar to the respective proportions in the general population. This confirms that Chinese women continue to be under pressure to marry at a young age, despite the interruption of migration. On the other hand, migration's effect on men's age of marriage was stronger, as the marriage rates of male migrants in age groups 20–24, 25–29, and 30–34 (13.38 percent, 63.48 percent, and 85.86 percent respectively) were significantly lower than the respective proportions for all Chinese men (30.53 percent, 79.97 percent, and 91.13 percent respectively).

Marriage has traditionally had a bigger negative effect on women's mobility than men's. The peaks in migrants' sex ratio at 25–29 for interprovincial migrants and 30–34 for intraprovincial migrants (Figure 15.2) are compelling evidence for this effect. The age differential in the peaks reflects differentials in age of marriage between intraprovincial and interprovincial migrants. Among female interprovincial migrants between the ages of 20 and 24, 57.7 percent were married, compared with 50.95 percent among their intraprovincial counterparts, suggesting that the former's average age of marriage was younger, hence an earlier peak in migrants' sex ratio.

Another factor in the rise in migrants' sex ratio during the late 20s and early 30s is the more concentrated age selectivity among female migrants. The proportion of female migrants aged below 25 was 56.70 percent, compared with 53.13 percent for their male counterparts. Marriage and industry/business were the two leading reasons for migration among Chinese women (see also Table 15.3 and note 9). The majority (53.01 percent and 64.15 percent respectively) of them completed these two types of migration before age 25. The leading reasons for male migrants were industry/business and job transfer, and the majority of male migrants who moved for these reasons (64.67 percent and 80.29 percent respectively) were 25 years or older. In other words, differentials in the age distribution of female and male migrants also contributed to a rise of sex ratio after the age of 25.

The second peak in sex ratio occurred between the 50–54 and 60–64 age groups (Figure 15.2). Unlike the young adult ages, migrants in these age groups accounted for relatively small proportions of total migrants (see Figure 15.1). Nonetheless, the high sex ratios reflect important life cycle and migration differentials between men and women. The leading reasons for migration in these age groups were, in ranked order, retirement, industry/business, and job transfer, all related to employment. Between the ages of 50 and 64, men accounted for more than 80 percent of the migrants for these three reasons, reflecting existing gender discrepancies in labor force participation and men's greater access to labor market opportunities.

Spatial Patterns of Migration

Table 15.2 shows intraprovincial and interprovincial migration rates by region, using the regional schemes popularized by the Seventh Five-Year Plan (see also Figures 15.3 and 15.4). Since the census definition of migrants excluded those aged below 5, migration rate is defined as the number of migrants per 100 population aged 5 or above. As expected, for both men and women, intraprovincial migration rates were higher than interprovincial migration rates. Intraprovincial moves accounted for respectively 65.85 percent and 69.4 percent of male and female migration. The higher proportion of female intraprovincial migration indicates that women were more likely than men to move short distances. But the data on interprovincial migration suggests otherwise, which will be discussed below. Intraprovincial migration

TABLE 15.2   Male and Female Intraprovincial, Interprovincial, and Interregional Migration

	Male				Female			
	Eastern	*Central*	*Western*	*Sum*	*Eastern*	*Central*	*Western*	*Sum*
Migration Rate								
Intraprovincial	2.62	2.39	2.16	2.43	2.10	2.11	2.21	2.13
Interprovincial								
Out-migration	1.33	1.20	1.23	1.26	0.79	0.90	1.28	0.94
In-migration	1.78	0.88	0.90	1.26	1.36	0.63	0.64	0.94

Regional Distribution of Interprovincial Migrants

	Origin				Destination			
Eastern	28.85[a]	11.41[b]	3.82[b]	44.07	24.39[a]	7.81[b]	3.07[b]	35.27
Central	21.80[b]	8.48[a]	3.78[b]	34.06	21.76[b]	8.92[a]	3.49[b]	34.16
Western	8.42[b]	4.98[b]	8.47[a]	21.87	14.58[b]	7.29[b]	8.71[a]	30.57
Sum	59.06	24.86	16.08	100.00	60.72	24.01	15.27	100.00

NOTE: Migration rate = migrants per 100 population aged 5 or above.
[a]intraregional move
[b]interregional move
SOURCE: 1990 Census One-Percent Sample.

rates were quite uniform across regions for female migrants, but for male migrants showed a significant downward gradient from the eastern (2.62 percent) toward the western (2.16 percent) region, indicating that men's migration propensity was highest in the eastern region.

In terms of interprovincial migration, men had the highest out-migration rate (1.33 percent) and also the highest in-migration rate (1.78 percent) in the eastern region. While women also had the highest in-migration rate in the eastern region (1.36 percent), their out-migration rate was the highest in the western region (1.28 percent). These differences hint at gender differentials in the paths of interprovincial migration. Regional distribution of interprovincial migrants shows that migration within the eastern region was the most popular and accounted for respectively 28.85 percent and 24.39 percent of male and female interprovincial migrants. But a smaller female proportion for eastern-eastern moves suggests that women were more prone to moves across regions. Specifically, interregional moves in total accounted for 44.8 percent of female interprovincial migrants, and 41 percent of male interprovincial migrants. If the notion that women move shorter distances were valid, then one would have expected a higher proportion of male migrants undertaking interregional moves.

Perhaps the biggest difference in interregional flows is one from the west to the east, which accounted for 14.58 percent of female interprovincial migrants but only 8.42 percent of male interprovincial migrants. Western-central moves also accounted for a bigger proportion of female interprovincial migrants (7.29 percent) than male interprovincial migrants (4.98 percent). These statistics indicate a significant eastward

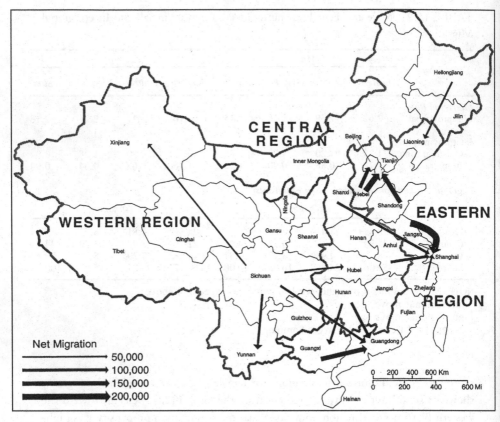

FIGURE 15.3   Male Net Interprovincial Migration, 1985–1990
SOURCE: 1990 Census One-Percent Sample.
NOTE: Only the 15 largest net flows are shown.

regional movement among female interprovincial migrants, more so than their male counterparts. These moves, across regions, involved longer distances than average intraregional moves. Contrary to the widely held view that women move shorter distances than men, the interprovincial migration statistics reviewed above suggest that women who moved across provinces were more likely than their male counterparts to move long distances (see also Li 1993).

Although available data do not permit analyses of the exact paths of migration,[7] it is possible to examine interprovincial migration flows. Figures 15.3 and 15.4 illustrate the spatial patterns of male and female interprovincial migration via the fifteen largest net interprovincial migration flows. A general west-to-east direction characterizes both male and female migration (Fan 1996), but the eastward component is clearly more dominant among female interprovincial migrants. Eleven of the fifteen largest female net flows were from western to eastern, central to eastern, or western to central regions, compared with seven for male net flows.[8] Fourteen of the fifteen

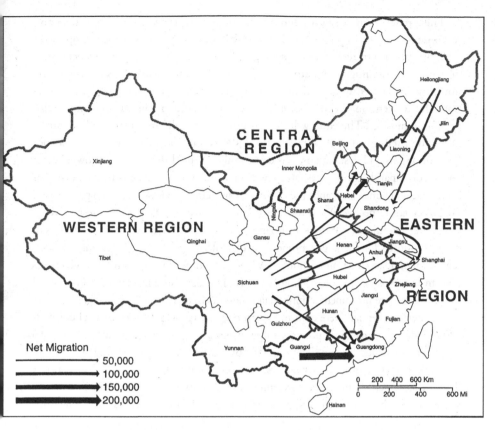

FIGURE 15.4   Female Net Interprovincial Migration, 1985–1990
SOURCE: 1990 Census One-Percent Sample.
NOTE: Only the 15 largest net flows are shown.

destinations among female net flows were in the eastern region, compared with twelve for male net flows. Perhaps the strongest testament of the propensity for female migrants to undertake long-distance moves is the number of intervening provinces between origins and destinations. Six of the fifteen female net flows, compared with only one male net flow, crossed two or more intervening provinces; and only five female net flows, compared with nine male net flows, were between adjacent provinces. A clustering of destinations for male migrants in Beijing, Tianjin, Shanghai, and Guangdong suggests the existence of regional "migration fields" where migrants were drawn from neighboring and nearer provinces. Female migration, on the other hand, was predominantly eastward, that is, to coastal provinces, which involves relatively longer distances. Although Sichuan was a popular origin for both male and female migration, it sent out female migrants primarily to the eastern region and male migrants to all directions, again reinforcing the notion that long-distance eastward movements were prevalent among female interprovincial migrants.

The predominant eastward movement of female migrants suggests an active response by women in poorer provinces to the concentration of economic opportunities in the eastern region, and a stronger relationship between regional economic development and migration paths than their male counterparts. This interpretation not only challenges the notion that women move shorter distances but also questions the conventional wisdom that women are less prone than men to migrate for economic motives. The latter will be examined more fully in the next subsection.

At the rural-urban level, although the 1990 census did not standardize residence types of origins and destinations, it is commonly accepted that cities and towns represent predominantly urban areas, and townships and counties represent rural areas. For both male and female migrants, the majority moved from townships (57.76 percent and 66.24 percent respectively) and to cities (58.4 percent and 54.23 percent respectively), suggesting a dominance of rural-to-urban moves. Specifically, moves from townships to cities accounted for the largest proportion of male migrants (33.71 percent). Among female migrants, a larger proportion (33.91 percent) moved from townships to counties than from townships to cities (32.34 percent). These statistics indicate that urbanward moves were more prominent among male migrants than among female migrants.

Gender differentials were even bigger among interprovincial migrants. The urbanward tendency of male interprovincial migrants was strong; cities accounted for 60.13 percent of their destinations. The majority of female interprovincial migrants (55.32 percent), however, moved to counties. In fact, rural-to-rural migration is a better description of female interprovincial migration, as moves from townships to counties accounted for the largest proportion (41.25 percent) of female interprovincial migrants.

Reasons for Migration

Although existing macro-statistical data do not address directly the processes of migration, one variable in the 1990 census—reason for migration—sheds important light on the various ways migration took place. Migrants were asked to choose from nine "reasons" for migration (Table 15.3).[9] Although labeled reasons for migration, their interpretation includes motives of migration as well as what migrants plan to do and what means and processes accompany their moves. Existing research has classified reasons of migration into one or more of the following categories: planned or institutional, economic, life cycle, family, and social (Fan 1999; Li and Siu 1994; Rowland 1994; Shen and Tong 1992:202; Tang 1993; Zhai and Ma 1994). The economic and social dichotomies are most popular (Kuashiji de zhongguo . . . 1994:258; Shen and Tong 1992:218; Wang and Hu 1996:90–92). Specifically, the 1990 census options of job transfer, job assignment, industry/business, and study/training are generally considered economic reasons; and friends/relatives, retirement, joining family, and marriage are generally considered social reasons (e.g., Li 1993; L. Li 1994).

The most commonly cited reasons for migration selected by female migrants were marriage (28.34 percent), industry/business (15.43 percent), and joining family (15.03 percent). Marriage and joining family both suggest the importance of family networks, or social channels, for achieving mobility. Male migrants selected primarily industry/business (29.95 percent), job transfer (15.07 percent), and study/training (14.8 percent) reasons, reflecting their greater access to labor market opportunities and their greater likelihood to migrate through work-related channels.

The role of marriage in female migration deserves more scrutiny. As discussed earlier, marriage provides poor rural women limited access to new resources. This includes the potential for physical and social mobility and the possible opportunity to obtain *hukou* in a desired destination. On the contrary, most industry/business migrants are not given local *hukou* and thus are denied the benefits available to local residents. In light of the *hukou* institution, therefore, marriage is indeed a more favorable alternative for women, especially for those whose skills and education render them less competitive in the industrial and business sectors.

Among female marriage migrants, 81.29 percent engaged in the labor force, compared with 65.35 percent for all female migrants, suggesting that the former were not merely tied movers but were indeed active participants in the labor force. While 86.2 percent of female marriage migrants originated from townships, 70.04 percent of them migrated to counties, indicating that female marriage migrants were primarily rural-to-rural migrants. The vast majority of female marriage migrants engaged in agriculture (85.27 percent), again indicating that they may be less competitive in nonagricultural work.

Among female interprovincial marriage migrants, 48.57 percent originated from the western region and 61.55 percent moved to the eastern region, indicating a prominence of eastward and long-distance moves. By moving to more developed, albeit rural, areas, these female migrants sought to achieve both marriage and economic betterment. The strong economic consideration behind marriage migration at least partly explains why marriage accounted for 29.13 percent of female interprovincial migrants, a proportion higher than that of intraprovincial migration (27.99 percent). Although short-distance marriages (especially those involving intraprovincial moves) remain dominant in China (Yang 1991; Zhuang and Zhang 1996), the significant proportion of interprovincial west-to-east marriage migration supports the notion that potential and actual economic gains are important factors offsetting the impediment of distance and intervening obstacles for these "brides from afar."[10]

Industry/business was by far the most important migration reason in China; it was the leading reason for all migrants (23.56 percent) and for male migrants (29.95 percent), and the second leading reason for female migrants (15.43 percent) (Table 15.3). However, this important motivation for female migration has seldom been highlighted in the literature, primarily because of the perception that women are less likely to migrate for economic reasons. For both male and female migration, industry/business accounted for higher proportions of interprovincial migrants (38.24 percent and 16.62 percent respectively) than intraprovincial migrants (25.64 percent and 14.9 percent respectively), again supporting the notion that migration

TABLE 15.3   Gender Differentials in Migration Reason

	All			Male			Female		
	Intra-provincial	Inter-provincial	Sum	Intra-provincial	Inter-provincial	Sum	Intra-provincial	Inter-provincial	Sum
Job transfer	10.90	14.31	12.01	13.29	18.51	15.07	8.01	8.33	8.11
Job assignment	7.36	5.48	6.75	9.31	7.55	8.71	5.00	2.55	4.24
Industry/business	20.78	29.31	23.56	25.64	38.24	29.95	14.90	16.62	15.43
Study/Training	15.14	8.14	12.85	17.73	9.15	14.80	12.00	6.70	10.38
Friends/relatives	9.09	11.08	9.74	6.98	8.60	7.54	11.65	14.59	2.55
Retirement	1.62	1.60	1.61	2.53	2.08	2.37	0.52	0.92	0.65
Joining family	10.97	11.02	10.99	8.02	7.41	7.81	14.53	16.15	15.03
Marriage	14.07	13.24	13.79	2.55	2.05	2.38	27.99	29.13	28.34
Other	10.08	5.82	8.69	13.95	6.40	11.37	5.40	5.00	5.28
Sum	100.00	100.00	100.00	100.00	100.00	100.00	100.00	100.00	100.00

SOURCE: 1990 Census One-Percent Sample.

backed by strong economic rationale is likely to take place over longer distances. But it was the eastern region that most interprovincial industry/business migrants originated from and moved to. Among interprovincial industry/business migrants, 52.56 percent of men and 52.51 percent of women originated from the eastern region, and respectively 59.62 percent and 70.95 percent moved to another province in the eastern region. The prominence of the eastern region as both origin and destination suggests that potential migrants in the region were most willing and able to migrate to seek employment in the industrial and business sectors. Moreover, such employment opportunities have been heavily concentrated in the region.

Not surprisingly, industry/business migration exhibited a strong urbanward movement, as respectively 85.21 percent and 85.39 percent of male and female industry/business migrants came from townships, and respectively 61.48 percent and 61.07 percent moved to cities. Among female interprovincial industry/business migrants, however, 53.54 percent moved to counties, and only 46.46 percent moved to cities. This is likely attributable to the expansion of rural enterprises in coastal open zones and their strong pull on *dagongmei*, discussed earlier. Guangdong, in general, and the Pearl River Delta, in particular, epitomize the labor-intensive production system that has absorbed large numbers of *dagongmei*. Unlike most other provinces, where marriage was the leading reason for female in-migration, more than half (51.42 percent) of female migrants entering Guangdong were industry/business migrants, underscoring the pull of economic opportunities there (Fan 1996).

Both job transfer and study/training, the second and third leading migration reasons for men, are highly economically oriented, as the former involves change in employment and the latter involves improvement in human capital, which likely brings about future economic gains. The third leading reason for female migration is joining family, which by definition refers to migration following the job transfer of other family members. It accounted for 15.03 percent of female migration and only 7.81 percent of male migration, suggesting that men's employment continued to be considered a priority in the Chinese family.

Migrants' Occupations

The above analyses provide convincing evidence for the argument that economic motives are the key to explaining both male and female migration in China. Both male and female migrants are active participants in the labor force. In 1990, among nonstudent migrants between the ages of 15 and 60, 90.40 percent of men and 78.94 percent of women were in the labor force. The extent to which migrants' economic goals are achieved, and the gender differentiations in these achievements, depend largely on the work attained after migration.

Table 15.4 shows the distribution of male and female migrants and nonmigrants by occupation. The male-female ratio is again used here to evaluate differentials in men's and women's relative proportions in each occupation (columns 5 and 6). A comparison of the nonmigrants' and migrants' male-female ratios can shed light on whether migration has contributed to more or less balanced gender distribution by occupation.

Although the vast majority of male and female nonmigrants (68.19 percent and 76.03 percent respectively) engaged in agricultural work, the proportions of migrants in that category (14.5 percent and 39.66 percent respectively) were significantly lower. However, the decline in proportion for male migrants was substantially larger than that for female migrants, so much so that the male-female ratio for agricultural work decreased from 90 for nonmigrants to 37 for migrants. Inasmuch as a shift from agricultural to nonagricultural sectors is likely to improve one's standard of living, male migrants seemed to have undertaken that sectoral shift more successfully than female migrants.

A shift in the labor force from agricultural to industrial work clearly accompanied the migration process. The proportions of male and female nonmigrants engaging in industrial work were, respectively, 16.95 percent and 11.58 percent, compared with 50.72 percent and 29.36 percent for male and female migrants. Nonetheless, the male-female ratio for migrants (173) was higher than that for nonmigrants (146), suggesting that migration contributed to a more uneven distribution of female and male labor in industrial work. As suggested in the last subsection, male migrants in industrial work were more likely than their female counterparts to have moved to urban areas, where wage rates are higher. This, in conjunction with men's higher success rates of shifting from agricultural to industrial work, suggests that male migration, more so than female migration, is accompanied by significant economic gains.

An increase in male-female ratios also characterized occupations commonly associated with greater power and higher prestige, namely professional, government, and administrative/clerical occupations. Although the proportions of female migrants in these occupations were generally higher than their nonmigrant counterparts, men have garnered greater proportional gains than women as a result of migration, leading to more uneven distributions of male and female labor in these occupations. For example, the male-female ratio in government work was 613 among nonmigrants, indicating a very uneven gender distribution; a migrant male-female ratio of 779 suggests that migration contributed to a worsening of gender distribution in that

TABLE 15.4   Gender Differentials in Occupation

	Nonmigrants (%)		Migrants (%)		Male-Female Ratio		Temporary Migrants (%)	
	Male (1)	Female (2)	Male (3)	Female (4)	Non-migrants (5)[a]	Migrants (6)[b]	Male (7)	Female (8)
Professional	5.12	5.25	11.91	11.41	98	104	3.26	3.63
Government	2.84	0.46	3.61	0.46	613	779	1.86	0.22
Administrative/ clerical	2.12	0.97	6.30	2.02	218	312	2.02	0.73
Commerce	2.72	3.07	7.31	7.59	88	96	9.82	11.33
Services	2.03	2.60	5.53	9.42	78	59	7.41	13.20
Agricultural	68.19	76.03	14.50	39.66	90	37	11.92	30.38
Industrial	16.95	11.58	50.72	29.36	146	173	63.61	40.45

NOTE: This table refers to population aged 15 and above.
[a](5) = (1) / (2) ×100
[b](6) = (3) / (4) ×100
SOURCE: 1990 Census One-Percent Sample.

occupation. The representation of women in services, on the other hand, has increased due to migration. The male-female ratio was only 78 among nonmigrants but decreased further to 59 among migrants. These statistics suggest that migration has reinforced the funneling of women into less prestigious, low-paying jobs and the positioning of men in occupations of leadership and responsibility (Riley 1996).

The occupational distribution of temporary migrants (columns 7 and 8) underscores the importance of the *hukou* institution as well as gender. Temporary migrants were less highly represented in professional, government, and administrative/clerical occupations and more highly represented in commerce, services, and industrial work than migrants as a whole. However, the interaction of gender and *hukou* status is even more revealing. While the vast majority of male temporary migrants worked in the industrial sector (63.61 percent), female temporary migrants were more widely spread among industrial, agricultural, services, and commercial occupations, further demonstrating the gendered division of labor among migrants. It appears that the feminization of agriculture has taken place for both temporary migrants and migrants as a whole. Women were clearly in greater demand in services, characterized by low prestige and pay, as the Chinese consumer market expanded and created needs for nannies, maids, restaurant servers, and similar kinds of positions (Yang and Guo 1996).

The analysis suggests a process of labor market segmentation accelerated by migration. As households send out men to work in nonagricultural sectors, the women who remain constitute the dominant labor force in agriculture. Even if women manage to migrate, a large proportion of them remain in the agricultural sector in their new destinations. The feminization of agriculture reflects sustained discrepancies in women's and men's human capital and access to opportunities in the labor market (Lu 1997). The higher proportions of female migrants in agriculture and services and their lower proportions in more prestigious occupations, relative to their male counterparts, suggest that not only has migration reinforced labor market segmentation, but it has also increased men's social mobility relative to that of women.

Conclusion

The main objective of this chapter has been to highlight the differentials between male and female migration, a topic largely absent in the voluminous literature on Chinese migration. Through analyzing gender differences in migration, I have highlighted the complexities in the migration processes for both men and women in China and the important role of gender in these processes. I have argued that gender differences in migration reflect not only the age-old sociocultural tradition, which continues to undermine the status of women, but also structural factors unique to the Chinese transitional economy. These factors include spatial-economic changes, which opened up new opportunities and specific destinations for male and female migrants, and the *hukou* institution, a mechanism of state control.

The empirical analysis examines several widely held notions about gender differences in migration. It supports, in general, the notion that Chinese men have a higher migration propensity than women and shows that women's mobility is especially constrained after marriage. The finding about migration distance is less conclusive. Although women were more highly represented than men in intraprovincial migration, suggesting that the former moved shorter distances, female interprovincial migrants were more highly represented among long-distance interregional moves. Such long-distance migration also contradicts the notion that men are more likely to migrate for economic reasons and women for social reasons. Although women's leading reason for migration was marriage, the very focused eastward movements signaled a strong economic rationale behind this type of migration and highlighted the agency of women. Large waves of women migrating for industry/business employment further underscored the economic motives behind female migration. On the other hand, male migrants were more successful than their female counterparts in their efforts to move to urban areas, transfer their labor from agricultural to nonagricultural sectors, and find work in more prestigious positions. The interaction of gender and *hukou* further reinforced the segmentation of China's labor market so that despite migration, poor and uneducated peasant women continued to be entrapped in the lower socioeconomic niches of the economy.

This chapter's findings underscore the importance of examining both male and female migration, highlighting the differences between them, and mainstreaming the study of female migration. Not only is gender a key to explaining differences in migration, it also provides an important perspective, integrating the sociocultural, spatial-economic, and institutional factors for understanding the patterns and processes of migration in China.

Notes

This research was funded by the National Science Foundation (SBR-9618500; SES-0074261), the Luce Foundation, UCLA Academic Senate, and UCLA International Studies and Overseas Program. The author would like to thank Chase Langford for cartographic production and Youqin Huang for research assistance.

1. The definition of migration has changed between the 1990 and 2000 censuses. Specifically, (1) the minimum duration of stay was reduced from one year to six months, and (2) individuals who moved within county-level units were not considered migrants in the 1990 census but were considered migrants in the 2000 census. At the time of this chapter's writing, only summary statistics from the 2000 census were available to the public. Thus, the analysis in this chapter uses primarily data from the 1990 census.

2. Two exceptions are Yang and Guo (1996) and Huang (2001), who provided rigorous and comprehensive analyses of occupational attainment by gender and type of migration (permanent versus temporary).

3. *Dagongmei* is literally translated as "working girls" and refers specifically to young women who migrate from the countryside to work in industries and services.

4. Despite recent efforts to reform the *hukou* system, only the most successful peasant migrants have benefited from these changes, and only small cities and towns have granted urban *hukou* or variants of urban *hukou* to significant numbers of peasant migrants (Yu 2000).

5. *Hukou* transfer to urban areas through marriage remains strictly controlled.

6. The one-percent sample is a clustered sample containing information about every individual in all households of the sampled village-level units (villages, towns, and urban neighborhoods in cities), drawn from China's 1990 census and made available by the National Information Center. It has a total of 11,475,104 records.

7. The 1990 census specifies the origin provinces of migrants, but not their county-level origin.

8. If Guangxi were considered a central-region province, which in fact is a more accurate description of its level of economic development, there would have been twelve eastward net flows for women and eight for men. Since the Guangxi-Guangdong net flow (223,700) was the largest among female migrants, allocating Guangxi to the central region would have strengthened even more the argument that eastward movements dominated female interprovincial migration.

9. Definitions of these reasons of migration are as follows: job transfer—migration due to job change, including demobilization from the military; job assignment—migration due to assignment of jobs by the government after graduation and recruitment of graduates from different schools; industry/business—migration to seek work as laborers or in commercial or trade sectors; study/training—migration to attend schools or to enter training or apprentice programs organized by local work units; friends/relatives—migration to seek support of relatives or friends; retirement—cadres or workers leaving work due to retirement or resignation, including retired peasants in rural areas with retirement benefits; joining family—family members following the job transfer of cadres and workers; marriage—migration to live with spouse after marriage; and other—all other reasons (SSB 1993: 513–514, 558).

10. A popular saying in rural Zhejiang, one of the main destinations of female marriage migrants, describes the current prevalence of long-distance marriage migrants: "In the 1960s, wives were from Subei (the northern and poorer part of Jiangsu, an east coast province); in the 1970s, they were from the rustic countryside; and in the 1990s, they come from afar" (Xu and Ye 1992).

References

Bauer, J., F. Wang, N. E. Riley, and X. Zhao. 1992. Gender inequality in urban China. *Modern China* 18(3):333–370.

Bossen, L. 1994. Zhongguo nongcun funu: Shime yuanyin shi tamen liuzai nongtianli? (Chinese peasant women: What caused them to stay in the field?). In *Xingbie yu Zhongguo* (Gender and China), ed. X. Li, H. Zhu, and X. Dong, 128–154. Beijing: Sanlian Shudian.

Cai, F. 1997. Qianyi juece zhong de jiating jiaose he xingbie tezhen (The role of family and gender characteristics in migration decisionmaking). *Renkou Yanjiu* (Population Research) 2:7–21.

Chan, K. W. 1996. Post-Mao China: A two-class urban society in the making. *International Journal of Urban and Regional Research* 20:134–150.

Chan, K. W., and L. Zhang. 1999. The *hukou* system and rural-urban migration in China: Processes and changes. *China Quarterly* 160:818–855.

Cheng, T., and M. Selden. 1994. The origins and social consequences of China's *hukou* system. *China Quarterly* 139:644–668.

Chiang, N. 1999. Research on the floating population in China: Female migrant workers in Guangdong's township-village enterprises. In *Population, Urban and Regional Development in China*, ed. N. Chiang, and C. Song, 163–180. Taipei, Taiwan: Population Studies Center, National Taiwan University.

China reforms domicile system. 2001. *China Daily*. Available at: http://service.china.org.cn/link/wcm/Show_Text?info_id=19022&p_qry=2000%20and%20census, September 12.

Christiansen, F. 1990. Social division and peasant mobility in Mainland China: The implications of huk'ou system. *Issues and Studies* 26(4):78–91.

Croll, E. 1987. New peasant family forms in rural China. *Journal of Peasant Studies* 14(4): 469–499.

Davin, D. 1997. Migration, women and gender issues in contemporary China. In *Floating Population and Migration in China*, ed. T. Sharping. Hamburg: Institut für Asienkunde.

_____. 1998. Gender and migration in China. In *Village Inc.: Chinese Rural Society in the 1990s*, ed. F. Christiansen, and J. Zhang, 230–240. Surrey, Great Britain: Curzon Press.

_____. 1999. *Internal Migration in Contemporary China*. London: MacMillan Press.

Ding, J. H. 1994. Zhongguo renkou shengji xianyi de yuanyin bie liuchang tezhen taixi (Characteristics of cause-specific rates of interprovincial migration in China). *Renkou Yanjiu* (Population Research) 18(1):14–21.

Fan, C. C. 1995. Of belts and ladders: State policy and uneven regional development in post-Mao China. *Annals of the Association of American Geographers* 85(3):421–449.

_____. 1996. Economic opportunities and internal migration: A case study of Guangdong Province, China. *Professional Geographer* 48(1):28–45.

_____. 1999. Migration in a socialist transitional economy: Heterogeneity, socioeconomic and spatial characteristics of migrants in China and Guangdong province. *International Migration Review* 33(4):493–515.

_____. 2000. Migration and gender in China. In *China Review 2000*, ed. C. M. Lau and J. Shen, 217–248. Hong Kong: Chinese University Press.

_____. 2002. The elite, the natives, and the outsiders: Migration and labor market segmentation in urban China. *Annals of the Association of American Geographers* 92(1):103–124.

_____. 2003. Rural-urban migration and gender division of labor in China. *International Journal of Urban and Regional Research* 27(1):24–47.

Fan, C. C., and L. Li. 2002. Marriage and migration in transitional China: A field study of Gaozhou, Western Guangdong. *Environment and Planning A* 34(4):619–638.

Fan, C. C., and Y. Huang. 1998. Waves of rural brides: Female marriage migration in China. *Annals of the Association of American Geographers* 88(2):227–251.

Gao, X. X. 1994a. Dangdai zhongguo nongcun laodongli zhuanyi ji nongye nuxinghua qushi (Transference of rural labor force and the tendency of farming femalization in contemporary China). *Shehuixue Yanjiu* (Social Science Studies) 2–3:83–90, 82–89.

_____. 1994b. China's modernization and changes in the social status of rural women. In *Engendering China: Women, Culture, and the State*, ed. C. K. Gilmartin, G. Hershatter, L. Rofel, and T. White, 80–97. Cambridge: Harvard University Press.

Gu, S. Z., and X. H. Jian, eds. 1994. *Dangdai zhongguo renkou liudong yu chengzhenhua* (Population Movement and Urbanization in Contemporary China). Wuhan, China: Wuhan daxue chubanshe (Wuhan University Press).

Guowuyuan dui nongmin banli jizhen hukou zuochu guiding (State Council regulations for peasants seeking permission to live in towns). 1984. *Renmin Ribao* (People's Daily), December 22, 1–2.

Honig, E., and G. Hershatter. 1988. Marriage. In *Personal Voices: Chinese Women in the 1980s*, ed. E. Honig and G. Hershatter, 137–166. Stanford: Stanford University Press.

Huang, Y. 2001. Gender, hukou, and the occupational attainment of female migrants in China (1985–1990). *Environment and Planning A* 33(2):257–279.

Ji, D. S., and Q. Shao, eds. 1995. *Zhongguo renkou liudong taishe yu guanli* (The Migration Trend and Management of Population in China). Beijing, China: Zhongguo renkou chubanshe (China Population Press).

Knight, J., and L. Song. 1995. Towards a labour market in China. *Oxford Review of Economic Policy* 11(4):97–117.

Kuashiji de zhongguo renkou zong bianweihui. 1994. *Kuashiji de zhongguo renkou (zhonghe juan)* (The Population of China Toward the 21st Century [General Volume]). Beijing: Zhongguo tongji chubanshe (China Statistics Press).

Li, L. 1994. *Gender and Development: Inner-city and Suburban Women in China*. Guangzhou, China: Centre for Urban and Regional Studies, Zhongshan University. Working Paper No. 11.

Li, L. M., et al. 1991. Guanyu wailai nuxing renkou de sikao (Ponderation of female population coming in from outside). *Nanjing Renkou Guanli Ganbu Xueyuan Xuebao* (Journal of Nanjing Population Management Cadre Institute) 3:43–46.

Li, M. H. 1994. Wuguo renkou qianyi de liuxiang (The migration stream in China). *Renkou Yanjiu* (Population Research) 3:48–51.

Li, S. M., and Y. M. Siu. 1994. Population mobility. In *Guangdong: Survey of a Province Undergoing Rapid Changes*, ed. Y. M. Yeung and D. K. Y. Chu, 373–400. Hong Kong: Chinese University Press.

Li, S. Z. 1993. Bashi niandai zhongguo renkou qianxi de xingbie chayi yanjiu (Gender-difference research on population migration of China in the 1980s). *Renkou Xuekan* (Population Journal) 5:14–19.

Li, W. L., and Y. Li. 1995. Special characteristics of China's interprovincial migration. *Geographical Analysis* 27(2):137–151.

Lim, L. L. 1993. The structural determinants of female migration. In *Internal Migration of Women in Developing Countries*. United Nations Department for Economic and Social Information and Policy Analysis, 207–222. Proceedings of the United Nations Expert Meeting on the Feminization of Internal Migration. Aguascalientes, Mexico, 22–25 October 1991. New York: United Nations.

Lin, G. C. S. 1997. *Red Capitalism in South China: Growth and Development of the Pearl River Delta*. Vancouver: University of British Columbia Press.

Liu, X. 1990. Guanyui Xiaoshanshi "wailainu" zhuangkuang ji qi guan li wenti (The conditions and management problems of "women from outside" in Xiaoshan city). *Renkou Yanjiu* (Population Research) 6:31–36.

Lu, L. 1997. Funu jingji diwei yu funu renli ziben guanxi de shizhen yanjiu (Women's economic status and their human capital in China). *Renkou Yanjiu* (Population Research) 2:50–54.

Mallee, H. 1996. In defense of migration: Recent Chinese studies on rural population mobility. *China Information* 10(3–4):108–140.

Maurer-Fazio, M., T. G. Rawski, and W. Zhang. 1997. Gender wage gaps in China's labor market: Size, structure, trends. Unpublished manuscript.

McGee, T. G. 1991. The emergence of *desakota* regions in Asia: Expanding a hypothesis. In *The Extended Metropolis: Settlement Transition in Asia*, ed. N. Ginsburg, B. Koppel, and T. G. McGee, 3–25. Honolulu: University of Hawaii Press.

Park, K. A. 1992. *Women and Revolution in China: The Sources of Constraints on Women's Emancipation*. Michigan State University, Franklin and Marshall College, Working Paper No. 230.

Pedraza, S. 1991. Women and migration: The social consequences of gender. *Annual Review of Sociology* 17:303–325.

Population Census Office and National Bureau of Statistics. 2002. *Zhongguo 2000 nian renkou pucha ziliao* (Tabulation on the 2000 Population Census of the People's Republic of China). Beijing: China Statistical Press.

Riley, N. E. 1996. China's "missing girls": Prospects and policy. *Population Today* 24(2):4–5.

Roberts, K. D. 1997. China's "tidal wave" of migrant labor: What can we learn from Mexican undocumented migration to the United States? *International Migration Review* 31(2):249–293.

Rowland, D. T. 1994. Family characteristics of the migrants. In *Migration and Urbanization in China*, ed. L. H. Day and X. Ma, 129–154. Armonk, N.Y.: M. E. Sharpe.

Shen, Y. M., and C. Z. Tong. 1992. *Zhongguo renkou qianyi* (China's Population Migration). Beijing: Zhongguo Tongji Chubanshe (China Statistics Press).

Smith, C. 1996. Migration as an agent of change in contemporary China. *Chinese Environment and Development* 7(1–2):14–55.

Solinger, D. J. 1995. The floating population in the cities: Chances for assimilation? In *Urban Spaces in Contemporary China*, ed. D. Davis, R. Kraus, B. Naughton, and E. J. Perry, 113–139. Cambridge: Cambridge University Press.

State Statistical Bureau (SSB). 1993. *Zhonguo 1990 nian renkou pucha ziliao* (Tabulation on the 1990 Population Census of the People's Republic of China), vol. 6. Beijing: Zhongguo Tongji Chubanshe (China Statistics Press).

Tang, X. M. 1993. Beijing shi renkou qianyi he renkou liudong (Population migration and mobility in Beijing). *Renkou Yanjiu* (Population Research) 4:52–55.

Wang, F. 2000. Gendered migration and the migration of genders in contemporary China. In *Re-Drawing Boundaries: Work, Household, and Gender in China*, ed. B. Entwisle and G. Henderson, 231–242. Berkeley: University of California Press.

Wang, J. M., and Q. Hu. 1996. *Zhongguo liudong renkou* (Floating Population in China). Shanghai: Shanghai caijing daxue chubanshe (Shanghai Finance University Press).

Wei, J. S. 1995. Gaige kaifang yilai woguo nongcun di renkou yidong (Population movements in China's villages since the reforms and opening). In *Zhongguo renkou liudong taishe yu guanli* (The Migration Trend and Management of Population in China), ed. D. S. Ji and Q. Shao, 104–119. Beijing, China: Zhongguo renkou chubanshe (China Population Press).

West, L. A., and Y. Zhao. 2000. *Rural Labor Flows in China*. Berkeley: Institute of East Asian Studies.

Xu, A. G., and X. Y. Zhen, eds. 1992. Zhejiangsheng qianru nuxing renkou hunyin shengyu diaocha (Survey of the marriage and reproduction of Zhejiang's moved-in female population). *Zhongguo Renkou Kexue* (China Population Studies) 4:49–55.

Xu, T. Q., and Z. D. Ye. 1992. Zhejiang wailai nuxing renkou tanxi (An analysis on female out-migrants in Zhejiang province). *Renkou Xuekan* (Population Journal) 2:45–48.

Yang, Q. F. 1991. Nannu beijia xianxiang ji qi libi qianxi (The phenomenon of southern women marrying to the north and its advantages and disadvantages). *Renkou Xuekan* (Population Journal) 5:51–55.

Yang, Q., and F. Guo. 1996. Occupational attainment of rural to urban temporary economic migrants in China, 1985–1990. *International Migration Review* 30(3):771–787.

Yang, X., and F. Guo. 1999. Gender differences in determinants of temporary labor migration in China: A multilevel analysis. *International Migration Review* 33(4):929–953.

Yang, Y. Y. 1994. *Zhongguo renkou qianyi yu fazhan di changqi zhanlue* (Long-Term Strategy for China's Population Migration and Development). Wuhan, China: Wuhan Chubanshe (Wuhan Press).

Yu, D. 2002. *Chengxiang shehui: Cong geli zouxiang kaifang—zhongguo huji zhidu yu hujifa yanjiu* (City and Countryside Societies: From Segregation to Opening—Research on China's Household Registration System and Laws). Jinan, China: Shandong renmin chubanshe (Shandong People's Press).

Yu, X., and L. H. Day. 1994. Demographic characteristics of the migrants. In *Migration and Urbanization in China,* ed. L. H. Day and X. Ma, 103–128. Armonk, N.Y.: M. E. Sharpe.

Zhai, J. Y., and J. Ma. 1994. Woguo guangdong sheng renkou qianyi wenti tantao (Migration in Guangdong Province). *Renkou Yanjiu* (Population Research) 2:18–24.

Zhang, Q. W. 1995. Dangqian zhongguo liudong renkou zhuangkuang he duice yanjiu (A study on the current situation of population migration in China and its policy implications). In *Zhongguo renkou liudong taishe yu guanli* (The Migration Trend and Management of Population in China), ed. D. S. Ji and Q. Shao, 56–64. Beijing, China: Zhongguo renkou chubanshe (China Population Press).

Zhu, B. S. 1994. Zhongguo shengji chengxiang renkou qianyi liutai moshi tantao (Interprovincial urban-rural migration in China). *Renkou Yanjiu* (Population Research) 6:3–9.

Zhuang, S., and M. Zhang. 1996. Chabie renkou qianyi yu xingbie goucheng diqu chayi de kuodahua (Migration between different regions and the enlargement of regional sex composition). *Renkou Xuekan* (Population Journal) 1:3–10.

16

Engendering Industrialization in China Under Reform

Carolyn Cartier

China's program of economic reform is a set of national modernization and development strategies in the broadest sense. The nature of the reform program encompasses both domestic perspectives and internationalized development strategies, especially in the arena of macroeconomic policy. This chapter examines the geographical characteristics of major reforms to show how, as development strategies, the reforms have borne gendered values and produced gendered results. While this survey focuses on the reform period, many gendered characteristics of reform policies and impacts cannot be simply attributed to the Deng Xiaoping and Jiang Zemin eras. The gendered characteristics of the contemporary reform program carry amalgamated histories of gendered social structures and gender-specific policy perspectives from earlier periods. It is the complex interaction of combined contemporary forces and historic social and economic conditions that creates gendered geographies in China under reform.

Changes in social and economic conditions as a result of reform—including population growth, household formation, education, migration, agricultural production, and manufacturing labor—have affected men and women differently all across China. At the aggregate scale, the macroeconomic reform policies restructuring the organization of production have promoted new patterns of gendered labor as well as heightened awareness of gender issues. Among more specific reforms, trends produced by the complex interactions among five initiatives have particularly structured gendered geographies: the open-door policy and the emphasis on export-oriented industrialization, which depends on low-wage women's labor for manufacturing production; the birth planning policy emphasizing one child, which has tipped the birth rate in favor of males; the rural household responsibility system, which has re-anchored many rural women's productive and reproductive activities in the household economy; the loosening up of the *hukou* or household registration system, which has given rise to gendered migration patterns; and the restructuring of state-owned enterprises (SOEs) resulting in significant layoffs, which has particularly limited women's employment options in a climate of high unemployment. In a 1997 survey conducted by the women's department of the All-China Federation of Trade Unions, over 70 percent of the 413 managers in fourteen provinces and cities said they would not hire women, even if they were better qualified than men (Liu and Zhang 1997). In Shanghai, Liaoning, and Guangdong, half of the qualified female workers seeking employment could not find jobs. The survey results also reported that over three-quarters of women employees worried about their lives and preferred jobs in state-owned companies. The large excess labor force in China presents a difficult problem in all arenas. Most recently the state has announced a new policy of compulsory maternity leave up to five years as a way to cut the urban industrial workforce (Wu and Xiang 1998), which would derail women's career paths and increasingly circumscribe women's domain within the domestic sphere. The All-China Women's Federation has protested the new policy. The more problematic gendered effects of the reform program are registering in rural areas, where women have increasingly formed the bulk of the labor force. Results of a recent international study reported that the highest suicide rate in the world is among rural Chinese women (MacLeod 1998; Wu and Mok 1998).

China under reform is a society experiencing the clashing forces of rapid economic restructuring, the arrival of globalized cultural and economic practices, and the resurgence of traditional cultural values, often in new forms. Enduring characteristics of traditional Chinese society inherited from Confucianism continue to influence contemporary society and how reforms are interpreted and gendered. As people and institutions interpret reform policies through traditional social practices, the implementation of ostensibly gender-neutral policies carries and transfers gender bias. The one-child policy, in addition to affecting the male-female birth ratio, has fostered new discourses about sex selection and the worthiness of males compared to females. Income generation problems in rural areas are also registering as gender-based decisions at the household level in removing girls from school for household and farmwork. These and other complex changes in society under reform have given rise to new debates on topics once considered to be subjects of comparatively private concern: issues about gender, the status of women, sex, and sexuality have unprecedentedly surged into the public sphere and popular discussion in the People's Republic in the past decade (Gilmartin et al. 1994; Evans 1995). In the context of domestic debates on gender and sexuality, taking gender issues seriously in China becomes an intellectual enterprise that derives fundamentally from the complexity of dynamic changes sweeping across China under reform.[1]

Amidst the transformation in society and economy, many men and women have found new personal and professional options. However, research on gender under the reform era has concluded that these conditions—the transhistorical conditions of traditional society combined with new reform era social and economic structures—have coalesced to promote distinct negative trends where the status of women is concerned. No matter the urban or rural circumstances, recent work (cf. Davin 1991; Ho 1995; Judd 1994; Kelkar and Wang 1997; Lin 1995; Meng and Miller 1995; Summerfield 1994a,b; 1997) on the effects of reform on women substantiates the position, taken earliest and most visibly by Croll (1978; 1983) and Stacey (1983), that the position of women in the social hierarchy has generally reverted to more traditional role–based expectations.

Engendering Industrialization

The analytical definition of gender is the socialization of male and female characteristics and the social structures based on those characteristics—not biological sex. Feminist theorist Gerda Lerner (1986: 238) has assessed the confusion over the term gender: "Unfortunately, the term is used both in academic discourse and in the media as interchangeable with 'sex.' In fact, its widespread public use probably is due to it sounding a bit more 'refined' than the plain word 'sex' with its 'nasty' connotations. Such usage is unfortunate, because it hides and mystifies the difference between the biological given—sex—and the socially constructed—gender." Most current work on gender and development utilizes the approach of the social construction of gender to demonstrate how men's and women's experiences are different and socially produced in the context of state and society under varying circumstances of family, sexuality, age, class background, educational level, and employment situation (e.g., Momsen and Kinnaird 1993). Anthropologist Henrietta Moore (1988) has reminded us that gender and gender relations are simultaneously both symbolic and material: How individuals think about and value aspects of femininity and masculinity influence the ways in which men and women act and react in situated contexts. Geographical perspectives on gender analyze the localized manifestations of gendered characteristics, how spatial and place-based processes inform the constitution of gendered identities, and larger scale influences of state and society in the production of gendered social characteristics with spatial patterns. In these ways geographers theorize gender relations, in both directly observable and symbolic forms, as spatially variable phenomena in everyday places and spaces, and across different levels of spatial scale from the household to the village community or towns and cities, to the national level of the state and society, and in international contexts of networked gender-related organizations (see Nicholson 1995).

In the theoretical terms of gender analysis, the reform policies promoting industrialization represent what Elson terms "male bias in the development process," which "operates in favor of men as a gender, and against women as a gender" (1991: 3). This is not to adopt essentialist categories, that is, that all men would be biased against all women, but that the development process tends to favor the accumulation of male power, privilege, and wealth. Male bias in the development process is produced in different and complex ways, through the practices and traditions of everyday life, applications of economic theory, and public policy that regularly embodies male bias of traditional practices and gender-insensitive theory. It is well known that normative economic theory regularly treats individuals as "rational actors" and does not incorporate variables about gender or domestic labor roles. In combination with traditional patriarchal practices, economic development policy regularly and unconsciously transfers male bias from both contemporary theoretical and traditional cultural modes of thought.

Despite the significance of situated gendered characteristics in China's reform process, and with the exception of important work on women and migration by Fan and Huang (1998), geographical research in China has yet to incorporate a gender analysis as a regular component of investigation. Gender and women are commonly treated as special interest topics, or externalities—variables that are not central to the issues at stake because they are difficult to measure empirically, and so they are not incorporated into basic economic analysis. Researchers also regularly work on topics that do not require them to assess gender variables, and so they are influenced not to raise questions about gender issues. In this light, this chapter serves dual functions: an analysis of the geographical characteristics of the gendered reform process, and an argument for including gender analysis as a regular component of geographical research on China under reform. Regular assessment of the gendered conditions of reform should yield awareness of the patterns of gendered conditions in society and the economy, and ultimately of the need to alleviate pressures limiting women's options; downward trends in the status of women in China bode ill for the broad spectrum of standards of development in China generally.[2] Thus this analysis does not seek to treat gendered social and economic structures in China as fundamentally limiting women's opportunities but rather to assess and critique how the development policies of the patriarchal state and society have narrowed women's options in a state-defined era of excessive population, dramatic labor surplus in the rural sector, and surging urban unemployment.

This analysis also adopts the perspective of the status of women as a position from which to evaluate gendered geographies. The status of women may be defined as the position or rank of women in society and in relation to the status of men. In many countries, including China, law contributes to defining the status of population groups by sex, race, and other factors of group definition. Chinese law defines women as equal to men in matters of educational opportunities, employment, marriage, property inheritance, and other aspects of societal existence. But the rule of law has not existed in China as a primary structural force in society, and many laws are not reliably observed; local and regional practices often and unpredictably prevail over national policies and laws. For example, Chinese law intercedes in the custom of patrilocal marriage by granting women the right to reside in their natal village after marriage, but if a couple wants to reside in the wife's village they may be expected to pay a fee, up to 8,000 yuan (about $1,000) in areas of Zhejiang province (Kelkar and Wang 1997: 75). In such circumstances, national law at best offers symbolic reform (see Mazur 1995). The realities of women's experience and comparative status result from the aggregate set of complex gendered negotiations taking place in society.

Transhistorical Conditions

Traditional social practices and social organization inherited from Confucianism have structured men's and women's life paths in China. Under reform some aspects of these transhistorical patterns have resurged. Of course the strictures of the Confucian cultural complex have increasingly fractured through the twentieth century, yet what does it mean that some rural localities have revived celebrations of Virtuous Wives Day, in which village leaders honor women who have unselfishly devoted themselves to caring for ailing husbands, to their children when widowed (the chaste widows of Confucian ideology were held to symbolize proper social order), or, in the logic of reform, to husbands to help them make a fortune (Rai 1994b: 128)? Virtuous women was a category of administrative record, officially promoted by the imperial order and recorded in dynastic era gazetteers for hundreds of years across Han China (see Elvin 1984). The revival of social practices like these raises questions about how the state, through local officials, supports and reinterprets traditional gender roles. In the case of so-called virtuous wives, localities have rewarded women who have demonstrated devotion to the patriline. In critical terms, this revival celebrates women's important role in the accumulation of patriline privileges under reform and represents how the state and society exert power to impose morality and surveil women's real and symbolic behavior. In spatial terms, these practices locate ideas about state-sanctioned proprieties, through women's conduct, in the household.

The basis of the Confucian social system in the patriline has maintained as the transhistorical force of lineage structure in the social system. Husbands, fathers, and eldest sons headed households and carried out lineage duties at the family home and ancestral hall. Traditional life paths tied sons to natal place; daughters relocated upon marriage to support the patriline responsibilities of their husbands. Women were also tied to the hearth, but not as inheritors of land, power, or status, but as resources of labor or reproduction and always as wives in relation to the patriline. Historically, in well-to-do families women rarely ventured outside the family compound. (Through the Ming and Qing dynasties the increasingly widespread practice of footbinding also limited women's physical mobility.) The widespread confinement of women to the domestic sphere has led Goldstone (1996) to attribute China's relatively late industrialization to the structures of the Confucian family that restricted women's work outside the home: Could women's traditional confinement to the home have prevented China from embarking upon industrialization much earlier, after the pattern of Britain, the United States, and Japan, where low-cost women's labor served the basis of industrial development in mechanized factory production of textiles? This research on the historic impacts of the Confucian cultural complex would explain the notion of China's delayed industrialization in gendered terms.[3]

Women's life paths in traditional China have allowed some status mobility, especially as women work through the expected stages from natal home to husband's home, to mother of sons, and ultimately to mother-in-law. In this typified life path women gained power through their husbands, through their sons, and in their elder years. In the Maoist era, directives about women's emancipation in part diluted the strict Confucian order and worked some flexibility for women into the social system. Under communism the reorganization of production attempted to equalize women's roles in the workforce and credited their labor contributions through the *danwei,* or work unit. The *danwei* also allotted housing and other services related to the domestic sphere, and so organization of production and reproduction in the same social unit promoted greater gender equality (Stockman 1994). Policies like the Marriage Law (1950) and Labor Insurance Regulations (1951) instituted, respectively, the right to seek divorce and the principle of equal pay for equal work. But the revolution for women was, in Wolf's (1985) words, "postponed" in China. Theory did not translate into practice, as husbands, mothers-in-law, and party cadres prevented women from stepping outside the bounds of many social traditions, especially from initiating divorce (Croll 1978: 22). The conclusions of Andors (1983) and Stacey (1983) echo each other in confirming that social practices of traditional gendering continued through the Maoist era.

With the onset of reform, decollectivization folded women back into a household-based gendered division of labor. Yet the relocation of labor activity in the household does not in itself diminish the status of women. It is how society values that labor contribution, and how women are able to interpret their own social status and quality of life in the context of existing power relations in the family, community, and workplace that determines the status of women (e.g., Rofel 1993). Given the attempts to improve conditions for women by legal and other policy measures, and women's own continuing interests to benefit from such measures, a great deal of cultural, political, and economic power must be continually wielded to shore up gender inequity. As this analysis suggests, it is reasonable to conclude that neither Chinese culture nor Confucianism is sufficient to explain women's subordination, but that these transhistorical cultural conditions in combination with state-sanctioned development policies that affect men and women produce the problematic climate for the status of women. As a result of the Maoist era, and now that China's modernization program is under way, certainly Confucian social order has been breaking down. Yet the status of women has not risen widely in response. In what ways do the role of the state and the economic policies pursued in China under reform limit opportunities for women?

Gender and the Role of the State

Policies of the contemporary state in China are sometimes explicitly gendered, as in the case of the compulsory maternity leave policy, and in other cases are gendered through existing social structures that influence, or occur as a result of, policy implementation. In the latter case, the mores and values of society shape the social landscape under reform. In the case of the birth planning policy exhorting the one-child model, the ultimate goal of the policy is to achieve high levels of industrial development by decreasing demands on the resource base. The policy itself is not explicitly gendered, but its implementation implicates male-biased gender-based decisionmaking at all levels from the household level and sex selection to the national level and forms of birth control sanctioned by the state.

The following two accounts about government opportunities to promote the status of women demonstrate how the state often lacks a committed position to transform gendered inequities, especially when faced with potential challenges to political and economic stability.

The United Nations Conference on Women, Beijing 1995

China's experience in hosting the Fourth United Nations Conference on Women offers a contemporary context for introducing contradictory forces shaping issues of gender and development in Chinese society. In China the arrival of the UN Conference on Women prompted new discussions on the situation and problems of women in Chinese society and lent definition to the Chinese women's movement (Wang 1996). An international view of the political landscape of the conference and the decisionmaking over actual conference venues, though, suggests the state's priorities and limitations on integrating issues of debates over gender, sex, and social change during the reform process.

The Fourth United Nations World Conference on Women convened in Beijing in August 1995. A few months earlier, in March, the Chinese government had relocated the Non-governmental Organization (NGO) Forum of the conference to the suburban city of Huairou on the northwest fringe of the capital. China's leaders cited a lack of facilities in Beijing, which proved a questionable rationale since the Huairou site could accommodate only two-thirds of the intended 36,000 NGO participants. NGO leaders protested in response; the Huairou site was fifty miles outside the city, which limited interaction with the governmental portion of the conference held in the capital. Officials limited traffic along the transportation route from Beijing to Huairou, ostensibly to ease transport flow but effectively to limit Chinese people from access to the site. Minimal event infrastructure, including tents in outdoor staging grounds and dormitory-style accommodations, symbolized the lack of significance accorded the NGO Forum participants. Meteorological forces complicated logistics at the largely outdoor site by unleashing a deluge over the North China plain that August; rains prevailed throughout the forum and made the state's promise of fine weather appear an undeliverable palliative. The lack of preparedness at the site and the flimsiness of the facilities became the focus of frustration and derision—and the media. China watchers skeptically assessed the official line, and alternative explanations emerged. Most persuasively, the Hong Kong press explained how the mere suggestion of NGO groups publicly demonstrating in Tiananmen Square had to be prevented (Nickerson 1995). China could not condone democratic activity in the very place where the state had crushed a popular uprising in June 1989. Later, NGO groups learned that the UN conference agreement with China had never guaranteed that the NGO Forum would take place in the capital itself (*Beyond Beijing* 1996), a predictable policy position considering that China had just a few fledgling NGOs in 1995.

The new debates on gender and sexuality in China carry on in Beijing, Shanghai, and other major population centers, but the state has not led any discussion on these recently uncloseted issues. In the NGO Forum the state faced not only the unwelcome challenge of democratic practice but representation of different orientations in

sexual practice. While Beijing would not accredit human rights NGOs for attendance at the UN Conference, lesbian NGOs found their accreditation an unpre-
dictable achievement (Rosenbloom 1996). But as the capital prepared itself to
manage the meetings, the imminent arrival of the NGO contingent was reduced to
sexualized extremes: What would demonstrating lesbians do? Fear, combined with
the heightened state of security, spun off widely repeated speculations that Beijing
police had been issued large white sheets to deploy in the event of women protest-
ing *en déshabillé*. Apparent near hysteria about the potential problems of the meetings coalesced in the notion of demonstrating naked lesbians; in intellectualized
term in the body (see Grosz 1994), in women's bodies, ultimately in the representative, collective body politic of the NGO participants.

   If China was not ready to host the NGO Forum of a decennial United Nations
conference, what did China want? When China tendered the offer to host the UN
World Conference on Women, it sought primarily the international recognition that
comes with hosting such a high-profile international event—most definitely not the
potential disruption of NGO politics. NGO groups met at Huairou, but press reports
never covered the activities of the NGO meetings; Chinese television coverage featured
only views of conference entertainment. Media executives of the major U.S. and international television and news organizations at the NGO Forum met daily to determine
the contents of news coverage: the weather, the conditions of the facilities, and famous
women participants. The actual topics of the meetings—the complex problems of
patriarchy and gender relations around the world—were not considered to be news.[4]
This is how issues about the status of women continue to be socially constructed as
externalities, and by inference, unimportant, at best special interest topics.

Overturning Legalized Patriarchy in Hong Kong

Male bias in state ideology is a common and pan-cultural condition, which also prevailed under the legal system of British Hong Kong. In the New Territories of Hong
Kong, a Qing Dynasty imperial law that denied women property inheritance rights
was overturned only in the early 1990s. Traditional law held sway without decisive
public question in the New Territories until 1991 when Legislative Council member Christine Loh challenged the existing inheritance law in the case of a forty-three-
year-old woman who was being harassed by her brothers to leave their family home
after the parents had died (Gargan 1994). The brothers sought to sell the property
without compensation to the middle-aged sister, who would be rendered homeless
by the patriarchal inheritance scheme. Loh argued successfully in court to replace
Qing rule with the non-sex-biased inheritance law in force in the rest of Hong Kong.

   Although the case was successful, male villagers in the New Territories accused
Loh of destroying their culture and threatened her with physical violence. A few
members of the Legislative Council also protested her activities, because they, like
the villagers who threatened Loh, were members of the Heung Yee Kuk. The Heung
Yee Kuk existed as an organization of wealthy Kejia (Hakka) landowners in the New

Territories and had been recognized as an advisory body to the Hong Kong government since 1926 (Jones 1995). The threats of physical violence toward Councilor Loh made by this de facto agent of the state were as shocking for their explicit content—including rape—as for their brazen forum—on television—projected directly into the households of millions of viewers, and thus were a threat to women generally who would challenge fundamentally patriarchal authority and support alternative views of power that threaten the hegemony of traditional male-gendered forms.

The actual law prohibiting female inheritance was a 1910 British ordinance that supported preserving local ways of life, a measure that helped prevent local opposition to colonial rule. The struggle to change the gender-based discriminatory inheritance laws in the New Territories demonstrates how transhistorical dominant values, whether Qing or British, nationalist or imperialist, Hakka or Han, aligned to reproduce a system of inequalities that supported the mutual interests of male-dominated elites at the expense of female equity. The majority of Hong Kong residents supported overturning this remnant of Qing Dynasty law. After the repatriation of Hong Kong to Chinese sovereignty in 1997, the State Council in Beijing affirmed the rule of law on equal inheritance rights in the New Territories. Nevertheless, media reports on the Heung Yee Kuk have continued to suggest that dissension remains over the equal inheritance law in the New Territories (Pang 1998). Changing such a system of entrenched gendered inequities through legal reform often first registers as symbolic reform until society adapts to accept the provisions of the new legal measures.

Gendered Reform Policies

How do relationships between the state and the macroeconomic policies of market reform construct gendered inequities? In China the socialist state continues to retain considerable planning and policymaking power; the state can choose to criticize or discourage the effects of reform that result in gendered inequities. The state, however, has not moved to intervene in the results of economic restructuring that limit women's labor market opportunities and so has effectively reinforced male bias in the development process. Research by Shirin Rai (1994a,b) shows that women are particularly disadvantaged under reform—in both education and employment opportunities—because of the concept of "market rationality," which Rai, like Elson, sees as an element of male bias in economic development generally. The market rationality argument holds that women workers as a group are ultimately more costly to employ than men at the same wage because of gendered female role expectations assumed by the employer, such as interruption of work to care for children or the elderly, the cost of maternity leave benefits paid by the employer, and other factors of socially constructed household roles. Similarly, if men are earning more than women, the market rationality argument holds that women, or the lower-wage worker, will opt to remain in the household if one spouse must attend to child care and other domestic duties. Rai critically views the gender-neutral language of market economic reform as an ineffectual discourse where gender equity is concerned. Her analysis of employment data on women with tertiary education showed that among unemployed young people of both sexes, the proportion of women rose to 61.5 percent even in 1986, a high-growth year; in the first decade of reform, the growing female unemployment peaked in the years 1988–1989, the time of slowest economic growth (Rai 1994b: 125). The data suggest that barriers to women's entry into the labor market may be more closely related to how opportunities fluctuate with political economic conditions than any simple notion that educated women should be able to gain higher status through labor market participation (see also Bauer et al. 1992; Summerfield 1994b). Indeed, Emerson (in Robinson 1985: 36) has suggested that during periods of job scarcity in the Maoist era, men were given employment priority as a matter of Party policy and that this pattern appears to have returned in the early 1980s. Pursuing this issue, Robinson (1985: 36) found no evidence that employment discrimination is accepted Party policy, but that "it is likely that factory and enterprise managers believe the Party is not serious about equal employment, especially since there do not appear to be equal employment compliance mechanisms." Here, once again, situated interpretations of state policy positions contribute to gendered inequities.

State Birth Planning Policy

In 1980 China announced the one-child policy in order to promote rapid industrialization by reducing demands on the resource base and to attain zero population growth by the middle of the twenty-first century. State promotion of the policy stressed widespread benefits, including better health care and social conditions, increased work efficiency and greater political awareness, and ultimately, the weakening of the patriarchal family system and greater gender equality (Chow and Chen 1994). Analyses of the effects of the birth planning policy, however, demonstrate just how complex its implications have been in Chinese society. The one-child policy particularly evinces how reproduction—what otherwise might be considered an element of the private or domestic sphere—binds private and public realms in dialectical relations along women's life paths, in which women's bodies become the real medium of policy implementation (Wong 1997).

Although a national policy, the one-child policy has been geographically uneven in implementation, acceptance, and results. In reality, the state has held only urban families to the one-child restriction; rural farming families commonly have two, and in some cases, three children. Officially classified national minorities are allowed to have more than one child, and other exceptions are allowed as well. Given that the majority of the population has remained classified as rural, the one-child policy has not been implemented in the majority of households. In impoverished areas, families with three, four, and even five children are not uncommon, as the state tacitly admits that it lacks the resources to alleviate the worst of rural conditions and turns the proverbial blind eye as poor people rely on one of the most fundamental forms of social security—large families. On the other end of the economic spectrum, some families have become affluent enough that they are simply able to pay the fines incurred by out-of-plan births, and they have little need for the package of benefits offered to the one-child family.

Despite apparent policy failure in many areas, the one-child policy has lowered the birth rate. It has also resulted in lower than natural female births. Between 1982 and 1989, the number of boys born per 100 girls rose from 107 to 114, well above the normal level of 105 to 106 (see Johansson and Nygren 1991). Although the one-child policy has definitely slowed the birth rate, many families with a first-born daughter have attempted to have a son without officially reporting the daughter, which has created a new social problem in unregistered female births. Yabuki (1995: 15) has estimated this undocumented so-called black population to be as high as tens of millions. Situations of unwanted girls result in some girls being abandoned and left at orphanages, while sex-selective abortions have also increased with the availability of amniocentesis and sonogram technology. Based on briefings from the Chinese Ministry of Health, Coale and Banister (1994) have reported that ultrasound technology is now ubiquitous at the county level, and although it is illegal to use ultrasound machines for sex selection purposes, technicians have regularly revealed the sex of the child to the parents.

New complex social patterns will continue to arise from these conditions. The perceived and real shortage of females, especially as the population enters marriageable age, is prominent among them, and the prospects for locating marriage partners have already begun to shift. New possibilities to migrate have, on the one hand, allowed women options to relocate for marriage to more distant and prosperous areas (Fan and Huang 1998); on the other hand, in more impoverished rural areas, the reorganization of farm labor in the household responsibility system, combined with the perceived shortage of marriage partners, has resulted in the revival of child betrothal arrangements (Croll 1994: 169; Rai 1994b: 125). At worst, the shortage of marriageable women has spurred illegal trade in women as brides. In the so-called marriage market of rural Anhui, local bachelors have taken brides from Sichuan, Shaanxi, Gansu, Yunnan, and Henan (Han and Eades 1995). In many cases, the movement of women is a process of chain migration; in other cases, women have been deceived into an arrangement. Nicholas Kristof and Sheryl WuDunn (1994: 217–219), former Beijing-based correspondents for *The New York Times,* have summarized the commodity trade in women and children for 1989–1990 based on a 1990 Chinese government report, which listed 18,692 cases of the sale of women, and 65,236 people arrested for trafficking in women and children, mostly girls. Even in Beijing, in 1993, a single gang abducted 1,800 women from the Beijing pick-up labor market to be sold in rural Shanxi province. For the period 1991–1995, statistics from the Public Security Ministry reported that nationwide 143,000 people involved in 32,000 gangs were arrested for kidnapping more than 88,700 women and children (Huang and Yin 1998). In 1998 a court in Quanzhou, Fujian province convicted 52 gang members for kidnapping and selling 110 women and two children from Guizhou and Guangxi provinces between 1991 and 1995; seven of the convicted were given the death penalty (Huang and Yin 1998). These government statistics lead to the conclusion that the decreasing numbers of women relative to men under reform has registered in a tortured form of market entrepreneurialism in which gang members abuse patriarchal power to trade in women as productive and reproductive resources (see Hirschon 1984; Ocko 1991; Evans 1997).

It has become commonplace to consider the worst problems of gender inequity in China as features of impoverished rural areas, about which it is difficult if not impossible to gain systematic empirical documentation. However, systematic research on the urban impacts of the one-child policy has demonstrated that it has had a significant effect on the status of women countrywide. Chow and Chen (1994), in a Beijing University-supported study, have problematized the state's position on the policy in an attempt to answer the question, Has the one-child policy strengthened or weakened the patriarchal family in China? As they summarize, this argument has already raged through the literature, and the bulk of the debate has come down on the side of the increasingly eroding status of women and the further strengthening of gender inequality. They explain how the state acts as a socialist patriarchy by determining women's reproductive behaviors to promote the primary goal of the one-child policy—economic development. Further, they conclude, based on survey data from central and suburban Beijing, that the one-child imperative creates its own pressures, as women intensify care for the only child at the expense of work-related or gender egalitarian priorities. Robinson's (1985: 49) conclusions about day care trends suggest why: The role of the state in providing child-care facilities is clearly decreasing in some areas, and private day care—promoted by state organizations as a partial response to unemployment—is increasing.

Gender, Rural Reform, and the Feminization of the Agricultural Labor Force

The household responsibility system revived the household as the basic unit of agricultural production and allowed the rural populace to work beyond meeting state grain quotas, from growing vegetables for market to setting up business enterprises. During the first decade of reform, market farming on private plots alone tended to increase rural incomes, but returns from farming stagnated later in the first decade of reform and through the second decade. Households responded by reallocating labor to higher-income-earning enterprises. Given women's reproductive roles in the domestic sphere, women tended to remain in the household, while men increasingly sought wage labor in new township, village, and private enterprises (TVPs) or in urban areas, especially construction work. After the first decade of reform, half of all male farm laborers in China had been employed in nonfarm sectors, while less than one-fifth of female farm laborers had worked in nonfarm sectors (Zhang 1994: 83). This trend has increased to the point that in some provinces, such as Jiangsu and Shandong, women constitute 80 to 90 percent of the agricultural labor force (Carino 1995: 36), and on average women now perform over 70 percent of the farm labor (MacLeod 1998: 63). This is the most significant pattern of labor re-gendering under reform. In the long sweep of Chinese history this is a remarkable reversal; women in imperial China—unlike in Japan or Southeast Asia—did not widely engage in farming labor (Bray 1997: 5).

The contradictions of the household responsibility system—the possibilities of entrepreneurial sideline wage activities, combined with stagnating incomes in the rural sector—have meant that decisionmaking within households favors adding new income-generating activities. The continuance of the patrilocal marriage system, in which families lose daughters and gain daughters-in-law, favors maximization of the labor contributions of girls and young women. These conditions have generated a new gendered problem in removing girls from school to become household income earners. Since the reforms, the dropout rate for both primary and secondary school-age girls has been rising (Croll 1994: 165; Rai 1994b: 125). Boys continue to be educated, except in the most impoverished areas. So far, China has not provided a detailed national survey to explain the patterns of school attendance, but available regional reports and national aggregate data indicate the seriousness of the trend. A study undertaken by the All-China Women's Federation (1993) sought to identify how reform had changed the lives of rural women. Data was gathered at sites in southern Jiangsu and Sichuan (both relatively prosperous areas), and it was found that the greatest change was the clear decrease in the education level of girls. In 1995, Hubei province released a five-month study that confirmed that the majority of Chinese children not attending primary school were girls (Jia 1995). In 1996 over 10 million school-age children, six to fourteen years old, dropped out of school due to household financial problems, and two-thirds were girls (Shen and Wan 1997). The state has responded to these negative trends in female education by indirect means. Its primary form of support for girls' education has been a campaign to generate private donations intended to help impoverished families. In 1992 the state sanctioned the *Chunlei jihua* or Spring Bud Plan, a campaign that has been promoted both domestically and internationally. The Spring Bud Plan (1997) suggests several levels of donations: 300 yuan (US$36) for one girl for one year, 20,000 yuan (US$2,500) for a full class of fifty girls above the fourth grade level, or 500,000 yuan (US$60,000) to be used more generally within a specific school. Chinese professional organizations have also responded to the cause; in 1997 the press reported that a banking organization donated 7.8 million yuan (US$939,000) to fund girls' education (Shen and Wan 1997).

The nature of the solicitation approach suggests that the state considers girls' education a philanthropic project, not especially different from charity campaigns for unusually disadvantaged groups, rather than a basic right and state responsibility. Although some historic precedent exists in China for this kind of response to female education, some of the solutions to economic problems under reform, like this one, resemble prescriptions of the increasingly internationalized neoliberal economic regime promoted by the World Bank and other international economic institutions, which advocate increased privatization and a diminished role of the state in providing basic social services (see Sparr 1994).

In the dynamic growth sector of township, village, and private enterprises (TVPs), men have dominated new business enterprises (Entwisle et al. 1995). Women have not widely benefited from increased employment opportunities (Ho 1995), and when TVPs have employed women, there has been serious within-occupation gender-based wage discrimination (Meng and Miller 1995). Ellen Judd's (1994) research on the transformation of the rural political economy in Shandong confirms these trends. In a detailed monograph-length treatment of village enterprises and village administration, Judd explains how male village leaders have gone about the process of defining who can hold village jobs by sex and age. Except for the usual position of *funü zhuren*, or women's head, village leaders reserved high-level jobs for males. This explicit male gendering of village management and leadership positions systematically excludes women from high-ranking labor roles. She also argues that rural reform policies have been formulated and debated as though they were gender neutral (Judd 1994: 253), when in reality the household should be viewed as a local-level representation of the patriarchal state through the policies of the Communist Party. Ultimately, Judd (257) concluded, "There is no differentiating feature in Chinese life that is more profound, continuing, and asymmetrical than gender."

Clearly, restructuring in the rural sector has created massive upheaval in many people's lives. The suicide rate for rural women is stark evidence of serious problems. A major new study on suicide in China conducted jointly by the World Bank, the World Health Organization, and Harvard University has concluded that new social and economic pressures in the rural sector are at least partly responsible for the female suicide rate: 56 percent of all women worldwide who commit suicide are Chinese, and most of them are in rural areas (MacLeod 1998; Wu and Mok 1998). Although details of the study have not been formally released and researchers dispute the precise rate (MacLeod 1998: 63), they agree that the rural suicide rate is nearly three times the urban rate, and the rate among rural women is about 40 percent higher than among men. Often, the chosen means is one of the inputs of the modern agricultural system itself: pesticide. Much more research will have to be done to assess the problem, but early indicators suggest that a significant proportion of women who commit suicide are comparatively better educated and who despair in the knowledge of their highly limited positions in society and the economy.

Gender and the Export-Oriented Industrialization Regime

China's open-door policy laid the foundation for the industrialization drive in export-oriented development and its reliance on a strong manufacturing sector and low-wage labor to produce consumer goods for the world market. The special economic zones on the South China coast became the centers of export-oriented industrialization and the major centers of assembly line work for women in electronics and toy manufacture, sewing garment in apparel production, and mixed assembly and sewing in the footwear industry. For many young women, migration to urban areas and manufacturing zones has meant dependable wage labor, new standards of living, and personal options (Davin 1996a,b). Assembly work, however, does not offer long-term options for women. It is regularly assumed that women will leave assembly jobs, especially upon marriage, and supervisory jobs remain the domain of men (see Lee 1995).

Development of the manufacturing sector under reform has also resulted in new domestic markets and household consumption, as millions of Chinese households buy refrigerators, televisions, washing machines, VCRs, and other household goods for the first time. It is commonplace to consider the widespread acquisition of consumer durable goods as symbols of rising standards of living. Yet where women's household labor is concerned, Robinson (1985: 46) has questioned state promotion of the manufacturing sector at the apparent neglect of basic infrastructural provision, such as plumbing and electricity supply. It will be left to the new housing reforms to redress problems in the existing housing stock; in the meantime, in some areas women have continued to collect thermoses and buckets of water while televisions blare.

The restructuring of the Chinese economy to receive foreign direct investment has resulted in new transnational and transboundary production regimes, especially in Guangdong and Fujian provinces, where the bulk of foreign investment has spilled across borders, from Hong Kong and Taiwan, respectively. In the special economic zones, relationships between workers and employers represent both the immediate need of manufacturing plants for large quantities of low-wage laborers and the insecurities workers face in relocating long distances to live in factory dormitories. Ching Kwan Lee (1995) has shown, based on research in branches of the same electronics company in Shenzhen and Hong Kong, how gender relations between employers and workers can differ substantially depending on age, marital, and residency status. In Shenzhen, employers adopted a rigid patriarchal stance by restricting the characteristically young, single migrant workers to the factory compound and by tethering them to the assembly line and restricting movement on the factory floor itself during working hours. To resist and overcome such restrictions in a dislocated environment, workers formed informal assistance networks based on common dialect and area of origin. In Hong Kong, where manufacturing employment has declined substantially, the remaining women factory workers are characteristically older, and their local family ties transcend the boundaries of the workplace so that they are able to maintain their bases of support outside the factory.

In the regions of rapid industrial zone development in South China, new patterns of household formation have emerged that reflect the unusual demographics of the production regime. Some businessmen from Hong Kong and Taiwan who spend the workweek in China have formed relationships with women who work in the manufacturing zones. For many working-class men, finding a wife and setting up a household in China has been an otherwise unattainable economic opportunity, especially in high-cost Hong Kong. But many more of the relationships are affairs and so-called second marriages. In Hong Kong, the Chinese phrase *bao yilai* came into popular use to refer to the phenomenon. *Bao yilai* means "keeping second wives." Siumi Maria Tam (1996) records how such so-called second marriages became "the talk of the town" in Hong Kong in 1995. An estimated 300,000 Hong Kong men have established second households in Guangdong province, and their China-born children, people realized with some alarm, could claim the right to reside in Hong Kong (Tam 1996: 117, 120). But widespread discussion of the problem in the media actually reinforced patriarchal power and stereotypical gender roles within the marriage, Tam explains. The press reported the notion that peer pressure in China to take a mistress was so great that men who were uninterested in initiating affairs were pressured to do so and were sometimes left out of business deals if they demurred. Tam's class analysis also points out that in contrast to the situation of historical China, in which only the wealthy strata of society could afford to practice polygyny, this new form is nonclass specific and affords working-class men an unprecedentedly accessible form of social prestige. In Hong Kong the male-gendered media message was female blame: Wives had lost their feminine powers; they should make themselves more attractive and become more skillful in pleasing their husbands. The rate of divorce in Hong Kong is positively correlated with this problem.

In Taiwan, the practice was sensationalized in a book with a politically unfortunate title: *Yiguo liangqi,* or *One Country, Two Wives* (Lin 1994). The illustrations in Figures 16.1 and 16.2 from *Yiguo liangqi* represent some of the problems of popular notions about cross-strait "second relationships" between Taiwan businessmen and mainland Chinese women in the areas around the special economic zones. In each of the illustrations the mainland woman is represented as comparatively younger, thinner, and more cheerful than her Taiwan counterpart and, given the pig-tailed hairstyle, even child-like. Rather than questioning the practice of so-called second marriages, these images serve to construct and reinforce gender stereotypes. Both images represent stereotypes about desirable and undesirable female characteristics based on appearance, and in each, the Taiwan woman shows her disapproval of the situation but is apparently powerless to do anything about it. The behavior of the male, by contrast, is represented as relatively powerful and based on economically logical decisionmaking. His behavior is not called into question, reflecting unstated privileges of male bias.

Chinese data on marriages shows that Fujian, Guangdong, and Shanghai have the highest rates of mainlanders marrying non-mainland spouses and that the rate of such marriages increased four times over the first fifteen years of reform (Ye and Lin 1996). Fully 85 percent of these marriages took place in Fuzhou, Xiamen, Quanzhou, and Zhangzhou—the cities longest opened under economic reform and with the greatest degree of overseas and cross-strait connections.

Gender and Sexuality in the Public Sphere

New debates on issues of gender in China are transforming the terrain of gender research on gender identity, sexuality, and the roles of men and women. As Harriet Evans (1995: 157) points out, "Sex—in some form or another—has emerged from obscurity to occupy a position of unprecedented prominence in public life in the People's Republic." The All-China Women's Federation has sponsored much of this discussion, and yet the role of the state in bounding sexual mores has declined substantially under reform. Evans (387) points to the emergence of "a new sexual culture in China's urban centers," and Schein's (1996) example of the state-authorized video, "Exhibition about Sex: From Ignorance to Civilization," which showed in Zhongshan

FIGURE 16.1 The sign on the noodle stand to the right reads, "Taiwan noodles 30 yuan per bowl"; the sign to the left reads, "Mainland noodles 10 yuan per bowl." The male consumer reads the Taiwan sign and remarks, "These are too expensive." He turns to the mainland sign and concludes, "This side is good— cheaper and more." SOURCE: Lin 1994:63.

FIGURE 16.2 The Taiwan woman stands and fumes while the businessman (note the collared jacket) says to his young mainland companion, "Because I have been busy in Taiwan recently," to which she bashfully and flirtatiously replies, "I don't like it, you've been gone so long." SOURCE: Lin 1994:53.

park in Beijing, suggests just how much things have changed. Compared to the Maoist era, the role of the state and the efficacy of the All-China Women's Federation in defining women's issues have declined, in large part because the Women's Federation remains linked to state perspectives rather than changing in tandem with the new social and economic problems women face (Zhang 1994). In rural areas, the Women's Federation has been displaced and supplemented at the village level by the rise of informal organizations, especially in areas where women dominate the agricultural labor force. Rai and Zhang (1994) analyzed the All-China Women's Federation campaign "Competing and Learning," a response to the increasing feminization of the rural labor force, and concluded that the state's position in this campaign has remained oriented toward the production regime rather than the new problems women face as a result of economic restructuring and so has not challenged the sexual division of labor fundamental to the status of women in rural China. Although the Women's Federation struggles to meet the changing needs of women and state priorities, it has faced oppo-sition by other state offices such as the All-China Federation of Trade Unions (Howell 1996), especially under the current conditions of high unemployment.

Conclusion

Industrialization is generally held up as the path to social and economic moderniza-tion, but where gender is concerned, the institutionalized cultures of development itself, and the historic models on which development is based, represent forms of male bias. Higher household incomes—the quantitative result sought by economists—have been achieved in China under reform, but higher incomes in rural areas have been achieved through the reinforcement of the gendered characteristics of the patri-archal family and the valuation of sons over daughters. Agricultural labor has rapidly become gendered female as women remain in the lowest-paid sectors, while men have tended to move into higher waged labor opportunities. The future implications of women bearing the responsibility for producing China's food will be important to understand at several scales, from the local scene to the global level of world food sup-ply. In urban areas, the new emphasis on rationalizing production through greater operating efficiency has resulted in widespread layoffs and fewer job possibilities for women, which the state has implicitly sanctioned even though existing laws explicitly guarantee women equal employment opportunities. In the export-oriented sector, low-cost women's labor has been the critical factor of production on which China's foreign-invested manufacturing production depends. In all these sectors of develop-ment, women's labor roles are less centrally examined contributions to economic growth. One can reasonably assume why this is the case: In addition to the theoreti-cal problem of externalities, focusing on the exploitation of women and women's labor is unpleasant business. It raises difficult questions that could undermine otherwise successful development trajectories and positive economic growth trends.

The recent six-part World Bank report, *China 2020,* exemplifies the absence of gender analysis in economic development. Two of the reports, *At China's Table: Food Security Options* (World Bank 1997a) and *China Engaged: Integration with the Global Economy* (World Bank 1997b), sport covers that show women engaged in work activ-ity. The cover of the report on food security features a young peasant woman shoul-dering a carrying pole laden with bundled chaff. The report, though, includes no information whatsoever on the feminization of agriculture in China under reform. This is an astonishing but all too predictable limitation of basic development analy-sis. The report on China's integration with the world economy focuses on foreign trade and global capital markets but nowhere engages with the subject suggested by its cover, a young woman with short permed hair intently soldering a circuit board.

As China continues to plan rapid industrialization, contemporary Chinese debates over meanings of gender reflect the clashing forces of rapid industrialization and the resurgence of traditional practices in an atmosphere of high unemployment.

Although women, like men, may have new economic opportunities in China since the introduction of economic reforms and market-based economic principles, they are also facing more discrimination. After two decades of reform, policies have given rise to complex patterns of social and economic change. From the resurgence of traditional celebrations honoring virtuous wives to new options for women to leave difficult rural circumstances by migrating to more prosperous regions, women face both the revival of traditional forms of patriarchal control as well as relatively uncharted opportunities. The new debates on sex and gender and the rise of new ad hoc women's groups at the local level suggest how the problem of the diminishing status of women may be solved: not by the state or as a result of the economic development process per se, but by women themselves as they negotiate the avenues of social and economic flexibility opened up by the reform process.

Notes

1. Jonathan Mirsky (1994), in a review article, has attempted to assert that the scholarship of gender analysis is an American liberal feminist project that has no indigenous place in China.

2. The United Nations has formulated the Gender-related Development Index (GDI) based on the major UN indicator of human development, the human development index (HDI). The GDI incorporates inequalities in living standards between men and women, based on categories of income, life expectancy, adult literacy, education, and other factors. Industrialized countries are the top-ranking countries in the GDI, just as they are in the HDI. In the GDI list of 130 countries, China ranks 72 (United Nations 1995).

3. Other analyses of the apparent inside/outside dualism do not view women's household position in such limited terms. Dorothy Ko (1994) has argued that conditions for women in traditional China should be assessed relative to the era and cultural context and not by comparison to modern perspectives about domestic confinement as a form of oppression.

4. Dr. Janice Engsberg, Visiting Professor of Journalism at Xiamen University and veteran of the 1984 UN Conference on Women at Nairobi, attended daily the press planning meetings of the conference and was informed that media leaders covering the conference sanctioned non-conference topics as the primary reporting subjects. Personal communication, Xiamen, August 1995.

References

All-China Women's Federation. 1993. *The Impact of Economic Development on Rural Women in China*. Tokyo: United Nations University.

Andors, P. 1983. *The Unfinished Liberation of Chinese Women, 1949–1980*. Bloomington: Indiana University Press.

Bauer, J., F. Wang, N. E. Riley, and X. Zhao. 1992. Gender Inequality in Urban China: Education and Employment. *Modern China* 18(3):333–370.

Bray, F. 1997. *Technology and Gender: Fabrics of Power in Late Imperial China*. Berkeley: University of California Press.

Beyond Beijing: NGO Participation at the UN Fourth World Conference on Women: Report on Barriers to Access, with Recommendations for Change. October 1996. Compiled by Amnesty International, Human Rights in China, Human Rights Watch, the International Human Rights Law Group, and the Robert F. Kennedy Memorial Center of Human Rights. http://www.igc.org/beijing/barriers.html.

Carino, T. C. 1995. Women of Henan. *China Currents* 6(1):35–39.

Coale, A. J., and J. Banister. 1994. Five Decades of Missing Females in China. *Demography* 31(3):459–479.

Chow, E. N., and K. Chen. 1994. The Impact of the One-Child Policy on Women and the Patriarchal Family in the People's Republic of China. In *Women, the Family, and Policy: A Global Perspective*, ed. E. N. Chow and C. W. White, pp. 71–98. Albany: State University of New York Press.

Croll, E. 1978. *Feminism and Socialism in China*. London: Routledge & Kegan Paul.

_____. 1983. *Chinese Women Since Mao*. London: Zed Books.

_____. 1994. *From Heaven to Earth: Images and Experiences of Development in China*. London and New York: Routledge.

Davin, D. 1991. Women, Work and Property in the Chinese Peasant Household of the 1980s. In *Male Bias in the Development Process*, ed. D. Elson, pp. 29–50. Manchester: Manchester University Press.

_____. 1996a. Gender and Rural-Urban Migration in China. *Gender and Development* 4(1):24–30.

_____. 1996b. Migration and Rural Women in China: A Look at the Gendered Impact of Large-Scale Migration. *Journal of International Development* 8(5):655–665.

Elson, D., ed. 1991. *Male Bias in the Development Process*. Manchester and New York: Manchester University Press.

Elvin, M. 1984. Female Virtue and the State in China. *Past and Present* 104: 111–152.

Entwisle, B., G. E. Henderson, S. E. Short, J. Bouma, and F. Zhai. 1995. Gender and Family Businesses in Rural China. *American Sociological Review* 60(February):36–57.

Evans, H. 1995. Defining Difference: The "Scientific" Construction of Sexuality and Gender in the People's Republic of China. *Signs* 20(2):357–394.

_____. 1997. *Women and Sexuality in China: Female Sexuality and Gender Since 1949*. New York: Continuum.

Fan, C. C., and Y. Huang. 1998. Waves of Rural Brides: Female Marriage Migration in China. *Annals of the Association of American Geographers* 88(2):227–251.

Gargan, E. A. 1994. Swept Away in 1911, Manchus Still Haunt Women. *New York Times*, April 17, p. A6.

Gilmartin, C. K., G. Hershatter, L. Rofel, and T. White, eds. 1994. *Engendering China: Women, Culture, and the State*. Cambridge: Harvard University Press.

Goldstone, J. A. 1996. Gender, Work, and Culture: Why the Industrial Revolution Came Early to England but Late to China. *Sociological Perspectives* 39(1):1–21.

Grosz, E. 1994. *Volatile Bodies: Toward a Corporeal Feminism*. Bloomington: Indiana University Press.

Han, M., and J. S. Eades. 1995. Brides, Bachelors, and Brokers: The Marriage Market in Rural Anhui in an Era of Economic Reform. *Modern Asian Studies* 29(4):841–869.

Hirschon, R., ed. 1984. *Women and Property—Women as Property*. London: Croom Helm; New York: St. Martin's Press.

Ho, S.P.S. 1995. Rural Non-agricultural Development in Post-Reform China: Growth, Development, Patterns, and Issues. *Pacific Affairs* 68(3):360–391.

Howell, J. 1996. The Struggle for Survival: Prospects for the Women's Federation in Post-Mao China. *World Development* 24(1):129–143.

Huang, J., and D. A. Yin. 1998. Court in Quanzhou Sentences 52 People for Trafficking in Human Beings. *China News Digest*, April 8, http://www.cnd.org/CND-Global/.

Jia, D. 1995. Majority of Chinese Illiterates Are Women/Girls. *China News Digest*, September 6, http://www.cnd.org/CND-Global/.

Johansson, S., and O. Nygren, 1991. The Missing Girls of China: A New Demographic Account. *Population and Development Review* 17(1):35–51.

Johnson, K. A. 1983. *Women, the Family, and Peasant Revolution in China*. Chicago: University of Chicago Press.

Jones, C. 1995. The New Territories Inheritance Law: Colonization and the Elites. In *Women in Hong Kong*, ed. V. Pearson and B.K.P. Leung, pp. 167–192. Hong Kong: Oxford University Press.

Judd, E. R. 1994. *Gender and Power in Rural North China*. Stanford: Stanford University Press.

Kelkar, G., and Y. Wang. 1997. Farmers, Women, and Economic Reform in China. *Bulletin of Concerned Asia Scholars* 29(4):69–77.

Ko, D. 1994. *Teachers of the Inner Chambers: Women and Culture in Seventeenth-Century China*. Stanford: Stanford University Press.

Kristof, N. D., and S. WuDunn. 1994. *China Wakes: The Struggle for the Soul of a Rising Power*. New York: Times Books.

Lee, C. K. 1995. Engendering the Worlds of Labor: Women Workers, Labor Markets, and Production Politics in the South China Economic Miracle. *American Sociological Review* 60(3):378–397.

Lerner, A. 1986. *The Creation of Patriarchy*. New York: Oxford University Press.

Lin, C. 1994. *Yiguo Liangqi* (One Country, Two Wives). Taipei: Jingmei chubanshe.

Lin, J. 1995. Women and Rural Development in China. *Canadian Journal of Development Studies* 16(2):229–240.

Liu, W., and R. Zhang. 1997. Chinese Women Face Discrimination in Job Search. *China Daily*, December 18.

MacLeod, L. 1998. The Dying Fields: Economic Pressures Have Spawned a Tragedy in Rural China: Women Are Killing Themselves at an Alarming Rate. *Far Eastern Economic Review* 161(17):62–63.

Mazur, A. G. 1995. *Gender Bias and the State: Symbolic Reform at Work in Fifth Republic France*. Pittsburgh and London: University of Pittsburgh Press.

Meng, X., and P. Miller. 1995. Occupational Segregation and Its Impact on Gender Wage Discrimination in China's Rural Industrial Sector. *Oxford Economic Papers* 47:136–155.

Mirsky, J. 1994. The Bottom of the Well. *New York Review of Books*, October 6.

Momsen, J. H., and Vivian Kinnaird, eds. 1993. *Different Places, Different Voices: Gender and Development in Africa, Asia, and Latin America*. London and New York: Routledge.

Moore, H. 1988. *Feminism and Anthropology*. Cambridge: Polity Press.

Nicholson, L. 1995. Interpreting Gender. In *Social Postmodernism: Beyond Identity Politics*, ed. L. Nicholson and S. Seidman, pp. 39–67. Cambridge: Cambridge University Press.

Nickerson, D. 1995. The Shape of Things to Come. *South China Morning Post*, August 21, p. 19.

Ocko, J. K. 1991. Women, Property, and Law in the People's Republic of China. In *Marriage and Inequality in Chinese Society*, ed. R. S. Watson and P. B. Ebrey, pp. 313–346. Berkeley: University of California Press.

Pang, J. 1998. Kuk Seeking Liberal Image to Win Votes. *South China Morning Post*, May 7.

Rai, S. M. 1994a. Gender Issues in China: A Survey. *China Report* 30(4):407–420.

_____. 1994b. Modernisation and Gender: Education and Employment in post-Mao China. *Gender and Education* 6(2):119–129.

Robinson, J. C. 1985. Of Women and Washing Machines: Employment, Housework, and the Reproduction of Motherhood in Socialist China. *China Quarterly* 101 (March):32–57.

Rofel, L. 1993. Where Feminism Lies: Field Encounters in China. *Frontiers* 13(3):33–52.

Rosenbloom, R. 1996. Beijing and Beyond: International Organizing and the Conference on Women. *Gay Community News* 22(1):10–13.

Schein, L. 1996. The Other Goes to Market: The State, the Nation, and Unruliness in Contemporary China. *Identities* 2(3):197–222.

Shen, S., and G. Wan. 1997. 10 Million Children Drop Out of Schools Due to Financial Problems. *China News Digest*, April 25, http://www.cnd.org/CND-Global/.

Sparr, P., ed. 1994. *Mortgaging Women's Lives: Feminist Critiques of Structural Adjustment.* London and New Jersey: Zed.

Spring Bud Plan. 1997. http://www.chinaplaza.com/springbud/.

Stacey, J. 1983. *Patriarchy and Socialist Revolution in China*. Berkeley: University of California Press.

Stockman, N. 1994. Gender Inequality and Social Structure in Urban China. *Sociology* 28(3):759–777.

Summerfield, G. 1994a. Chinese Women and the Post-Mao Economic Reforms. In *Women in the Age of Economic Transformation: Gender Impact of Reforms in Post-Socialist and Developing Countries*, ed. N. Aslanbeigui, S. Pressman, and G. Summerfield, pp. 113–128. London and New York: Routledge.

_____. 1994b. Economic Reform and the Employment of Chinese Women. *Journal of Economic Issues* 28(3):715–732.

_____. 1997. Economic Transition in China and Vietnam: Crossing the Poverty Line Is Just the First Step for Women and Their Families. *Review of Social Economy* 55(2):201–214.

Tam, S. M. 1996. Normalization of "Second Wives": Gender Contestation in Hong Kong. *Asian Journal of Women's Studies* 2:113–132.

United Nations. 1995. *Human Development Report.* New York: Oxford University Press.

Wang, J. 1996. What Are Chinese Women Faced with After Beijing? *Feminist Studies* 22(3):497–502.

Wolf, M. 1985. *Revolution Postponed: Women in Contemporary China.* Stanford: Stanford University Press.

Wong, Y. R. 1997. Dispersing the "Public" and the "Private": Gender and the State in the Birth Planning Policy of China. *Gender and Society* 11(4):509–525.

World Bank. 1997a. *At China's Table: Food Security Options.* Washington, D.C.: World Bank.

_____. 1997b. *China Engaged: Integration with the Global Economy.* Washington, D.C.: World Bank.

Wu, F., and C. Mok. 1998. High Suicide Rates in China. *China News Digest*, January 26, http://www.cnd.org/CND-Global/.

Wu, L., and B. Xiang. 1998. Women's Federation Warns Government Against Forced Maternity Leave. *China News Digest*, March 6, http://www.cnd.org/CND-Global/.

Yabuki, S. 1995. *China's New Political Economy: The Giant Awakes*, translated by Stephen M. Harner. Boulder: Westview Press.

Ye, W., and Q. Lin. 1996. Fujiansheng shewai hunyin zhuangkuang yanjiu (Research on marriage with non-mainlanders in Fujian province). *Renkou yu jingji* (Population and Economics) 2:21–27.

Zhang, J. 1994. Development in a Chinese Reality: Rural Women's Organizations in China. *Journal of Communist Studies and Transition Politics* 10(4):71–91.

17

Growth and Management of Large Cities

Yehua Dennis Wei

The growth of large cities has generated considerable government concern in developing countries and has attracted much scholarly attention (e.g., Ginsburg et al. 1991; Kasarda and Parnell 1993; Fuchs et al. 1994). This issue is important to governments, as it is associated with population distribution, uneven development, intergovernmental relations, and social problems. Many developing countries have had difficulty coping with rapid urbanization and large city growth. Troubled by urban problems, developing countries have attempted to control the speed of urbanization and the growth of large cities. On the other hand, large cities in developing countries, such as Shanghai, Bangkok, and Taipei, are striving to become "global cities." The growth of large cities is expected to continue in the twenty-first century, and the dramatically changing global economy and growing cities in developing countries have necessitated more urban research (UNDP 1991). Scholars have questioned the effectiveness of previous government policies and called for a reevaluation of urban policy in developing countries (Richardson 1989). However, existing theories explaining the growth of large cities remain fragmented (Krugman and Elizondo 1996), and sources of urban growth are still not well studied. While scholars have rediscovered the importance of institutions (North 1990), few have examined the impact of state policy on large cities. A better understanding of forces underlying large city growth is necessary to plan and manage urban growth in developing countries.

Like many other developing countries, China has implemented an urban policy to control the size of large cities, which are defined as cities with populations of more than one-half million nonagricultural residents (Zhao and Zhang 1995). During the Mao era, China's spatial policy attempted to restrict the growth of cities, but the industrialization policy favored major industrial cities to a certain extent, as cities, especially large cities, are centers of industrialization and administration. The growth of Chinese cities has accelerated during the post-Mao period of economic reform, which has made it difficult to implement China's urbanization policy. Indeed, millions of rural migrants have flooded into the cities. Chinese cities, large or small, are experiencing unprecedented growth, and China has become one of the world's most rapidly urbanizing countries. Although migrants have had positive contributions to urban development, they have also brought new problems to Chinese cities. Rapid growth and restructuring of large Chinese cities have challenged the planning and management approaches of cities (Wei 1995). A better understanding of the changing Chinese cities is necessary in order to cope with rapid urbanization and large city growth in China (Yusuf and Wu 1997). This chapter addresses the extent and underlying forces of the growth of large cities and the implications for China's urban planning and management.

Migration and the Growth of Large Cities

Socialist countries were very concerned with uneven distribution of population and industrial activities and implemented urban policies to control urbanization, but their industrialization policies tend to favor the growth of large cities (Wei 1994). Although the impact of these two conflicting policies on the change of city size distribution remains controversial (Fan 1988; Clayton and Richardson 1989; Wei 1997), it is generally agreed that large cities in socialist countries remained centers of economic development and their growth continued (Rowland 1983). Some scholars working on urban China tend to emphasize the impact of industrialization policy on China's urban development (Chan 1994). Others, however, focus on China's explicit urban policy (Buck 1981), which advocates control of the size of large cities and promotes the development of small cities. This policy was implemented during the Mao era and has continued into the reform period, although scholars disagree over the extent of implementation. The drive for industrialization and economic growth has greatly shaped China's urban and regional development (Naughton 1995). Indeed, the central and local governments often overrule urbanization policy when there is a pragmatic need for industrialization and economic growth, such as the recent expansion of Shanghai, the premier city of China.

Development policies in China during the past five decades were inconsistent and were influenced by changing domestic and international contexts. China's changing political ideologies and development policies have greatly influenced China's urban development. During the Mao era, policies implemented in the First Five-Year Plan (FYP) period, the Third Front period, and the early 1970s were quite different (Ma and Wei 1997). During the 1950s, Soviet principles of industrialization and industrial allocation were adopted by China, and the government implemented urban-centered industrialization policies. With little control over migration, Chinese cities experienced rapid growth. Later, however, China became concerned about this rapid urban growth, and with the failure of the Great Leap Forward and the intensification of the cold war, the government restricted the growth of cities, mainly through the registration system, work permission policies, food rationing, and service provisions.

China's spatial policy experienced further changes during the early and mid-1970s when China gradually opened its door to the outside world. Large-scale industrial equipment was imported from Western countries, and most of the equipment was allocated to a few coastal cities (Ma and Wei 1997). After the Cultural Revolution (1966–1976), the problem of shortages and the return of rusticated people placed tremendous pressure on Chinese cities. Although economic reforms were initiated in 1978, urban policy on large cities remained fundamentally unchanged. In 1980, China's urban policy was revised with the stated goals of strictly controlling the size of large cities, rationally developing medium-sized cities, and vigorously developing small cities. With slight changes, this policy has been incorporated into China's Urban Planning Law (DCP 1991). However, the dramatic changes that have accompanied reform have made the implementation of this policy difficult.

Three sources contribute to the growth of cities: natural increase, migration, and the expansion of city boundaries. During the 1950s, with the recovery of the Chinese economy and the implementation of industrialization programs, rural-urban migration was a major source of urban growth in China. Population in coastal cities recovered, and the allocation of key projects stimulated the growth of major industrial cities in the interior (for regions and provinces, see Figure 17.1). For example, population in Lanzhou, planned as one of the major interior industrial centers, blossomed from 373,600 in 1952 to 797,000 in 1957, an increase of 113.6 percent (SSB 1990). Beijing, China's new capital, also recorded dramatic population growth (Table 17.1). From the late 1950s to the end of the Cultural Revolution, however, rural-urban migration was strictly controlled and millions of urban residents were even sent to rural areas (Kirkby 1985). With strict control over migration, Chinese cities recorded slow or zero growth, and natural increase was the major source of the growth of cities. Some large cities even experienced population decline. For example, Shanghai's population declined from 6.1 million in 1957 to 5.78 million in 1958; after some improvement in the early 1960s under the economic recovery program, population in Shanghai decreased again from 6.43 million in 1965 to 5.8 million in 1970 then to 5.57 million in 1975 (SSB 1990). The importance of migration in large city growth reemerged in the late 1970s when many rusticated urban youth returned to cities, especially large coastal cities.

The growth rate of large Chinese cities has been accelerating since the mid-1980s, when China initiated urban reforms and broadened its open-door policy.

FIGURE 17.1   China's Province-Level Units and Three Autonomous Regions

During this time period, migration has become a major source of the growth of large cities in China. According to surveys conducted by the Shanghai Bureau of Public Security, the floating population in the city increased dramatically from 0.75 million in 1984 to 1.1 million in 1985, 1.334 million in 1986, 1.25 million in 1988, and 2.81 million in 1993. At the end of 1989, the floating population was estimated at 1.35 million in Beijing, while a survey conducted on November 10, 1994, revealed that the floating population had reached 3.295 million (*Beijing Daily* 1995). A study of twenty-eight million-plus cities in 1988 found that, on average, the floating population accounted for 18 percent of city population (Li and Hu

TABLE 17.1   Total Population of Selected Cities, 1952–1995 (in thousands)

	1952	1965	1978	1985	1995	1952–1978 Growth (%)	1978–1995 Growth (%)
Shanghai	5,058	6,431	5,574	6,983	8,567	91.1	53.7
Beijing	2,519	4,760	4,926	5,860	7,337	95.6	48.9
Lanzhou	374	1,036	1,285	1,354	1,652	243.6	28.6
Suzhou	448	552	564	709	1,057	25.9	87.4
Hangzhou	689	965	1,045	1,247	1,435	51.7	37.3
Guangzhou	1,622	2,430	2,831	3,289	3,854	74.5	36.1

SOURCE: SSB 1990, 1997.

1991). The share of the floating population in total city population has increased further in the 1990s, reaching 20 to 30 percent in many cities. By the mid-1990s, almost all Chinese cities had surpassed their population control objectives for the year 2000 that were established in the early 1980s. Although many large interior cities have recorded rapid growth, most large cities are still located in the coastal region (Figure 17.2).

Population control in Shanghai is supposed to be one of the most strict as it is the largest city in China. However, total population in Shanghai reached 13.3 million in 1990, surpassing the control objective of 13 million for the year 2000 that was established by the Shanghai Comprehensive Plan and sanctioned by the State Council in 1986. Ironically, in the late 1980s, development of the Pudong district of Shanghai was initiated to propel the economic growth of Shanghai and the Yangtze Valley. Under the motto of "Open Pudong, Develop Pudong," Pudong's building has been accelerated in the 1990s. The designated development area of Pudong covers 350 square kilometers, and eventually 2 million people will reside there. This alone will dramatically increase the population of Shanghai.

The population of Beijing grew rapidly from 9.2 million in 1982 to 10.5 million in 1993, a net increase of 1.3 million. The control objective of 10 million urban residents for the year 2000 was reached in 1988. In 1990, net migrants almost doubled the control objective for that year set up by the Seventh Five-Year Plan (1985–1990) of Beijing.

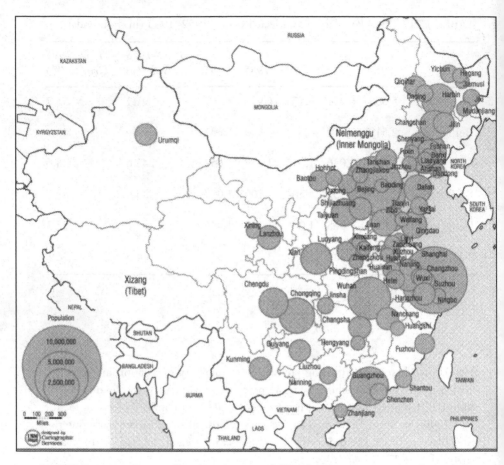

FIGURE 17.2   Geographic Distribution of China's Large Cities

The control objectives for the year 2000 established a residential population *(changzu renkou)* of 11.6 million and a floating population of 2 million (BICPD 1992) (for statistical definitions of urban population, see Chan 1994b). However, in 1995 Beijing's floating population reached 3.295 million *(Beijing Daily* 1995), which surpassed even the control objective for the year 2040 (a floating population of 3 million). It is evident that the growth of the migrant population and the urban residential population in Beijing is much greater than expected.

In Hangzhou, the control objective for urban population planned during the early 1980s for the year 2000 was 1.05 million. As one of the most important historical and tourist cities in China, the State Council stressed the strict control of population growth in Hangzhou. However, according to the 1990 census, the residential population had already reached 1.47 million. Urban statistics show that in 1995, the registered city population reached 1.44 million (Table 17.1). Moreover, the rapid development of new industrial districts is expected to increase the city size substantially. Many other coastal cities, as well as interior cities, are experiencing the rapid growth of both migrant and urban populations, although the speed of growth varies from city to city.

As Hamer (1990) points out, many migrants, who are considered temporary residents, have actually lived in cities for a long period of time. In 1988, in the above-mentioned twenty-eight million-plus cities, 28.67 percent of the floating population had been in residence for more than a year (Li and Hu 1991). According to the 1990 census, both Beijing and Shanghai had 0.49 million temporary migrants who had resided there for more than a year (SSB 1993). In 1993, 34.01 percent of Guangzhou's 1.7 million floating population had lived there for more than one year (Wu et al. 1995). Although scholars disagree over the effectiveness of China's urban policy, it is clear that large cities in China are experiencing rapid growth.

As I will elaborate in a later section, the unexpected growth of large cities has placed a tremendous pressure on them to improve urban planning and management. Government policy in China has had a substantial impact on its urban development, but multiple forces have contributed to the growth of large cities.

Economic Reforms, Growth, and Large Cities

Scholars working on China tend to agree that economic growth is the most important force driving urbanization and the growth of large cities. Indeed, scholars have long argued that more efficient production due to economies of scale and agglomeration have prompted the growth of large cities (Alonso 1971; Wheaton and Shishido 1981). Economies of scale and agglomeration are especially important in developing countries due to resource constraints and market segmentation. Higher productivity and returns in metropolises are due to their greater industrial mix, more specialized services, higher-quality labor force, better urban infrastructure, and greater access to international markets. The push-pull theory maintains that substantial rural-urban differentials in income and job opportunity drive the migration into large cities. While rapid urbanization forces some governments to restrict the growth of large cities, governments and private investors in developing countries often emphasize economic growth and the attractiveness of large cities (urban bias), contributing to the growth of large cities.

Although Mao advocated rural development and the control of large cities, the country's industrialization policy to a certain extent intensified rural-urban gaps and facilitated the growth of large cities. Since the launch of economic reforms in 1978, China's urban policy has encouraged the development of small towns, but the Chinese government has stressed economic growth and the importance of cities in economic development. Since the reform, the Chinese economy has grown rapidly (Table 17.2). Economic reforms and economic growth have stimulated an increase in the number of migrant workers and the expansion of large cities.

TABLE 17.2   Major Economic Indicators in China, 1980–1995

	1980	1990	1995	1980–1990 Average Annual Growth (%)	1990–1995 Average Annual Growth (%)
GDP (billion yuan)	452	1,855	5,826	10.2	12.0
Per Capita GDP (Yuan)	456	1,627	4,815	7.8	10.7
Fixed Assets Investment (billion yuan)	91	452	2,002	22.9	34.7
Foreign Investment (US$ billions)	–	3.5	37.5	–	36.1
Export (US$ billions)	18.1	62.1	148.8	15.1	19.1

NOTE: Data in current prices. Growth rates in comparable prices.
SOURCE: SSB 1996.

Rural reforms were initiated in 1978 with the elimination of the commune system and the adoption of the responsibility system. Under this new system, farmers have more flexibility in their economic activities. They are allowed to plant more lucrative cash crops on their land, to sell agricultural products in the market, and to engage in nonagricultural activities. The impacts of rural reforms on the growth of large cities are multifaceted. First, the dismantling of rural communes and improvements in farming efficiency have increased the amount of rural surplus laborers who are available to migrate to cities. Second, rural reforms, facilitated by price adjustments, decentralization, and the development of markets, have greatly fostered the growth of agriculture and rural industry, which provide agricultural products, capital, and rural markets for the urban economy and to a certain extent facilitate the development of industries and services in cities. Third, the rural reform diminishes the ability of rural administrative organizations to control employment and migration. These factors have contributed to the migration of rural workers to cities and the growth of the urban economy.

Rural-urban differentials have long existed in China. Agricultural products have traditionally been lower priced than industrial products, and rural residents have lower incomes and poorer living conditions. During the Mao era, they were kept in rural areas and were not allowed to take up nonagricultural activities or to move into cities. Rural reforms partially reduced rural-urban differentials in the early 1980s, but with the shift in the emphasis of reforms to cities in the mid-1980s, rural-urban differentials in many areas of China became larger. For example, from 1978 to 1985, per capita consumption of the agricultural population increased 9.6 percent annually, higher than that of the nonagricultural population (5.1 percent);

however, for the 1986–1993 period, the growth rate of the former was 4.3 percent annually, which was much slower than the latter (7.4 percent) (SSB 1994). The increase in the rural-urban gap stimulated the migration of rural workers to cities in the 1990s. Rural reforms and rural-urban gaps have acted as forces that "push" the out-migration of rural workers.

Economic reform shifted its focus to urban areas in 1984, and urban reforms have stimulated the growth of urban economies, particularly those in the coastal region (Xie and Costa 1991). The introduction of new systems in product pricing and exchange, taxation, investment, employment, and production management have provided urban enterprises more power in decisionmaking. Urban reforms have also given urban governments considerable autonomy in administration. Large city authorities have acquired more decisionmaking power as a consequence of decentralization and the development of markets. Because the advancement of local officials is largely dependent on their achievement of economic growth, urban governments have used their increasing power to mobilize resources for urban growth. In this sense, local governments are acting as development or entrepreneurial agencies that are keen to expand urban economic capacities and to stimulate urban economic development.

Cities are China's economic centers (Table 17.3), and it is often argued that large cities perform more efficiently than small cities due to economies of scale and agglomeration (Zhao and Zhang 1995). For example, in 1993, large cities, with 19.76 percent of the total urban population, produced 34.67 percent of the gross value of industrial output and 44.29 percent of pre-tax profits generated by all Chinese cities (SSB 1995). The government considers the development of large cities and their trickle-down effects necessary for economic development (Wei 1994). Large cities are expected to act as powerful centers of regional economies and strong engines to propel the growth of the national economy. Large cities are critical to China's economic development and are often emphasized by the Chinese government (Yeung and Hu 1992).

Some preferential policies have been granted to large cities, especially those in the coastal region. The development strategies of central cities and coastal cities have been implemented to take full advantage of the economic endowments of large cities. More state investment has been allocated to large coastal cities to strengthen their industrial bases and to improve their urban infrastructure. Especially with the decentralization of investment control, large cities are able to invest heavily in their economies and infrastructure. Reform policies have encouraged regions to organize production, circulation, consumption, and management around central cities. The improvement of transportation and communication networks centered on large cities has also facilitated the growth of these cities. Although large cities have suffered from more government control and from problems with state-owned enterprises (SOEs), their governments have bargained

TABLE 17.3   Major Indicators of Chinese Cities, 1995

	China	Urban Region		Urban District	
		Volume	Share (%)	Volume	Share (%)
Population (millions)	1,211	893	73.7	500	41.3
GDP (billion yuan)	5,826	5,274	90.5	3,998	68.6
Industrial Output (billion yuan)	9,189	7,883	85.8	6,284	68.4
Fixed Capital (billion yuan)	4,499	4,199	93.3	3,672	81.6
Profit and Tax (billion yuan)	505	487	96.4	422	83.6

SOURCE: SSB 1997.

intensely with the central government for more decentralized policies and for reduced taxes.

Large cities have also experienced dramatic structural changes. The Chinese economy, like many other economies under socialism, traditionally had shortages in urban services (Wei and Dutt 1995). With the drive to develop urban services, particularly since 1992, the service sector has boomed in many cities. While on the one hand, millions of urban workers have been laid off by SOEs, rural migrants are more willing to take jobs as nannies, repairers, vendors, and construction workers. Labor market segmentation exists in many cities and the need for rural workers is real, although city governments often attempt to limit the jobs available to rural migrants. In short, although urban policy attempts to control large city growth, economic reforms and growth have facilitated the expansion of large cities in China.

Globalization, Foreign Investment, and Large Cities

The globalization of economic activities has emerged as a major trend of the world economy, and a large amount of foreign investment has been flowing to the newly industrializing countries (NICs) in Asia. China has gradually opened up its economy to foreign investment and trade. As a result, foreign investment has increased since the late 1970s, particularly during the 1990s.

Foreign investment and trade in China were very limited before economic reforms due largely to China's policy of self-reliance and antagonism toward capitalism. During the 1950s, China's major foreign trade and investment partners were the Soviet Union and Eastern European countries. Loans from the Soviet Union were used for the construction of 156 major projects in the First Five-Year Plan period, and this stimulated the growth of the major interior cities. With the breakup of Sino-Soviet relations in the early 1960s, China, for the following decade, favored self-reliance and was isolated from the global market. During the early 1970s, China began to trade with Western countries and to import some advanced industrial equipment, mainly from Japan, Western Germany, and the United States. Several coastal cities, such as Nanjing, benefited from the allocation of industrial equipment there.

Since the late 1970s and the end of the cold war in the 1980s, China has dramatically restructured Mao's policy of self-reliance and favored opening up its domestic economy. Attracting foreign investment and promoting international trade are major elements of economic reforms. The government believes that such investment can be used to reduce the shortage of capital, advance technology, promote exports, increase foreign exchange, and stimulate economic growth (Grub and Lin 1991). China's opening up to the outside world has been facilitated by the globalization of economic activities, economic restructuring in Asian NICs, and overseas Chinese networks (Leung 1996). Besides the favorable open-door policy of the central government, Chinese cities have also initiated numerous local policies to attract foreign investment (Zhang 1994). Rapid economic growth, expanding consumption, and policy incentives have made investment in China attractive. China's large potential markets have attracted the interests of global investors, and volumes of foreign investment and trade have increased dramatically in China (Table 17.4).

The opening of China has been spatially uneven, with a strong focus on the coastal cities. The four special economic zones (SEZs) of Shenzhen, Zhuhai, Shantou, and Xiamen were opened in 1980, provided with various state incentives (Figure 17.1). Special economic zones soon became major destinations of foreign investment, and population in these cities has grown dramatically. Subsequent to opening the four SEZs, China established fourteen open coastal cities (OCCs) in 1984 and three delta areas (Zhujiang, Minjiang, and Changjiang deltas) in 1985 and 1986. A series of laws and incentive policies were introduced in the mid-1980s to encourage foreign investment and trade. Open areas have enjoyed higher foreign exchange retention rates, lower tariffs on imports, lower taxes on foreign investment, tax breaks for exports, and more decisionmaking power in management. China introduced the coastal development strategy and the strategy of the "grand international economic circle," and coastal cities were encouraged to actively participate in the international economy.

After the 1989 Tiananmen Square incident, China's economic reforms experienced a short period of retrenchment. Many foreign countries boycotted the crackdown by canceling or withholding investment projects and trade contracts. As a result, FDI in China decreased sharply during this period, and only in 1991 did signs of recovery begin to appear. In 1992, China deepened economic reforms by easing restrictions on foreign investment in the service sector, reducing tariffs, and further opening up domestic markets. Relations with South Korea, Taiwan, and Singapore were improved. More preferential polices were granted to Shanghai, especially the Pudong district, and many interior cities were opened up for foreign

TABLE 17.4   Foreign Capital in China, 1979–1995 (US$ millions)

Year	Total	FDI	Foreign Loans	Other
1979–1983	14,438	1,802	11,755	881
1984	2,705	1,258	1,286	161
1988	10,226	3,194	6,487	545
1990	10,289	3,487	6,534	268
1992	19,202	11,007	7,911	284
1995	48,133	37,521	10,327	285

SOURCE: SSB 1996.

investment. Moreover, local governments received considerable power from the central government and have actively lured foreign capital by providing policy incentives. By 1995, nearly half of Chinese cities and counties were open for foreign investment and trade. China's economic reforms and favorable investment conditions have attracted the interest of global investors, whose confidence has been growing as a result of their experience in previous years. FDI increased dramatically from US$4.4 billion in 1991 to US$37.5 billion in 1995 (Table 17.4).

Although all of the provinces have attracted foreign investment, the coastal region has captured the lion's share of FDI (Figure 17.3). China's coastal region has generally attracted about 90 percent of regional FDI. From 1983 to 1995, the central region's share of FDI increased from 1.1 percent to 9.2 percent, while the western region's share actually declined. Among China's provinces, Guangdong has attracted the largest amount of FDI, and investment generally has spread from Guangdong to other coastal provinces. By 1995, though Guangdong remained dominant, its share of FDI in China had decreased to 27.6 percent, while shares of FDI in other coastal provinces, particularly Jiangsu, Fujian, Shanghai, and Shandong had increased. The share of FDI in the interior region had increased only slightly, with most of the investment flowing to cities near the coastal region, such as cities in Hubei, Hunan, Henan, Anhui, and Jilin. This regional change, that is, from Guangdong province to other coastal provinces, reflects the increased business experience of foreign investors and the spatial unevenness of China's open-door policy and economic development.

Coastal provinces have long-established close ties with foreign countries and global investors; overseas Chinese and their networks have channeled much of the foreign investment in China. Foreign investment tends to concentrate in large cities, which have better-trained labor and facilities, larger markets, and more exposure to foreign countries. In 1993, large city regions, including central cities and suburban counties, attracted 39 percent of total foreign investment in China (SSB 1995).

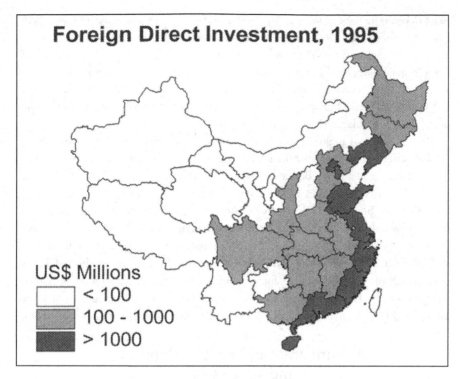

Foreign Direct Investment, 1995

US$ Millions
☐ < 100
▨ 100 - 1000
▪ > 1000

FIGURE 17.3   Foreign Direct Investment, 1995

Although very large cities, such as Shanghai and Beijing, have been very successful in attracting foreign investment, some of the relatively less populated large cities, such as Suzhou, saw their foreign investment soar in the 1990s. While many interior cities were opened up to foreign investment during the 1990s, foreign investment is still concentrated in coastal cities, including large coastal cities.

The economic policies of the SEZs and OCCs, all located in the coastal region, have been the most attractive to global investors. These cities with more preferential policies, active local governments, easy access to the international market, and good transportation facilities tend to attract more FDI than other cities. Between 1980 and 1984, SEZs attracted about half of the regional FDI. In 1984, when the fourteen OCCs were opened up, FDI in SEZs reached US$343 million, which accounted for 38.75 percent of regional FDI, and OCCs attracted 25.4 percent of the regional FDI. Their dominance in FDI was not challenged during the 1980s; only in the 1990s, did these cities' shares of FDI decline somewhat, partially due to rising costs and the opening up of competing cities for foreign investment. In 1993, SEZs and OCCs attracted 13.7 percent and 30.3 percent of FDI respectively.

Open districts, in particular, have become the areas where the bulk of foreign investment is concentrated. Open districts have been established primarily in large cities to promote foreign investment and trade. The size of these districts can be very large, and these districts are experiencing the fastest population growth. In Suzhou, a city whose population is supposed to be strictly controlled, four newly established development zones (including the Singapore Industrial Park) are planned to occupy an area of 139.81 square kilometers, more than doubling the current built-up area (59 square kilometers). The establishment of open districts has greatly increased the population size of large cities.

Recently, several major large cities have begun pursuing the goal of becoming "global" cities. More than ten cities have announced ambitious plans to reach such goals, including Shanghai, Beijing, Tianjin, Guangzhou, Xian, Wuhan, and Dalian. The development strategies of these cities include more aggressive expansion of administrative boundaries, the in-migration of educated people, and the construction of new industrial districts. As foreign investment tends to be attracted to places with more preferential policies and locational endowments, these efforts have further stimulated the growth of foreign investment in these cities. In short, China's open-door policy and the drive for economic development, facilitated by global restructuring and the globalization of economic activities, have attracted large amounts of foreign investment, which has contributed to the growth of large cities.

Administrative Changes, Planning, and Large Cities

China's large cities play important roles in economic development, in the political arena, and in the social spheres of culture and education. Hu and Meng (1987) argue that these multiple functions are the unique characteristics of China's large cities and are the fundamental forces driving the growth of these cities. Large cities are important political (administrative) centers and China's political centers have developed more rapidly than nonpolitical centers due to strong government involvement in the economy. Because the advancement of local officials is very much dependent on their achievement in economic development (Walder 1995), local governments have a strong motivation to stimulate economic growth. Marketization and decentralization have strengthened the administrative powers of large cities, whose officials have used these powers to promote urban growth.

A system of "city leading county" has been adopted in many large cities. This system is not new in China; it was implemented in some cities as early as the 1950s. In 1982 Secretary General Zhao Ziyang gave new impetus to this system, starting in the coastal provinces of Liaoning, Jiangsu, and Guangdong. Although not all of China's cities have adopted this plan, it has nonetheless expanded quickly from the coastal provinces to the interior provinces. By the end of 1995, 774 counties and 260 county-level cities, which accounted for 48 percent of cities and counties in China, were administrated by central cities (NERC 1997). The purpose of this reform is to reduce administrative barriers perceived to have hampered urban-rural interactions and to enable cities to play more roles in stimulating regional economic growth and gradually forming economic regions centering on major cities (Ma and Noble 1986). Officials of central cities argue that this centralized system of urban management is necessary for concentrating resources needed for the development of central cities, which are considered to be the driving engines of regional growth. Although it is difficult to evaluate the effectiveness of this system, it is clear that it has improved the interaction between cities and counties. However, the "city leading county" system has provided more power for large cities in centralizing capital, personnel, and production materials from the surrounding counties and in expanding their administrative boundaries, leading to the growth of urban population.

Another example of administrative reform is the establishment of autonomous planning systems *(jihua danlie)*, separate from the central government, in fourteen large cities (Chongqing, Wuhan, Shenyang, Ningbo, Dalian, Guangzhou, Xian, Harbin, Qingdao, Shenzhen, Xiamen, Changchun, Chengdu, and Nanjing). These cities now have provincial-level decisionmaking powers in industrial production, capital investment, transfer of commodities, and some other areas. These cities' fiscal authority to retain local revenue has been greatly expanded. SOEs previously managed by the central government have been largely decentralized and turned over to these cities, which has greatly strengthened their economic bases. In addition, in 1995, fifteen major cities and provincial capitals officially obtained the status of vice-provincial city *(fu sheng ji shi)*.

Modifications in urban planning have been carried out to facilitate economic development and to promote residential convenience. The rise of urban land rent, the adoption of land-use laws, and stronger local government administration have led to the restructuring of urban land-use patterns. Shortages in urban land and the rapid increase in urban population have prompted the expansion of large cities. In addition, Chinese planners have made considerable efforts to change the traditional monocentric spatial structure to a polycentric pattern. In many cities, Soviet-style self-sufficient work-unit districts have been revamped into more functionally specialized urban districts (Gaubatz 1995). The administrative areas of large cities have increased considerably to accommodate rural migrants and the relocation of industries previously located in downtown areas due to rapid urbanization and suburbanization. Western ideas of zoning and central business district organization have been introduced into planning strategies. It is clear that the restructuring of China's urban spaces, much of which has resulted from changes in planning and administration, has significantly contributed to increasing urban population.

The suburban areas (or urban periphery) of large cities, in particular, have been experiencing rapid growth and dramatic change. Pannell (1992) suggests that suburbs may be more suitable locations for development than the built-up areas of cities, as land acquisition, site preparation, building construction, and production preparation cost less in such locations. In addition, suburbs have loose government control and have become favorable places for migrants. Chinese authorities have

used suburban areas as the major areas for urban expansion, and parcels of agricultural land have been transformed into urban land for industrial, commercial,
and residential development. Suburban population, even the population of counties, which was previously counted as rural population, has become part of urban
population. For example, Wuchang county in Hubei province was annexed to the
city of Wuhan in 1995 and has become a district of the city, the Jiangxia district.
It is evident that rapid growth in the suburbs has also contributed to the growth
of large cities.

In short, the recent growth of large cities is related to the increasing rural surplus
labor and to job opportunities provided by cities as a result of economic reform and
increasing foreign investment. Administrative changes have also opened the door
wider for migration into large cities. The increase in migrants and the growth of
large cities are greatly challenging the management of large cities in China.

Implications for Urban Planning and Management

Migration, as a major cause of large city growth, has had a positive impact on
Chinese cities. Rural-urban migrants help to solve the problem of labor shortages in
cities due to rapid economic growth, economic restructuring, and declining fertility
rates. Migrants are playing important roles in the sectors that have difficulty hiring
urban residents, such as sanitation, construction, packaging and hauling, and hotel
and restaurant services (Solinger 1995). Petty vendors and nannies are also largely
rural migrants. They have significantly contributed to the development of urban
services and have helped to improve the living conditions of cities. Rural workers, as
cheap laborers, are particularly important to SEZs and coastal cities where foreign
investment concentrates. Moreover, migrants themselves are better-off working in
cities. They have also facilitated the interchange between cities and the countryside
and among different regions of China.

However, migrants have brought new challenges to large cities. During the
Maoist period, under the influence of the Soviet model of development that emphasized production/accumulation and neglected services, China's cities were overwhelmingly industry-oriented, and urban services lagged far behind. Since the
initiation of economic reform, the improvement of urban services has captured the
attention of the Chinese leaders. However, to a large extent, China's large cities still
suffer from significant problems in the areas of sewage facilities, housing, solid waste
disposal, air pollution, and traffic congestion. Pannell has asserted that "the cities of
China are now beginning to experience the kinds of urban problems found in many
third world countries, with the emergence of slum and squatter settlements and an
urban underclass of marginally or temporarily employed persons who have little or
no access to health-care facilities, adequate housing, education opportunity, or
social-welfare benefits" (Pannell 1992: 34).

The lack of adequate urban infrastructure has long plagued large cities in China. Although the Chinese government has emphasized the improvement of transportation, the transport sector is still far from well developed. The recent increase in migrants has put tremendous pressure on the transportation sector, and severe congestion has become a serious problem in large cities. The situation is even more critical during the annual Spring Festival. Large cities also have poorer housing conditions than other urban settlements, especially in coastal cities (Kim 1991). Housing rural migrants is a serious problem. Migrants have placed great pressures on urban services, which were already overburdened. A substantial increase in the fiscal support for urban infrastructure is necessary to improve the conditions in large cities. Overemphasis on capital investment in the industrial sector remains a characteristic of China's fiscal policy. Great potential exists in the development of market instruments and the use of foreign capital to improve urban infrastructure.

Because large cities are having so much difficulty coping with the dramatic increase in migrants, many have implemented strict policies to control rural-urban migration. The mid-1990s experienced tougher restrictions with the announcement by central and city governments of policies to control housing, employment, and resident status of rural migrants. Employment and residence permits are now required for rural workers to move into cities. These permits are issued only to those with higher levels of education and needed skills, and the jobs must be temporary in nature and incapable of being filled by city residents. However, the implementation of these policies does not solve the problem of rural surplus labor or the shortage of urban services. Although some scholars have argued that the rural-urban gap is diminishing (Wang 1997), the "wall" between the cities and the countryside still exists in many regions of China. Rural workers and migrants continue to be treated less favorably than urban workers, creating a two-class society within Chinese cities (Chan 1996). The complexity of rural-urban relations and the increase in rural-urban migration demand new approaches to managing employment, urban growth, and rural-urban relations.

China's urbanization policy is largely based on the size of cities and provides little guidance for urban development. The policy of controlling the growth of large cities is a defensive policy responding to rapid urbanization and growth of cities. It does not fundamentally solve the problems underlying urbanization and urban growth. As economic reforms continue, especially decentralization, marketization, and the dismantling of the household registration system, the ability of the Chinese government to control large city growth is likely to decline further. On the other hand, the roles of markets, localities, and global investors are increasing, which will certainly affect urban development and policy implementation. Although the emphasis of urban policy is shifting from control to planning and management, the size of cities will remain an important issue for China, as well as other developing countries. This is largely due to the fact that developing countries usually lack the financial resources and expertise to manage large cities. China's urban policies, with regard to economic development, property rights, quality of life, and social justice, need thorough review and renewal.

The complexity and rapid change of large cities necessitate the reform and strengthening of urban planning and management. Chinese urban planning was interrupted for many years, particularly during the Cultural Revolution, when urban planning departments were eliminated. Although great improvement has been made since the late 1970s through increases in personnel, funding, and decisionmaking power, the current urban planning approach still largely follows the Soviet mode, emphasizing physical planning and project allocation. Urban planning is often isolated from planning implementation and urban management, necessitating the constant revision of plans. Planning has not provided effective guidance for urban development and growth management. Greater consideration of social and economic conditions and more attention to local factors should be incorporated into the planning process. The reform of the urban planning system and the improvement of urban management remain urgent issues in China.

The root of the problems surrounding migration and large city growth lies in the dualism inherent in Chinese society and its concomitant power structure. China is still largely a segregated society, expressed in the dualism of city and countryside, urban population and rural population, and the affluent coastal region and the neglected interior regions. Mao failed to reduce this dualism, and post-Mao policies have intensified certain dimensions of it. Rural migrants are a disadvantaged class and are sometimes treated as second-class citizens. Urban governments welcome cheaper rural laborers but are less interested in providing services, including housing, education, health care, and employment services, to migrant workers. National policy has a role to play in solving these problems.

China is also characterized by the concentration of power in cities, particularly provincial and national capitals. Although decentralization has been introduced to stimulate local economic growth, re-centralization took place in the mid-1990s as a response to the lack of effective adjustment mechanisms. Efforts must be made to improve the management of migration and large cities, to develop the labor market, to reduce rural-urban dualism and regional inequality, to improve the welfare of rural people, and to balance power in the society. This implies that economic reform alone cannot solve all of China's social problems; further reforms in urban administration and public finance are necessary to improve the management of large cities in China.

References

Alonso, W. 1971. The economics of urban size. *Papers of Regional Science Association* 26:67–83.
Beijing Daily. 1995. Statistical reports on national economy and social development in Beijing in 1994. February 21.

BICDD (Beijing Institute of City Planning and Design). 1992. *Beijing chengshi zongti gui-hua, 1991–2000* (City Master Plan of Beijing, 1991–2000).

Buck, D. D. 1981. Policies favoring the growth of smaller urban places in the People's Republic of China, 1949–1979. In *Urban Development in Modern China,* ed. L.J.C. Ma and E. W. Hanten. Boulder: Westview Press.

Chan, K. W. 1994. *Cities with Invisible Walls.* New York: Oxford University Press.

_____. 1996. Post-Mao China: A two-class urban society in the making. *International Journal of Urban and Regional Research* 20(1):134–150.

Clayton, E., and T. Richardson. 1989. Soviet control of city size. *Economic Development and Cultural Change* 38(1):155–165.

DCP (Department of City Planning, Ministry of Construction). 1991. *The Urban Planning Law of the People's Republic of China.*

Fan, C. C. 1988. The temporal and spatial dynamics of city-size distribution in China. *Population Research and Policy Review* 7:123–157.

Fuchs, R. J., E. Brennan, J. Chamie, F. Lo, and J. I. Uitto, eds. 1994. *Mega-City Growth and the Future.* Tokyo: United Nations University Press.

Gaubatz, P. R. 1995. Urban transformation in post-Mao China. In *Urban Spaces in Contemporary China,* ed. D. S. Davis, R. K. Kraus, B. Naughton, and E. J. Perry, 28–60. New York: Woodrow Wilson Center Press and Cambridge University Press.

Ginsburg, N., B. Koppel, and T. G. McGee, eds. 1991. *The Extended Metropolis.* Honolulu: University of Hawaii Press.

Grub, P. D., and J. H. Lin. 1991. *Foreign Direct Investment in China.* New York: Quorum Books.

Hamer, A. 1990. Four hypotheses concerning contemporary Chinese urbanization. In *Chinese Urban Reform,* ed. R.Y.W. Kwok, W. L. Parish, A.G.O. Yeh, and X. Xu, 233–242. Armonk, N.Y.: M. E. Sharpe.

Hu, Z., and X. Meng. 1987. Socioeconomic and political background of the growth of large cities in China. *Asian Geographer* 6(1):24–28.

Kasarda, J. D., and A. M. Parnell, eds. 1993. *Third World Cities.* Newbury Park: Sage.

Kim, W. B. 1991. The role and structure of metropolises in China's urban economy. *Third World Planning Review* 13(2):155–177.

Kirkby, R.J.R. 1985. *Urbanization in China.* London: Croom Helm.

Krugman, P., and R. Livas Elizondo. 1996. Trade policy and the Third World metropolis. *Journal of Development Economics* 49:137–150.

Leung, C. K. 1996. Foreign manufacturing investment and regional industrial growth in Guangdong province. *Environment and Planning A* 28:513–536.

Li, M. B., and X. Hu. 1991. *Liudong renkou du dachengshi fazhan de yingxiang ji duice* (Impact of Floating Population on the Development of Large Cities and Solutions). Beijing: Economic Daily Press.

Ma, L.J.C., and A. G. Noble. 1986. Chinese cities. *Urban Geography* 7(4):279–290.

Ma, L.J.C., and Y. Wei. 1997. Determinants of state investment in China, 1953–1990. *Tijdschrift voor Economische en Sociale Geografie* (Journal of Economic and Social Geography) 88(3):211–225.

NERC (National Economic Reforms Committee). 1997. *Zhongguo jinji tizhi gaige nianjian* (China's Economic Reform Yearbook, 1996). Beijing: Reform Press.

Naughton, B. 1995. Cities in the Chinese economic system. In *Urban Spaces in Contemporary China,* ed. D. S. Davis, R. K. Kraus, B. Naughton, and E. J. Perry, 61–89. New York: Woodrow Wilson Center Press and Cambridge University Press.

North, D. 1990. *Institutions, Institutional Change, and Economic Performance.* Cambridge: Cambridge University Press.

Pannell, C. W. 1992. The role of great cities in China. In *Urbanizing China*, ed. G. R. Guldin, 11–39. New York: Greenwood Press.

Richardson, H. 1989. The big, bad city. *Third World Planning Review* 11(4):355–372.

Rowland, R. H. 1983. The growth of large cities in the USSR. *Urban Geography* 4(3):258–279.

Solinger, D. 1995. The floating population in the cities. In *Urban Spaces in Contemporary China*, ed. D. S. Davis, R. K. Kraus, B. Naughton, and E. J. Perry, 113–139. New York: Woodrow Wilson Center Press and Cambridge University Press.

SSB (State Statistical Bureau). 1990. *Zhongguo chengshi sishi nian* (China: Forty Years of Urban Development). Beijing: China Statistical Information and Consultancy Service Center.

———. 1993. *Zhongguo renkou tongji* (Population Statistics of China). Beijing: China Statistics Press.

———. 1994, 1996. *Zhongguo tongji nianjian* (Statistical Yearbook of China). Beijing: China Statistics Press.

———. 1995, 1997. *Zhongguo chengshi tongji nianjian* (Urban Statistical Yearbook of China). Beijing: China Statistics Press.

UNDP (United Nations Development Program). 1991. *Cities, People, and Poverty*. New York: UNDP.

Walder, A. G. 1995. Local governments as industrial firms. *American Journal of Sociology* 101(2):263–301.

Wang, M.Y.L. 1997. The disappearing rural-urban boundary: Rural socioeconomic transformation in the Shenyang-Dalian region of China. *Third World Planning Review* 19(3):229–250.

Wei, Y. 1994. Urban policy, economic policy, and the growth of large cities in China. *Habitat International* 18(4):53–65.

———. 1995. Chinese large cities: A review and a research agenda. *Asian Geographer* 14(1):1–13.

———. 1997. State policy and urban systems: The case of China. *Journal of Chinese Geography* 7(1):1–10.

———. 1998. Regional inequality of industrial output in China. *Geografiska Annaler B* 80(1):1–15.

Wei, Y., and A. K. Dutt. 1995. Spatial patterns of sectoral shifts in China. *Asian Profile* 23(4):273–288.

Wei, Y., and L.J.C. Ma. 1996. Changing patterns of spatial inequality in China. *Third World Planning Review* 18(2):177–191.

Wheaton, W. C., and H. Shishido. 1981. Urban concentration, agglomeration economies, and the level of economic development. *Economic Development and Cultural Change* 30:17–30.

Wu, Y. W., C. M. Wang, and D. N. Cheng. 1995. A study of the administration of floating population in the city of Guangzhou. *Redai dili* (Tropical Geography) 15(1):49–55.

Xie, Y. C., and F. J. Costa. 1991. The impact of economic reforms on the urban economy of the People's Republic of China. *Professional Geographer* 43(3):318–335.

Yeung, Y., and X. Hu. 1992. *China's Coastal Cities*. Honolulu: University of Hawaii Press.

Yusuf, S., and W. Wu. 1997. *The Dynamics of Urban Growth in Three Chinese Cities*. New York: Oxford University Press.

Zhang, L. Y. 1994. Location-specific advantages and manufacturing direct foreign investment in south China. *World Development* 22(1):45–53.

Zhao, X. B., and L. Zhang. 1995. Urban performance and the control of urban size in China. *Urban Studies* 32(4-5):813–845.

18

Suburbanization in Beijing

Yi-Xing Zhou and Fahui Wang

Empirical studies of urban population distribution in developed countries have been abundant since the classic study by Clark (1951), but much less research has been undertaken on developing countries due to less reliable data available (Mills and Tan 1980). On cities in a planned economy, such as in China's, very little has been reported. What are the spatial patterns of population distribution in Chinese cities? How have they changed over time? How are they different from cities in Western countries?

Suburbanization started as early as the 1920s and became trendy in the 1950s and 1960s in America (Muller 1981). Since the economic reform in 1978, China has undergone dramatic changes toward a market economy. The experimental urban land-use management and housing reforms have had significant impacts on the urban structure in China. Have Chinese cities experienced suburbanization? Can China's suburbanization be attributed to the same causes of suburbanization in America? Are there any differences?

In the literature, there are few studies on the change in spatial patterns of population distribution in Chinese cities (see Yan 1995 for a review). Recent research (e.g., Lo 1994; Ning and Yan 1995; Yeh et al. 1995) has benefited from the data sources in the 1982 and 1990 national population censuses. This chapter examines Beijing, the national capital city, as a case study. In the existing literature on Beijing, Sit (1995) describes its basic population distribution patterns. Hu and Foggin (1994b) and Zhou (1996) present the various growth rates of urban population across Beijing and raise the issue of suburbanization. This chapter is based on the existing literature and the results from our recent research on modeling Beijing's urban density patterns (Wang and Zhou 1999).

Central City and Suburbs in Beijing

As described in Sit (1985, pp. 52–53), the administrative organization of Chinese cities is intended to integrate urban and rural, industry and agriculture, and maintain urban self-sufficiency in food. A city usually has a large territory of rural counties *(xian)* in addition to the municipality *(shi)*. In Beijing, there are eight rural counties surrounding the municipality and under its administration: Huairou, Miyun, Yanqing, Pinggu, Changping, Shunyi, Tongxian, and Daxing. The municipality consists of four city districts *(cheng qu:* Dongcheng, Xicheng, Chongwen, and Xuanwu), four inner suburban districts *(jing jiao qu:* Chaoyang, Fengtai, Haidian, and Shijingshan), and two outer suburban districts *(yuan jiao qu:* Mentougou and Fangshan). Since the outer suburban districts are mostly rural like the rural counties, they are usually classified together as "remote districts and counties" *(yuan jiao qu xian)*. See Figures 18.1 and 18.2.

The whole region of Beijing can be broken down into three rings: the inner ring of city districts, the middle ring of inner suburban districts, and the outer ring of remote districts and counties (Hu and Foggin 1994b; Zhou 1996), shown in Figure 18.1. Population densities vary significantly across the three rings (see Table 18.1). The inner ring *(cheng qu,* 34 mi^2) is approximately the built-up area before the 1949 Revolution, including the Old City *(Laocheng,* 24 mi^2) bounded by the old city walls (now replaced by the Second Ring Road) and Guanxiang outside of the Old City. This is an area comparable to the concept of central city in America. In 1990, the central city had a population (all urban) of more than 2.3 million and an average density of 68,882 persons per square mile. The middle ring *(jing jiao qu,* 496 mi^2) had a population of nearly 4 million and a moderate density averaging 8,028 persons per square mile. In 1990, this area was mostly urbanized with an urban population of 3.128 million, or 78.6 percent of its total population of 3.978 million. The total urban population in Beijing was 5.445 million in 1990. The outer ring *(yuan jiao qu xian,* 5,963 mi^2) is mostly rural with a low density of 753 persons per square mile. The inner ring's density is about ten times the middle ring's, and the middle ring's density is about ten times the outer ring's. Among the ten remote districts and counties, six (Mentougou, Fangshan, Changping, Shunyi, Tongxian, and Daxing) have close ties to the central city, and together with the four city districts and four inner suburban districts they form an area comparable to the concept of metropolitan area in America.

Our study area is the urbanized portion of Beijing, approximately the combination of the central city districts and the inner suburban districts *(cheng jing jiao qu)*, that is, the inner and middle rings in Figure 18.1, illustrated in Figure 18.2. Even within the inner suburbs, two tracts remained rural in 1990: the tract at the map's northwest corner in Haidian District and the large tract extending south-north in Chaoyang District (see Figure 18.3). This rural portion is composed of forty-one villages *(xiang)* and is excluded from the analysis, since our focus is on the urbanized area.

In most cities in China (including Beijing), a subdistrict *(jiedao)* is designated whenever the area is dominated by urban population. Subdistricts on the edge of the

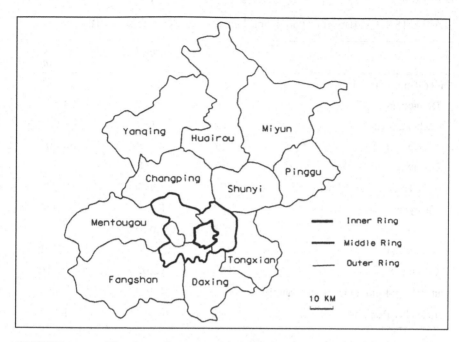

FIGURE 18.1 The City of Beijing and Its Three Rings

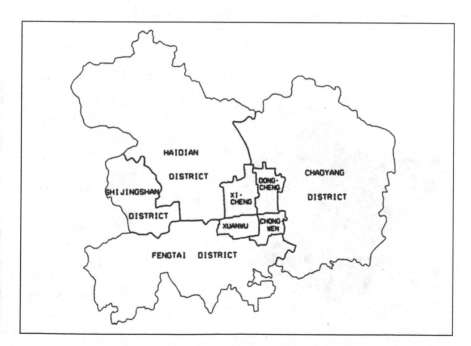

FIGURE 18.2 Districts in Beijing's Central City and Suburbs

TABLE 18.1   Variation of Population Densities in Beijing's Three Rings

	Area (mile²)	1990 Population ('000s)	Density (person/mile²)
Inner Ring: City Districts	34	2,318	68,882
Dongcheng	10	606	63,539
Xicheng	12	756	65,224
Chongwen	6	418	68,002
Xuanwu	6	538	84,377
Middle Ring: Inner Suburbs	496	3,978	8,028
Chaoyang	182	1,449	7,966
Haidian	118	1,030	8,769
Shijingshan	32	298	9,421
Fengtai	165	1,105	6,716
Outer Ring: Remote Districts/Counties	5,963	4,492	753
Whole City Region	6,492	10,791	1,662

SOURCES: Sit 1995, p. 116, and Zhou 1996, p. 199.

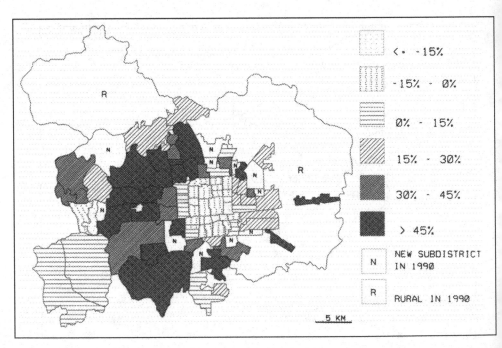

FIGURE 18.3   Urban Population Density Change in Beijing, 1982–1990

municipality (e.g., the three large subdistricts on the city's southwest corner in Fengtai District) may contain a small portion of rural residents. The area covered by the subdistricts is a good approximation for the urbanized areas in Beijing. In the study area, there were 82 subdistricts in 1982 and 93 subdistricts in 1990. Four other designated subdistricts (Qinghe Farm, Shoudou Airport, Shougang Mine, and Laoshan) are not included since they are very far from the continuously built-up area. Between 1982 and 1990, 12 new subdistricts were designated. Excluding Laoshan, 11 are in the study area (see Figure 18.3).

Censuses count population according to their residence rather than their registered status used by annual statistics reports. Data from the censuses provide a better measurement of true residents in an area. For a variety of reasons, there is a significant portion of population in Beijing who live in one place but hold the registered permanent residence status (RPRS) of another place. Socialist countries, including China and the Soviet Union, had stringent migration control through the methods of population registration and job control. Since the 1978 economic reforms, which gave people more freedom to migrate, people's registered residence often does not match their true residence. For example, some central city residents have moved to suburbs but are reluctant to give up the RPRS of the central city. Many rural residents work and live in the city but have not obtained the city RPRS. Population in our analysis includes longtime registered permanent residents and temporary residents. Temporary residents include (1) those who have lived in the subdistrict for more than one year but have RPRS for elsewhere; (2) those who have lived in the subdistrict for less than one year but have been away from the place where they hold RPRS for more than one year; and (3) those with unknown RPRS.

Evidence of Suburbanization in Beijing

We use data aggregated at both district and subdistrict levels and density contour maps to present the evidence of suburbanization in Beijing. We define suburbanization as the moving of central city residents to suburbs, which leads to negative growth in the central city and positive growth in the suburbs. In a broad definition, suburbanization starts when the growth rate in the central city is lower than in its suburbs. Although the spatial limits of most central cities are static, urbanization always leads to the expansion of suburbs and usually faster growth in suburbs than in central cities. The broad definition is less meaningful in the sense that the stage of suburbanization is not distinguished from the general trend of urbanization.

Evidence at the District Level

Based on the data from the 1964 (second), 1982 (third), and 1990 (fourth) censuses, the population changes in Beijing at the district level are summarized in Table 18.2. Because the areas remained unchanged, the rate of change in population is equivalent

TABLE 18.2   Changes in Population and Densities in Beijing

	1964-1982		1982-1990		Population Density (person/mi²)		
	Total Growth %	Annual Growth %	Total Growth %	Annual Growth %	1964	1982	1990
City Districts	3.00	0.16	-3.38	-0.43	69,247	71,326	68,882
Dongcheng	4.80	0.26	-6.96	-0.90	65,160	68,287	63,539
Xicheng	3.70	0.20	-1.10	-0.14	63,593	65,947	65,224
Chongwen	8.43	0.45	-5.14	-0.66	66,117	71,691	68,002
Xuanwu	-3.80	-0.21	-0.94	-0.12	88,669	85,298	84,377
Inner Suburbs	31.13	1.52	40.46	4.34	4,349	5,703	8,028
Chaoyang	41.88	1.96	41.68	4.45	3,961	5,620	7,966
Haidian	23.48	1.18	44.62	4.72	4,911	6,063	8,769
Shijingshan	32.75	1.58	31.39	3.47	5,274	7,000	9,421
Fengtai	27.15	1.34	34.89	3.81	3,917	4,978	6,716
Remote Districts and Counties	27.64	1.36	13.12	1.55	523	665	753
Whole City Region	21.50	1.09	17.20	2.00	1,170	1,416	1,662

SOURCE: Zhou, 1996, p.199.

to the rate of change in density. The following observations can be made about Beijing's population in the period 1964 to 1990.

1. During 1964–1982, the population in the central city districts increased slowly, except for a small drop in population in Xuanwu District. In other words, suburbanization had not started.

2. During 1982–1990, all four central city districts experienced negative growth. Given a natural population growth rate of 4 per thousand in the central city, the negative growth of 1982–1990 is certainly due to the out-migration of residents.

3. Both in 1964–1982 and in 1982–1990, suburbs (including inner suburbs and remote districts and counties) grew faster than the central city; outside of the central city, inner suburbs grew faster than remote districts and counties.

4. The inner suburbs had an annual growth of 4.34 percent in 1982–1990, significantly higher than the 1.52 percent of 1964–1982. This is attributed to in-migration from both the central city and remote districts and counties.

TABLE 18.3 Populations of the Subdistricts in Beijing's Central City, 1982 and 1990

	1982	1990	Change (%)		1982	1990	Change (%)
Dongcheng Qu				Xicheng Qu			
Chaoyangmen	51,840	43,058	-16.9	Fengsheng	59,597	550,633	-15.0
Jianguomen	87,622	73,007	-16.7	Xichanganjie	92,564	78,867	-14.8
Andingmen	57,828	49,702	-14.1	Fusuijing	86,204	74,184	-13.9
Jingshan	52,074	44,966	-13.6	Xinjiekou	77,626	71,064	-8.5
Beixinqiao	82,399	72,290	-12.3	Changqiao	82,992	77,378	-6.8
Dongsi	51,577	45,339	-12.1	Erlonglu	74,905	72,568	-3.1
Jiaodaokou	54,694	48,827	-10.7	Fuwai	59,147	60,849	2.9
Donghuamen	86,598	81,204	-6.2	Zhanlanlu	66,548	71,627	7.6
Hepingli	86,897	99,474	14.5	Yuetan	103,605	116,999	12.9
Dongzhimen	39,991	48,336	20.9	Dewai	61,003	81,644	33.8
Xuanwu Qu				Chongwen Qu			
Dashila	67,483	50,776	-24.8	Qianmen	59,358	46,685	-21.4
Chunshu	59,359	51,804	-12.7	Chongwenmen	56,121	46,653	-16.9
Taoranting	71,848	66,971	-6.8	Donghuashi	67,599	62,007	-8.3
Niujie	55,721	52,098	-6.5	Tiyuguanlu	55,8472	53,895	-7.8
Tianqiao	55,516	52,284	-5.8	Tiantan	74,316	68,968	-7.2
Guangnei	94,835	93,246	-1.7	Longtan	60,040	60,919	1.5
Baizhifang	75,679	78,691	4.0	Yongdingmen	64,395	78,524	21.9
Guangwai	63,200	91,890	45.4				

Evidence at the Subdistrict Level

Based on the data aggregated at a smaller geographical unit—the subdistrict (*jieda*) level—we see more details of the spatial variation of population change. Table 18.3 presents the population sizes of subdistricts in 1982 and 1990 (1964 data are not available at the subdistrict level) and their change percentages in the central city. Subdistricts had an average population size of 53,780 in 1982 and 58,552 in 1990. The eleven new subdistricts designated after 1982 have no data for 1982. We have grouped the population growth rates into six categories and show the spatial variation in Figure 18.3. Findings are summarized as follows.

1. Most subdistricts in the central city experienced population loss during 1982–1990. Among the 82 subdistricts in 1982, 28 subdistricts lost population. Out of these 28 subdistricts, 25 were in the central city, which account for 71 percent of the total 35 central city subdistricts. Six subdistricts

(Dashila, Qianmen, Chaoyangmen, Chongwenmen, Jianguomen, and Feng-sheng) near the city center or the Second Ring Road lost 15 percent or more of their population.

2. Only 10 subdistricts on the edge of the central city gained population and that gain was slight. Two of those (Longtan and Baizhifang) were at the corners of the Old City with relatively low densities in 1982. The remaining 8 subdistricts were all in Guanxiang, outside of the Old City.

3. Almost all suburban subdistricts gained population in 1982–1990, excluding 3 subdistricts (Yongdinglu, Gucheng, and Beixin'an). The fastest growth occurred in those subdistricts with proximity to the central city, which were the favorite destinations of out-moving central city residents. The newly designated 11 subdistricts indicate the expansion of the urbanized area.

4. Among the suburban subdistricts with population loss, Yongdinglu of Haidian District was due to the relocation of a large company. Employees were relocated at the same time, because most housing in Chinese cities is supplied by employers *(danwei)*. Similarly, in Gucheng and Beixin'an in the western suburb of Shijingshan District, many residents moved to nearby subdistricts where new housing units were built and supplied by employers.

Data at the subdistrict level suggest that suburbanization in Beijing is most active in suburbs close to the central city and slower in remote areas.

Evidence from Density Contour Maps

Using GIS surface modeling techniques, we have drawn density contour maps for 1982 and 1990 (Figures 18.4 and 18.5). In the maps, the density contours are in an interval of 12,950 persons per square mile, and the contour lines of 25,900, 77,700, 129,500, and 181,300 persons per square mile (10,000, 30,000, 50,000 and 70,000 persons/km^2 respectively) are highlighted in bold.

In the 1982 density contour map, we identified four peaks with a density greater than or equal to 129,500 persons per square mile: three in the central city, and one in the western suburb of Haidian District. The peak in the west was attributed to the high density in Yongdinglu subdistrict before the large company relocated. In the central city, the highest peak appears in Xuanwu, the second in Chongwen, and the third in Dongcheng. The Forbidden City and emperor's gardens (now museums and public parks) occupy a large area of land in the central city, forming a density trough in the middle of a U-shaped ridge.

The 1990 density contour map shows similar patterns, with changes summarized in the following.

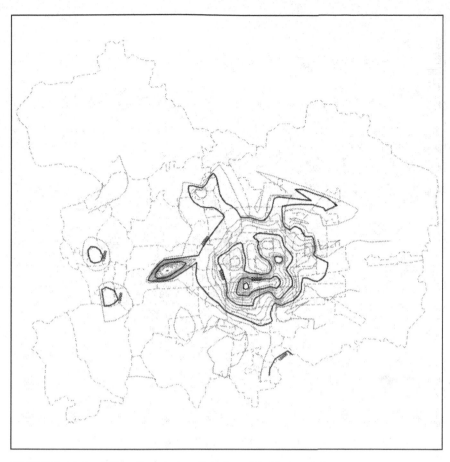

FIGURE 18.4   Urban Population Density of Beijing, 1982

1. Density peaks are flattened. Densities decrease significantly around the peaks. The 50,000 (129,500 persons/mi²) contour lines across the four central city districts in 1982 shrink to contain a smaller area in Xuanwu District in 1990. The 70,000 (181,300 persons/mi²) contour line disappears in the 1990 map. The density peak in the western suburb is no longer outstanding, and its density has dropped to less than 30,000 persons per square kilometer (77,700 persons/mi²) in 1990.

2. Areas of moderate density have expanded. The 10,000 contour lines extend outward. The expansions toward the north and northeast are most noticeable, as a result of the development of the Asian Olympic Village (*ya yun cun*) in the north and developments such as the master-planned Wangjing residential district along the Shoudou Airport Highway.

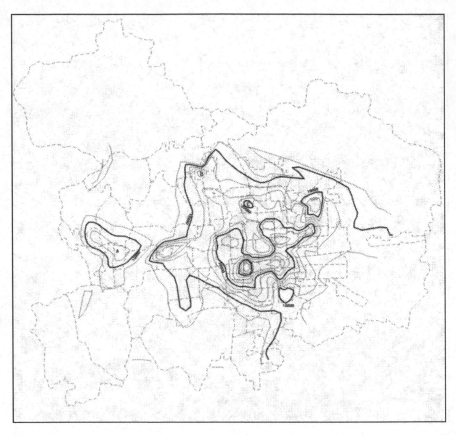

FIGURE 18.5   Urban Population Density of Beijing, 1990

3. Some suburban centers start to emerge (see Figure 18.5). Center A is where
   the Shijingshan Iron and Steel Company (one of the largest companies in
   China) is located. The company has grown significantly in the 1980s and
   has built many new housing units for its employees. Center B reflects the
   fast growth of high-tech industries centered in the Zhongguancun area—
   Beijing's university and intellectual area. Center C emerges due to the
   large-scale housing development project in Shuangyushu, the largest master-planned residential district in Beijing.

Figures 18.6 and 18.7 further depict the population patterns in 1982 and 1990
respectively in three-dimensional displays. All the above evidence suggests the beginning of suburbanization in Beijing since 1982.

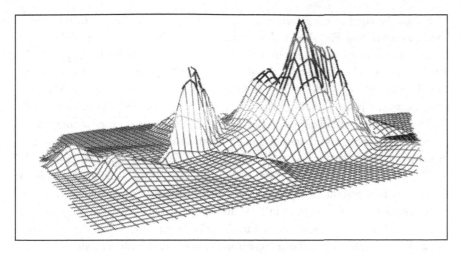

FIGURE 18.6   Urban Population Density of Beijing, 1982

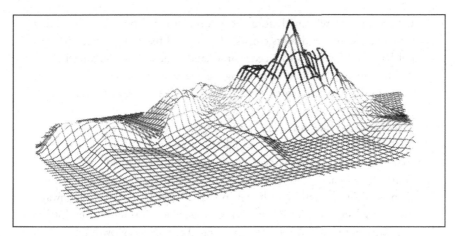

FIGURE 18.7   Urban Population Density of Beijing, 1990

Causes of Suburbanization in Beijing

Intensive suburbanization in America occurred in the 1950s and 1960s. The reasons are complex "but include transportation improvements, growing affluence, the rapid rate of urban growth, and growing dissatisfaction with the city itself . . . . The prospect of lower tax rates in the urban fringe also encouraged outward movement, as did the possibility for whites to escape from an increasingly minority-dominated central city" (Hartshorn 1992, p. 458).

Except the rapid urban growth, none of the above seems to apply to the suburbanization of Beijing. First, "the widespread adoption of the private automobile helped launch the age of mass-scale suburbanization" in America (Muller 1981, p. 38). In 1993, of the 1 million privately owned vehicles in China, only 50,000 were family cars. The commuting modes in Beijing in 1990 consisted of the following: 57.8 percent, bicycle; 31 percent, public bus; 4 percent, subway; and 5.9 percent, automobile (Li 1995). Beijing's suburbanization can hardly be attributed to private automobiles.

Second, American suburbanization was accompanied by growing middle-class families. In China, this class is very weak. The 1990 GDP per capita in China was only $317 (World Resource Institute 1994, p. 256). In addition, the urbanization ratio in China was 26.4 percent in 1990 (Hu and Foggin 1994a, p. 3), whereas the U.S. urbanization ratio was 64 percent in 1950 and 70 percent in 1960 (Hall 1977, p. 11).

Finally, although Americans moved to the suburbs for better schooling, tax benefits, and racial segregation, none of these is an incentive for suburbanization in Beijing. So what are the causes of suburbanization in Beijing?

1. Urban land-use reform. Before the reform, urban land values in China were not assessed, and rents consisted of a small fixed rate. After differential land rents were introduced, some land uses such as manufacturing, warehousing, and residential housing were replaced by more profitable tertiary businesses (retail stores, banks, hotels) in the central city of Beijing.

2. Rapid growth of the intrametropolitan highway network. In America, intraurban freeways built in the 1950s and 1960s accelerated suburbanization. In recent years, Beijing has widened and improved the Second-, Third-, and Fourth-Ring Roads and other urban arteries and constructed many new interchanges. Beijing's highways, though not as efficient as American highways, have played similar roles in improving accessibility to suburbs. Subdistricts near these beltways have all experienced rapid population growth.

3. Housing reform and central city renovation. Before the 1978 reform, housing in Beijing was quite poor. Residential areas in the central city were crowded, with an average living space of 10–16 square feet per person. In the 1980s, Beijing conducted many renovation projects by relocating central city residents to the suburbs, where new and more spacious housing units were supplied. About 20 percent of the residents involved in the central city renovation projects moved to suburbs (Lu 1994).

4. Improvement of suburban infrastructures and services. In the new political environment, suburban infrastructures (roadways, sewage, water, utility supply, and retail services) have been improved. The suburbs have become more livable.

TABLE 18.4   Comparison of Suburbanization in China and Western Countries

	Western Countries	China
Macro Background	Began in the 1920s and peaked in the 1950s and 1960s	Began in the 1980s
	Free market economy	Transition from a central planned economy to free market economy
	Growing affluence	Increasing per capita income, but much lower than in Western countries at the outset of suburbanization
	Urbanization level > 60%	Urbanization level < 30%
	Improvement of transportation	Improving transportation, but premature highway systems
Micro Impetus	Dissatisfaction with social and environmental problems in central cities	Central cities are still attractive, but their renovation projects lead to population relocation
	Widespread adoption of private automobiles	Just beginning use of privately owned automobiles
	Spacious houses in suburbs	Improved living space in suburbs to some extent
	Industrial and office parks were attracted by cheaper suburban land	Land-use reform forces central city industries to relocate, and service and financial businesses pay higher rents
Consequences	Middle-class and wealthy families moved out first	Rich can afford the housing in central cities and stay; working class moves to suburbs
	Residential decline and financial crisis in central cities	More rents and taxes are generated by service and financial businesses in central cities
	Transition from single to multiple functions in suburbs (from bedroom community to suburban downtown)	Improving infrastructures in suburbs, but still backward.

SOURCE: Zhou and Meng 1997.

The overlapping forces of out-migration of central city residents and in-migration of rural residents to suburbs has accelerated suburbanization in Beijing. Suburbanization has also been observed in other Chinese cities, including Shanghai (Ning and Yan 1995, p. 590) and Shenyang (Zhou and Meng 1997), with similar causes. Zhou and Meng (1997) have compared suburbanization in the West and in China and summarized the differences (see Table 18.4).

Conclusion

Based on data from the third and fourth censuses conducted in 1982 and 1990 respectively, suburbanization has indeed taken place in Beijing. Central city residents have started moving to the suburbs. Inner suburbs with proximity to the central city are the most favorable destination, where housing is more spacious and residents still have good access to central city services and jobs. Suburbanization in more remote suburbs is still rather moderate.

Although China shares some similarities with America's suburbanization experience, such as rapid urban growth, transportation improvement, and an increase in people's living standards, there are several differences. China's urban land-use reform has introduced differentiated land rents to the land management system and has driven old manufacturing and residential housing from central cities to the suburbs. Housing reform has provided more attractive housing developments in the suburbs. Economic reform has shifted the investment focus from defense-oriented heavy industry to consumption-related light industry and services, which has improved the suburban living environment. It is the new political and economic environments that have opened the path to suburbanization in China's large cities.

References

Clark, C. 1951. Urban population densities. *Journal of Royal Statistics Society* 114:490–494.

Hall, P. 1977. *The World Cities.* London: Weidenfeld and Nicolson.

Hartshorn, T. A. 1992. *Interpreting the City: An Urban Geography.* New York: John Wiley and Sons.

Hu, Z. L., and P. Foggin. 1994a. Comparison of urban population from two statistical measurements. *Urban Problems (Cheng Shi Wen Ti,* in Chinese) 13 (1):2–4.

_____. 1994b. Ringed variations of Beijing's population. *Urban Problems (Cheng Shi Wen Ti,* in Chinese) 13 (4):42–45.

Li, A. 1995. Cheer for Chinese Automobiles. *Beijing Youth Newspaper (Zhong Guo Qing Nian Bao,* in Chinese), February 8.

Lo, C. P. 1994. Economic reforms and socialist city structure: A case study of Guangzhou, China. *Urban Geography* 15:128–149.

Lu, X. 1994. Housing rehabilitation in Beijing. *Urban Planning (Cheng Shi Gui Hua,* in Chinese) 4:8–12.

Mills, E. S., and J. P. Tan. 1980. A comparison of urban population density functions in developed and developing countries. *Urban Studies* 17:313–321.

Muller, P. O. 1981. *Contemporary Suburban America.* Englewood Cliffs, N.J.: Prentice Hall.

Ning, Y., and Z. Yan. 1995. The changing industrial and spatial structure in Shanghai. *Urban Geography* 16:577–594.

Sit, V. F. S. 1985. Introduction: Urbanization and city development in the People's Republic of China. In *Chinese Cities: The Growth of the Metropolis Since 1949,* ed. V. F. S. Sit, pp. 1–66. Oxford: Oxford University Press.

_____. 1995. *Beijing: The Nature and Planning of a Chinese Capital City.* New York: John Wiley and Sons.

Wang, F., and Y. X. Zhou. 1999. Modeling urban population densities in Beijing, 1982–1990: Suburbanization and causes. *Urban Studies* 36 (2): 271–287.

World Resources Institute. 1994. *World Resources, 1994–1995.* Oxford: Oxford University Press.

Yan, X. 1995. Chinese urban geography since the late 1970s. *Urban Geography* 16:469–492.

Yeh, A. G., X. Xu, and H. Hu. 1995. The social space of Guangzhou City, China. *Urban Geography* 16:595–621.

Zhou, Y. X. 1996. Suburbanization in Beijing and some recommendations. *Scientia Geographica Sinica* (*Di Li Ke Xue,* in Chinese) 16:198–208.

Zhou, Y. X., and Y. C. Meng. 1997. Shenyang's suburbanization: Suburbanization comparison between China and the Western countries. *Acta Geographica Sinica* (*Di Li Xue Bao,* in Chinese) 52:289–299.

19

Village Transformation in Taiwan and Fujian

Ronald G. Knapp

Chinese rural settlements traditionally took shape over a lengthy period of time as a result of population growth, small-scale agricultural development, and popular beliefs concerning *fengshui*. Over the past half-century, however, more formal planning, top-down political decisions, and the broad economic policies of the State have played increasing roles in reshaping the Chinese countryside. On both Taiwan and the coastal mainland since 1949, quite different political and economic systems have acted to alter the geometry and morphology of villages. The appearance and function of the hybrid rural settlements that have emerged echo traditional village forms but are often neither rural, nor urban, nor suburban settlements. Using field observation, published survey data, and press reports, this chapter examines rural habitat changes across space and through time as well as the role of the State in reshaping villages in the People's Republic of China (PRC) and Republic of China (ROC). While information concerning rural housing and village environs has been collected annually on Taiwan since 1950, relevant and comparable survey data on the mainland began to appear only in the early 1980s.

Governmental policies in both Taiwan and the mainland immediately after 1949 included compulsory land reform programs that significantly decreased tenancy and increased the number of owner-cultivators. Although the implementation of land reform programs in the ROC and PRC differed, each undergirded the distinct socioeconomic development policies that followed. The intent of both land reform programs was improving rural living conditions through altering rural economic structure. However, living conditions actually worsened in many villages throughout much of the mainland over the next three decades, whereas they generally

improved dramatically in Taiwan. Rapid economic development in many areas of
the mainland after 1978, however, brought a building boom to the countryside,
transforming many dilapidated and once quiescent villages into vibrant settlements.
Indeed, as the twenty-first century begins, one observes a greater similarity in the
appearance of Taiwan and mainland villages than at any time since 1949. This seeming convergence has come about after decades of seeming divergence, but village
transformation, unfortunately, has not always been for the better.

Fujian and the Mainland

Several policies guided the physical changes of village habitats in the China mainland after 1949. Of signal importance was the rigid household registration *(hukou)*
system that bound villagers to their localities. Originally implemented to deter the
"blind movement" of peasants into China's cities and to facilitate centralized planning, the system actually strained life in the villages. As population grew and households subdivided, existing village dwellings were subdivided to provide minimal
shelter for increasing numbers of people. Makeshift sheds for kitchens and other
uses multiplied, contributing to crowded and unkempt villages during the 1950s.
Sometimes indirectly, often directly, political movements fostered these changes.
The "differentiation of class status in rural areas" that accompanied land reform led
to the classification of rural residents as landlords, rich peasants, middle peasants, or
poor peasants. Not only land but also houses and tools were confiscated from landlords and rich peasants and then redistributed to middle and poor peasants.
According to the Land Reform Law, ancestral shrines, temples, and the houses of
landlords "should not be damaged," but the "surplus houses of landlords . . . not
suitable for the use of peasants" were to be refitted into buildings for "public use"
(*Land Reform* 1950). Even so, many fine old residences, temples, lineage halls, and
ancestral shrines were destroyed. Representing sociocultural realities then in dis-
grace, many structures of this type were simply emptied—becoming mute shells—
as they were stripped of the traditional meaning and activity that had given them
life. Others were modified to serve such socialist utilitarian purposes as brigade
headquarters, storage, farm machinery repair, workshops, or schools (Knapp and
Shen 1991, 1992). Few Chinese villages grew in size or altered their appearance
because of the addition of new buildings between 1949 and 1957, a period of collectivization of agriculture, as little attention was paid to the construction of new
housing or even public structures. There is no doubt some peasants were able to
build new houses, but most simply repaired what had become ramshackled because
of decades of neglect. There was no focused national or local concern for the
improvement of ventilation, lighting, or sanitation of dwellings, or even the need to
reduce land occupied by housing in order to expand agricultural production. Rather
than a larger area of housing or functional improvements of occupied space, the
emphasis was on increasing grain production and the availability of basic consumer

items such as thermos bottles and enamel products that were the primary indexes of living standard advancement.

The Patriotic Sanitation Movement *(Aiguo weisheng yundong)*, begun in the early 1950s, was directed at the elimination of "the four pests" but was expanded in the 1960s to embrace environmental sanitation and the prevention of disease through the proper maintenance of manure pits and the elimination of unclean ditches and ponds. Attention to the cleanliness of village drinking water sources and toilets accompanied efforts to control rodents and mosquitoes. Model villages—where wells were capped, latrines and animal-manure pits covered, lanes swept, and pigs and poultry penned and separated from residential space—were highlighted in the press. The norm, however, was only a modest upgrading of village sanitary conditions. The frenzy of the Great Leap Forward, beginning in 1958, led to an accelerated destruction of old residences, lineage halls, ancestral shrines, and temples. This assault on structures was aimed, as earlier in the decade, at rooting out "feudal superstitions" and also, in the case of Fujian and other southern provinces, eliminating the power of local lineages by expunging architectural elements that expressed such domination. Lineages in many single-surname villages in Fujian had been weakened in the early 1950s when their landholdings were confiscated, while ritual practices had often been transferred to private homes (Huang 1989, 29).

To increase the efficiency of collectivized agriculture and military-style organization under the recently instituted people's communes, a variety of "modern" and purely functional structures were sometimes built. Using bricks and timber from demolished dwellings and other buildings, communal dining halls and dormitory-style accommodations were constructed in villages in northern China, but they were less common in Fujian and other areas of the southeast. Multipurpose facilities for assembly, lectures, recreation, and study sessions, however, were constructed in villages in Fujian just as elsewhere in China. Initially during this period, because emphasis was placed on agricultural production rather than consumption, villagers were explicitly dissuaded from investing in improving their housing stock. In recent years in China, it has been common to look back on the commune period as a time of excess and error, and indeed much waste accompanied overzealous efforts to meet unrealistic goals. However, much positive construction occurred during this time, especially the building of water conservancy facilities, bridges, and roads. Nonetheless, province-wide in Fujian, per capita cash incomes actually decreased by 22.6 percent to 104.43 *yuan* between 1966 and 1976 (Dangdai Zhongguo 1991, 130). This decrease from an already low base indicates the stress on meeting only villagers' basic needs, leading to a persisting deterioration of living conditions.

The popularization of Dazhai as a model for village development spurred changes in some Fujian villages just as it did elsewhere in China. In Fujian, well into the 1970s and 1980s, some second- and third-generation Dazhai-type villages further spread the message of egalitarianism and self-reliance in village construction. A common element of Dazhai-type villages was a geometrically regular village grid with consolidated placement of public/collective facilities such as assembly hall, pig sties, school, grain-drying areas, repair stations, and administration within a production brigade center. To conserve materials and space, new dwellings were usually barracks-style, with the same low height, depth, and length. Reflecting the prevailing restrictions on private plots, courtyards were especially small. Information on Dazhai-type villages in Fujian is extremely limited (Knapp and Shen 1991, 56–63).

With the death of Chairman Mao in the autumn of 1976, new political and economic options began to be considered. In Fujian, officials acknowledged the critical state of rural housing and the need to rebuild dilapidated villages. Echoing national concerns, new policies allowed villagers to invest their own labor and capital in house construction. The production responsibility system and, subsequently, the promotion of a commodity economy were accompanied by substantial increases in rural per capita incomes, reaching 171.75 *yuan* in 1980 before escalating to 396.45 *yuan* in 1985 (Dangdai Zhongguo 1991, 132). These numbers of course disguise village-to-village disparities.

As cash incomes increased, farmers quickly began to repair, renovate, expand, and even rebuild deteriorated dwellings. Between 1980 and 1985 in Fujian, one-third of all rural households built new houses. Most of these new houses were built adjacent to old houses or on original building lots, with little attention to village planning. Compared with the 1980 figures, per capita building area in 1985 increased by 61.7 percent to 181.2 square feet (16.86 square meters), 155.7 square feet of which was living space. One-third of new dwellings were multistoried, and 62 percent were of fired brick and wood (Dangdai Zhongguo 1991, 137). Locally available cut granite blocks and fired red bricks were the material of choice for single-story traditional-style houses in southern Fujian in the 1980s. Prefabricated cement building components—pre-stressed panels used as floor and roof sections, window and door frames, as well as stairs, lintels, and purlins—increasingly became common. Between 1978 and 1992 countrywide, the share of housing expenditures as a percentage of overall consumer expenditures more than tripled, while those of food, clothing, and fuel decreased (Yan 1994, 17–18).

Many new village houses in Fujian preserve elements of traditional dwellings, even as they employ modern materials and layouts. At ground level, dwellings are often U-shaped with a centered entryway through a gate. Termed *sanhe yuan,* this type usually has a three-bay *(jian)* central structure facing a courtyard. The middle *jian* is the traditional ceremonial focus of the dwelling, doubling today as a living room, with bedrooms on both sides. Duplicated above this traditional element usually is another, second and more modern set of rooms. The upstairs central room has become in many dwellings the equivalent of a family room, with television and other entertainment equipment. Upstairs bedrooms are breezier and generally cooler than the traditional downstairs bedrooms. On ground level and perpendicular to the three-*jian* core of *sanhe yuan*-type houses are two wings that provide space for a kitchen, bathroom, and storage. Stairs normally reach to the flat roofs of these wings and are used for household tasks such as drying vegetables. As is typical in many areas of China, there is a preoccupation with bigness in new houses in rural Fujian, as the scale of construction clearly goes beyond the current needs of households. Ground floor space in many houses is used for entrepreneurial activities, such as a small workshop, restaurant, or retail shop. Large and expansive dwellings not only serve to "store" wealth, possibly for future living space for as yet unmarried sons, they also declare a household's changing status in the community.

Because rural dwellings are often mere copies of nearby dwellings, there is a striking monotony of building styles. Piles of building materials and dwellings in various stages of completion give an unfinished and chaotic appearance to villages throughout Fujian province. All too often, as architects and engineers note, scarce building materials are wasted because of the poor skill levels of construction workers who have had little experience with the materials. Inexperience has led to overbuilding, the wasteful use of costly materials. Brick and reinforced concrete represented some 82 percent of building materials used in 1991 in rural housing construction (*Fujian jingji* 1992, 235). The standards for kitchens and indoor toilet facilities remain backward throughout the province.

Beginning in the early 1980s, there was growing concern for the amount of cultivable land being consumed by building. The need to gain control over the unregulated use of land and an awareness of deteriorating environmental quality prompted some important yet often crude preliminary planning exercises to be carried out in many rural towns and villages. Of 1,383 central villages *(zhongxin cunzhen)* in which master plans *(zongti guihua)* and construction plans *(jianshe guihua)* were completed, authorities noted the quality of those of Haixing village in Changleyingqian *zhen* and Yangtan village of Qiujiang *xiang*, Sha *xian* (Dangdai Zhongguo 1991, 138). Similar exercises carried out elsewhere involved surveying and mapping, inventorying of facilities, assessing building quality, and recommending planned development. By the end of 1984, 39 percent of Fujian villages reported having completed a planning exercise with the year 2000 as the target for implementation (Yuan 1987, 191).

For most of the 1980s, however, neither policies, planners, nor those who enforce regulations were able to keep abreast of the illegal and unapproved upsurge of rural construction. Architects and others participated in design competitions to improve the layout, ventilation, and lighting, popularized the use of appropriate materials, and guided attention to village infrastructure projects such as drainage, running water, and transportation facilities, as well as amenities such as cultural centers. All too often, unfortunately, there was a neglect of community facilities as villages underwent a transformation from a redistributive to an increasingly market-oriented economy. Nee and Su noted that those Fujian communities that were relatively affluent were able to employ income from village collective enterprises, public welfare funds, and public accumulation funds to continue to support and build schools, health care and recreational facilities, expand drinking water and electric service and less visible village improvements, as well as invest in agriculture and industry (Nee and Su 1990, 18–23). Poorer villages, on the other hand, frequently failed to initiate village improvements, even as they let the fruits of collective actions during the commune period deteriorate.

In general, efforts at village planning were frustrated by the proliferation of small-scale factories and workshops that multiplied in the wake of the economic boom throughout coastal southeast China. Often using old equipment and obsolete technologies, many township and village enterprises (TVEs) came to strain existing infrastructure and pollute groundwater and the air with their effluents. From 1988 to the present in China, the cooling of an overheated national economy has helped restrain general consumption. Furthermore, increasing costs of building materials and the ability to enforce land-use regulations have contributed to the reduction of "building fever" *(jianfang re)*, as it was termed in the Chinese press. This slowdown has provided the hiatus needed for planners to catch up with and perhaps correct recent blatant excesses. Environmental issues in and around Fujian's villages, as elsewhere in China, must be addressed so as to avoid the poisoning of the countryside, a subject all too well understood in Taiwan, just ninety miles across the Straits.

Taiwan

The retreat of the Nationalist government to the island of Taiwan in 1949 set in train pragmatic policies that transformed the island in less than half a century from a developing country with a predominantly agricultural economy to a developed one with a mature industrial sector and a prominent role in the global economy. Among the policies facilitating this transition were a three-stage land reform program, broad improvements in agricultural production, a strategy of import substitution, and finally the sustained development of export-oriented, capital- and technology-intensive industrialization. Between 1953 and 1990, agriculture's share of the net domestic product decreased from 34 percent to 4 percent, and the percent of farm households fell from 51 percent to less than 20 percent. In the early 1950s, land reform was accompanied by the reduction of rents and the sale of farmland to former tenants. The net effect of these policies was a redistribution of income, new opportunities for mobilizing domestic savings, and expanded off-farm employment opportunities, as agricultural output by owner cultivators became increasingly diversified, intensive, efficient, and highly productive (Thorbecke 1992). With these profound shifts, substantial improvements occurred in overall rural and urban standards of living, including a transformation of housing and village environments (Fields 1992). Not all of the changes have been for the better, however.

An especially important facilitator of the integrated approach to rural development was the Chinese-American Joint Commission on Rural Reconstruction (JCRR). The development focus of JCRR included agricultural extension work, financing, agricultural research, farm management, crop and livestock improvement, soil conservation, development of water resources, promotion of farm mechanization, rural health improvement, agricultural cooperativization, and organizing the successful farmers' associations (Shen 1970). In the decades following land reform, the size of farms in Taiwan remained quite small, averaging only 1.1 hectares, thus limiting the income potential from agriculture for farm households. During the 1960s and 1970s, as industrialization quickened, substantial migration occurred from the countryside to the cities as farm labor increasingly became redundant. Countering these flows during the late 1960s, however, industry began to disperse to the countryside to tap the underemployed village labor. Off-farm part-time employment in factories and workshops within villages and small towns grew throughout the 1970s and 1980s. This sectoral shift contributed significantly to raising farm household incomes for villagers who increasingly became part-time farmers while maintaining their residence in the countryside. The bilateral nature of JCRR was terminated in 1979 when diplomatic relations between the United States and the Republic of China were broken. Subsequently, JCRR was reorganized into the Council for Agricultural Planning and Development (CAPD) and in 1984 into the Council of Agriculture.

An Intensive Village Improvement Program, sponsored by JCRR, was initiated in five areas in 1957 and then expanded to fifty-seven demonstration villages in 1960. Employing successful demonstration villages as models, this comprehensive grassroots rural development program set out to improve living conditions, including sanitation, nutrition, and health, education and culture, household practices, transportation and communications, as well as recreation and social welfare (Chinese-American JCRR 1960). An important index of improvements in village living standards was the presence of concrete areas used to dry paddy rice so that it could be safely stored. Popularized also were techniques for small-scale processing and preservation of agricultural products such as vegetables, fruits, and meat. Among various sideline activities promoted were embroidery, basket-making, artificial flower making, duck raising, and the weaving of straw bags and shoes. Narrow roads were widened and culverts and bridges were built. Nurseries and kindergartens were set up to provide child care for women who were active in the local farm and nonfarm economy.

Within each household, special attention was paid to improving sanitation and personal hygiene. Among the unsatisfactory conditions that had been widespread were "towels and toothbrushes used in common by the family members, windows too small or even no window at all, pigsty connected with kitchen, cattle pen adjacent to bedrooms, uncovered dirty latrine, fowls excreting everywhere, garbage in corners, drainage ditches full of dirty water, and darkness, dampness, mosquitoes and flies over the whole building." Among problems noted in correcting these living conditions were beliefs that "towels can be hung only on lucky days. The installation of a stove door to save fuel will insult the kitchen god, and the nailing of the stove door may cause abortion in a pregnant woman" (Chinese-American JCRR 1960, 28–29).

Correcting these unsatisfactory behaviors was facilitated by the training of local housewives as change agents. Larger-scale projects, such as water storage tanks, deep wells, public and private latrines, drainage ditches, improved kitchens, and community compost shelters were coordinated by JCRR and completed with village labor as community projects. Rural activity centers *(nongmin huodong zhongxin)* were constructed to facilitate cultural, educational, welfare, and recreational activities. Comprehensive rural development efforts based on model demonstration villages were key elements in the transformation of the lives of Taiwan's villagers.

Ten years after the completion of the Land-to-the-Tiller program, an extensive study assessed the effects of land reform on people's lives (Yang 1970, 261–276). Questions were directed to former tenant-farmers, current tenant-farmers, original-owner farmers, farm laborers, former landlords, and nonfarm people. Detailed information about changing patterns of food consumption, clothing, communications, health care, potable water, and housing were collected. Although two-thirds of the former tenants expressed betterment in overall living conditions, all other groups reported less improvement, with former landlords, predictably, declaring an overall deterioration.

The repair, remodeling, or rebuilding of deteriorated housing stock, much of which was near collapse, was substantial by 1964 but varied according to household status. Of former-tenant (now owner-cultivator) households, 39 percent carried out repairs with 38 percent building new dwellings—a clear indication not only of the sorry state of existing houses but also the newfound wherewithal of villagers to invest in changes. Sixty-two percent of current tenants rebuilt or repaired their houses. Repair work principally involved fixing roofs, often replacing straw with fired roof tiles. Some 40 percent of original owners built new houses, with the remainder carrying out substantial repairs. Few former landlords reported any change in housing since they generally continued to live in their original well-appointed brick and tile houses. For other groups, new dwellings were generally in the traditional single-story red brick and roof tile style. Little improvement generally was made in the interior walls; most remained bare without being whitewashed or covered with paper. Some new two-story dwellings and even a small number of thatched adobe dwellings were reported built during this period. Rural households stated that new construction was prompted by the fact that old dwellings were near collapse, the need to meet the requirements of growing children, or massive destruction by typhoons.

Observers of rural Taiwan in the 1950s and 1960s focused on the environments within and around dwellings as the indexes of village modernization and economic well-being. Tamped earth floors of dwellings decreased only slightly from 88 percent of all sampled farm households in 1952 to 76 percent in 1958. Brick or cement floors were a qualitative yet quite costly improvement that was adopted slowly. Inside privies doubled from only 5 to 10 percent in the same period, being far outweighed by outside privies and inside chamber pots and night soil pits. While nearly one in five rural households kept farm animals within the dwelling in 1952, this was more than cut in half by 1958 (Kirby 1960, 150, 152–153). Yang observed that "a bright, clean home, neatly arranged, with decoration are signs (sic) of the upper [socioeconomic-cultural] levels, whereas, dark, unclean, disorderly, or bare rooms reflect low standards." Yet, he continued, there is

> a kind of irrational agrarian axiom that was in direct contradiction to this general rule. It holds that the truly prosperous farm home will have a yard full of chicken droppings, pig manure, and all kinds of refuse, and inside the house there will be articles, utensils, etc., scattered or piled up everywhere, without any order or arrangement. This disarray was supposed to imply that the people of the household were so busy with farm work and cooking that they hardly had time to do any cleaning and arranging. It is only a poor family, which does not have a good farm to work on, does not have pigs and chickens to keep and feed, nor many children to raise and discipline, that can afford the time, energy and thought to cleaning and decorating. Thus, in an inverse way, cleanliness implies a scarcity of the things that a normal, or prosperous farm household ought to have. Should this tradition still prevail in any rural district, then a dirty, disorderly, undecorated home there may not necessarily mean a low-living standard. In general, especially nowadays, however, the expected rule holds. (Yang 1970, 294–295)

Streams, small ponds, and deep wells traditionally were the sources of water for Taiwan's villagers. With land reform, more hygienic sources began to be employed, such as artesian wells or village reservoirs connected to village dwellings with crude bamboo "pipes" and later plastic hoses. In some villages, public faucets were a temporary improvement as villagers continued to carry water from source to home. In the early 1960s, some 61 percent of former-tenant households still depended upon "primitive" means to obtain water (Yang 1970, 289–293). Township governments were important in initiating the delivery of potable running water.

Yang observed that "the kitchen was the most gloomy and disorganized place" in rural dwellings. Surprisingly, after a decade of educational extension work, only 30 percent of rural households reported kitchen improvements. Only 1 percent of rural dwellings in 1962 even had screens on doors and windows (Burke 1962, 11). Chiang Ching-kuo's several visits to the countryside in the early 1970s, as well as his widely reported comments in 1979 and 1980, underscored the continuing erratic pace of change.

In 1979, some 12 percent of Taiwan's remaining housing stock had been built before 1945, but by 1988 this percentage had been reduced to only 5.2 percent. As recently as 1982, some 24 percent of dwelling units island-wide were single-story red brick farmhouses. Traditional dwellings are most common in central and southern Taiwan, such as in Yunlin and Tainan prefectures, where in 1982 they still accounted for 57 percent and 49 percent, respectively, of all dwellings (Council for Economic Planning and Development 1982, 70). Here, too, as elsewhere throughout Taiwan, row houses and semidetached dwellings came to contribute to the architectural diversification of village landscapes. Whether old-style or new-style, most residences at the end of the 1980s dated only from the period 1972–1981 (Council for Economic Planning and Development 1982, 102). Traditional building materials such as clay, bamboo, and wood have been universally replaced by fired bricks as well as mixed brick and reinforced concrete construction in Taiwan, yet even within the highly urbanized core of Taipei municipality one still can encounter an occasional relict farmhouse.

Since the 1960s, the dispersal of small-scale workshops and businesses into the villages has spurred off-farm employment and led to rising rural incomes. Continuing even to the present, this dispersal has dramatically affected the physical appearance of Taiwan's countryside. Not only have traditional single-story dwellings been modified to meet entrepreneurial needs, they have been supplemented with multistory residences, countless retail shops, small factories and workshops, recreational structures, and religious buildings—all encroaching on adjacent arable land in a kind of village sprawl. In Hsin Hsing village in Changhua County in the early 1990s, for example, some 90 percent of household income was derived from industrial and commercial sources rather than agriculture. Dilapidated dwellings, sometimes abandoned or rarely used by households who had moved to cities, stand neglected. Modern consumer amenities—telephones, color televisions, refrigerators, VCRs, washing machines, propane gas stoves—proliferate in village houses. Crude animal pens and compost pits, once common in Taiwan's villages and the source of farmyard smells and uncleanliness, have all but disappeared. Today, the pollution is from the ephemera of manufacturing—plastic, paper, and glass refuse:

> Piles of materials used in the production process are heaped on porches of dwellings. Lean-tos, constructed of wood, tin, or tarpaulin, conceal parts of homes; their contents, which range from decaying bamboo baskets to manufacturing materials and from unused bicycles to motorcycles, often spill out into the areas they adjoin, adding to the disorder. Farm implements, such as plows and harrows, abandoned to the weather, lie rusting in alcoves and courtyards. Cartons packed with products destined for local or foreign markets stand in rows waiting for the trucks that will carry them to their destinations. Motorcycles, cars, and trucks—all signs of relative affluence—block the pathways and yards in front of the buildings. (Gallin and Gallin 1992, 290)

Much pollution, however, is not simply a relatively benign, incidental deterioration of a rural environment's aesthetic character. Farms and factories, both legal and illegal, have become competing neighbors over the past three decades. A survey commissioned in 1987 by Taiwan's Environmental Protection Administration revealed that 15 percent of farmland island-wide was polluted by heavy metals washed into the soil by wastewater, much derived from the decentralized nature of electroplating workshops and other small factories dispersed in the countryside in the 1960s and 1970s. Using nickel, chrome, and zinc, small electroplating workshops add an important step in giving strength and luster to a wide array of items from belt buckles to computer components. There is also evidence that some of Taiwan's pig farmers have added zinc and copper to animal feed to promote growth, in the process contributing toxicity to the animal excrement used for fertilizer. These insidious types of pollution are relatively invisible but extremely dangerous, sometimes even leading to the abandonment of farming because of the fear of growing and then eating "cadmium rice and vegetables." Ironically, the deliberate polluting of village groundwater sometimes led to the poisoning of paddy fields in order to have the land rezoned for building purposes that command a higher market price than farmland alone would. Small rural factories—whether because of undercapitalization and thus an inability to observe existing regulations or because of a fixation on massing wealth—have been major polluters in rural Taiwan, fouling not only productive fields but also village residential environments, including sources of fresh drinking water and adjacent fields.

Attention to upgrading village sanitation and householder hygiene continues to be popularized by government programs and in widely available handbooks such as *Taiwan nongjia yaolan* (Farmer's Guide). Running water and electricity are found in virtually all rural dwellings throughout Taiwan today. Although there has been improvement in toilet facilities, rudimentary outdoor toilet facilities and kitchens are found throughout the countryside. Still, into the 1980s, criticisms of standards of ventilation and lighting, of outdoor toilets, as well as bathing facilities in villages persisted (Luo 1984, 31). As with rural industries, residential wastewater and sewage throughout the countryside generally drain untreated directly into canals and rivers, seeping eventually into the groundwater from which drinking water is drawn.

From the vantage point of formal planning mechanisms and enacted statutes, Taiwan appears to be a model for preventing environmental problems and balancing economic growth with environmental protection. Yet, the reality has been quite different. Taiwan does not lack for plans, whether island-wide, such as the current Six-Year National Development Plan, called Challenge 2008, regional plans, such as those for the east, south, west, or north, or local schemes, such as those of individual counties. The urbanization of the island over the past half-century has led to intense real estate speculation as village farmland has been transformed into industrial, residential, and recreational needs for the mushrooming urban population. There is a continuing disregard for building codes and plans all over the island. Construction consistently outruns whatever policies and plans have been implemented to regulate development.

Even with elevated levels of public consciousness, battles must still be waged even in cases where the issues seem all too obvious. A prime example in the late 1980s and early 1990s was the development of Taipei's last remaining natural but environmentally fragile ecosystem, the Kuantu plain. Only twenty kilometers from downtown Taipei, the 1,000 hectares of the Kuantu plain offered an uncommonly rare opportunity to develop the area while paying attention to conservation. The scope of the project includes not only a national park and nature preserve but also substantial public and private housing as well as sports facilities. As so often in the past, government planning was carried out without the involvement of local residents, who viewed the emerging project not only with suspicion but also dissatisfaction with the likely amount of compensation the government would make for appropriated land. Yet, what appeared to be a rather straightforward conflict between "bad" developers and "good" residents took on a different color when it became clear that local farmers themselves had literally poisoned the agricultural environment with their abusive overuse of chemical pesticides (Lee and So 1999).

Fujian and Taiwan at the Threshold of a New Century

Whatever similarities there had been in the traditional rural landscapes of Fujian and Taiwan, they were dramatically altered in the three decades after 1949. The transformation of villages on the mainland resulted primarily from campaigns that erased extensive portions of the architectural and spatial heritage that had accumulated over hundreds of years. Very few new structures built in China's villages during this time added anything of substantial form or meaning to the landscape. Furthermore, the swelling of the rural population so increased the pressure on living and working space that buildings throughout the countryside markedly deteriorated. On Taiwan, at the same time, houses and village patterns also changed due to government policies, but the policies did not directly "attack" traditional structures in order to obliterate them as on the mainland. Rather, economic policies led to a decentralization of labor-intensive industrial activity and the intrusion of small-scale workshops into once primarily agricultural settlements. Newly built multistory dwellings often were placed adjacent to traditional red-brick farmhouses, resulting in a village-scape that mixed the past with the present in a jarring fashion. Increases in village wealth often led to the refurbishing or rebuilding of village temples and shrines, elevating rather than diminishing traditional landscape features. The past two decades of mainland economic reform, on the other hand, has strikingly altered the face of rural Fujian. Today, as Taiwan and Fujian stand on the threshold of the twenty-first century, the dwellings and other structures found in their villages are decidedly more similar than they have been in recent times.

Farmhouses in both Fujian and Taiwan continue to be owned by the households occupying them. House plots in Taiwan are also owned, while on the mainland the land occupied by a dwelling remains in collective, usually village, ownership. Renting of residences is not common in either. For individual households in both places, the dwelling provides a vehicle that can be used to store wealth as well as provide opportunities for earning cash incomes outside of agriculture. On the mainland, where ownership of farmland is not an option for investment by farm households, the dwelling structure with its manifold potential uses offers a ready locus for entrepreneurial investment.

There has been an evolution in building materials in both Taiwan and Fujian over the decades. Households with growing incomes purchase higher-quality traditional materials such as fired brick and tile rather than adobe brick or bamboo and also use cement, pre-stressed concrete, asbestos, glass, and metal for walls, posts, floor, ceilings, roofs, as well as window and door frames. Specialists continue to observe in both Taiwan and Fujian that materials are often used wastefully because of lack of experience and poor design. Multilevel dwellings throughout China continue to be built without the benefit of architectural or engineering expertise. Village low-rise buildings up to three or four stories, it must be admitted, are remarkably forgiving of shortcomings in their construction, a fact that has fueled their erection, yet news of structural flaws leading to dangerous deformation and even collapse frequently appear in both the Taiwan and Fujian press.

Throughout the 1990s, the Taiwan media uncovered a disturbing trend in the use of dangerous building materials: radioactive steel, sea gravel, and unwashed coastal sand. While tests revealed radiation contamination affecting health only in large apartment buildings where radioactive steel was used in pre-stressed concrete, more widespread has been the use of sea gravel and unwashed sand throughout urban and rural Taiwan. Sea gravel and coastal sand have a high salt content that is said to be sufficiently corrosive to cause concrete to disintegrate and crumble, resulting in the collapse of multistory structures. Industry sources themselves estimate that perhaps 40 percent of Taiwan's construction up to the mid-1990s utilized sea gravel, while some civil engineers even claim that the number may be as high as 80 percent (Chan 1994, 4). Paradoxically, the illegal sale and use of sea gravel and sand surfaced as environmentalists were successful in limiting the excavation of riverbank gravel. It is likely that the irresponsible and illegal use of unsatisfactory building materials in Taiwan foreshadows similar practices on the mainland, where even today the quest for building materials far outstrips supply. As disposable incomes rise, villagers on the mainland and in Taiwan are able to identify new standards of building quality, yet the reality of regulations and enforcement suggests that much construction will continue to be of poor—even dangerous—quality. As a population long starved for decent habitation sets out to meet its expanding needs—pursuing ever rising expectations about building materials, water, lighting, hygiene, furnishings, and leisure—all too often, as admittedly is true even in the West, limited resources are frequently squandered and gross mistakes continue to be made.

Within all too many new and old Chinese dwellings, the differentiation of space according to use is not clear. In many rooms, farm equipment and stored materials lie aside the family's furniture. This condition is often true even when the household has sufficient space to separate working from living space. Indeed, new rural dwellings in Taiwan and Fujian are usually generous in size, significantly larger than a household would appear need. The flexible use of space, of course, was a traditional value and virtue of farm dwellings, developing out of necessity because of limitations in resources, but today lingers more out of habit than design. Building multistoried dwellings as residences in the countryside generally has meant that the ground floor—which must be entered first—often is the most cluttered portion of the dwelling, a mixture of attic and garage at the dwelling's front door. Overcoming such problems will come about as a result of improvements in design as well as changes occurring in people's subjective sense of aesthetics.

Changes in household composition in both Fujian and Taiwan have led to a reduction in the number of dwellings housing several generations under one roof. Where once there might have been a ramified structure with wings and stoves added to meet the needs of a growing extended family, today the pattern is separate space for each nuclear family as the total number of such households increases in both Taiwan and Fujian. Between 1964 and 1990, average rural household size in Taiwan contracted dramatically from 7.67 to 4.66 persons (Directorate-General of Budget 1990, 30). In Fujian, there was an increase from 4.69 persons in 1964 to 5.24 in 1973, with a subsequent decrease to 4.67 persons per farm household in 1992 (Chuan and Chen 1990, 270–271; Guojia 1993, 204). Today, rural household size in Taiwan and Fujian is quite similar, both approximately 3.6 persons, and households in both areas are essentially nuclear in form. The overall scale of dwellings necessary to house such relatively small nuclear families, by contrast, continues to expand in both areas as family expectations and resources swell.

The pollution of the countryside in Taiwan and Fujian includes not only the relatively benign visual pollution of clutter and disorder but also the widespread contamination of groundwater and soils because of the dispersal of polluting industrial and agricultural activities. The environmental and social consequences of accelerating industrialization especially are now being acknowledged in Taiwan as a result of the actions of a growing and active environmental movement. The farmer's "rebellion" against Dupont in 1986–1987 in Lukang in Changhua prefecture was a turning point in public awareness of toxic issues in Taiwan, and such efforts continue to increase. No such raised consciousness is apparent in rural Fujian, where the pace of industrialization indeed is accelerating with marked environmental deterioration, including the pollution of farmland and groundwater. It is ironic, but not necessarily surprising, that many of the small workshops in Fujian that are polluting groundwater and air with toxic effluents are subsidiary outposts of Taiwan concerns that must look beyond the island in order to avoid the vigilance of environmental monitors. Still, the "anarchy of development" continues in Taiwan, even as its excesses are exported across the Straits to Fujian and other southeast mainland provinces (Arrigo 1994).

Laws and regulations to control air and water pollution as well as solid and hazardous waste disposal continue to be enacted in Taiwan and Fujian but with little overall effect. There is some evidence indeed that many citizens in Taiwan "have no faith in the government's ability to solve the problems" and are taking issues into their own hands (Huang 1996, 6). In the mid-1990s, for example, rural residents in the periurban areas of Taipei and Kaohsiung in Taiwan sealed off landfills near their villages in order to bar industrial and household waste from being dumped in their environs. The multiplying waste stream of the island's increasingly affluent urban population had brought unsatisfactory levels of secondary pollution to rural environments because of seepage from poorly managed landfills. Some believed that the long-awaited implementation of the Environmental Impact Assessment Law in 1995 in Taiwan has offered some hope that unbridled and insensitive development could be constrained and that sustainable development was now possible. More recent laws and regulations affecting soil, groundwater, toxic chemicals, and waste disposal continue to offer hope. Across the Straits on the mainland, "a new criminal code issued on October 1, 1997, classified destruction of the environment as a crime to be seriously punished," but this threat is largely a hollow one inasmuch as environmental protection throughout most of China remains at the stage of admonition (Ding 1998, 12).

Villages in Taiwan and Fujian are no longer simply nucleated or dispersed clusters of peasant dwellings. Almost universally in Taiwan and increasingly in Fujian, villages are transitional settlement forms in which residents are both farmers and factory workers. The well-developed transportation system in Taiwan and proximity of the countryside to major urban centers accelerated these changes. In Fujian, the change process started later and is less pervasive, but the trend is inescapably observable—the morphology and function of villages are transforming. In both areas, the resulting environmental damage and stress underscore the underdeveloped nature of rural infrastructure and the ease with which those who wish to can operate beyond the control of the State. Governments at several levels in both Taiwan and the mainland are attempting to guide village development by using planning principles and regulations, but these efforts have only been partially successful. Regulations and education have up to now had only limited impact because of poor enforcement and low levels of community concern. Jack F. Williams's observation about Taiwan a decade ago applies to recent conditions in Fujian as well:

> The environmental crisis cannot be put entirely on the shoulders of the government or business. The people have been collaborators, knowingly or unknowingly, in the process of environmental degradation. In their blind pursuit of wealth, in their fixation on the welfare of their families, at the expense of public welfare and social good, individuals in Taiwan are the guilty parties in creating the mess with which the island now struggles. (Williams 1992, 196)

All too few residents in the villages and towns of both Taiwan and Fujian express any awareness of the escalating destruction of any part of their architectural heritage. Driven by increasing wealth and expanding economic opportunities for their residents, many villages in both areas have become patchworks of intruding structures and activities that collectively extinguish whatever bucolic character there was to the traditional rural settlements they are replacing. One must guard against romanticizing or sentimentalizing traditional village environments, but one must also recognize an obligation not to summarily destroy what took centuries to take form.

Conflicting national policies, lack of coordination between public and private efforts, as well as contradictory and ill-enforced regulations at the local level together contribute to the continuing reckless deterioration of Chinese village environments.

Reasonable land-use decisionmaking has been less the focus of attention in Taiwan than has been unbridled land speculation. This tendency has all too clearly emerged recently in Fujian in the wake of economic reforms and openness to Taiwan's commercial interests. While segments of the public in Taiwan have become increasingly vocal about the need for balanced development with only limited results, such voices in Fujian unfortunately are still largely silent. The illegal building of polluting factories, the mining of agriculture land to make bricks, the spilling of untreated industrial and domestic wastewater and by-products, the spewing of noxious gases and noises, the drawing down of groundwater resources, the erecting of new residences, among many other components of "development," have not only impacted agricultural production but also altered the nature and quality of rural habitats. Some villagers are now aware of the public health and ecological costs of practices thought necessary in the past for progress and are awakening to issues once ignored. Observations of future scenarios range from cautious optimism to the dire foreboding of unavoidable and cataclysmic harm. If there is a lesson from Taiwan's experience with improving rural living environments that is relevant to developments in Fujian, it is that shortsighted decisions made today are not so easily corrected tomorrow.

Note

This is an updated version of an article that appeared in *The China Quarterly* 147 (September 1996) and appears with permission granted by Oxford University Press. Research for this paper was supported by a grant from the East Asia Program, Cornell University (Peter Halpern Associates Fund).

References

Arrigo, L. G. 1994. The Environmental Nightmare of the Economic Miracle: Land Abuse and Land Struggles in Taiwan. *Bulletin of Concerned Asian Scholars,* Vol. 26, No. 1.

Burke, J. T. 1962. *A Study of Existing Social Conditions in the Eight Townships of the Shihmen Reservoir Area.* Taipei: Chinese-American Joint Commission on Rural Reconstruction.

Chan, V. 1994. Fixing Flaws in Building Sector. *Free China Journal,* Vol. 11, No. 21.

Chinese-American Joint Commission on Rural Reconstruction. 1960. *Taiwan sheng jiceng minsheng jianshe shiyan nongcun* (Intensive village improvement in Taiwan). Taipei: Chinese-American Joint Commission on Rural Reconstruction.

Chuan, Z., and J. Chen, eds. 1990. *Zhongguo renkou—Fujian fence* (The population of China—Fujian volume). Beijing: China Urban Economics Press (Zhongguo shizheng jingji chubanshe).

Council for Economic Planning and Development. 1982. *Dushi ji quyu fazhan tongji huibian* (Urban and regional development statistics). Taipei: Executive Yuan, Council for Economic Planning and Development, Housing and Urban Development Department.

Dangdai Zhongguo Congshu Bianjibu (Contemporary China Series Editorial Board). 1991. *Dangdai Zhongguo de Fujian, Shang, Xia* (Contemporary Fujian, Parts I and II). Beijing: Contemporary China Press (Dangdai Zhongguo chubanshe).

Ding, M. 1998. The Menace of Environmental Pollution. *China Today,* Vol. 47, No. 8:10–13.

Directorate-General of Budget, Accounting, and Statistics. 1991. *Report on the Survey of Personal Income Distribution in Taiwan Area of the Republic of China, 1990.* Taipei: Executive Yuan, Directorate-General of Budget, Accounting & Statistics.

Fields, G. S. 1992. Living Standards, Labor Markets, and Human Resources in Taiwan. In *Taiwan: From Developing to Mature Economy,* ed. G. Ranis. Boulder: Westview Press.

Fujian jingji nianjian, 1992 (Almanac of Fujian's economy, 1992). Fuzhou: Fujian People's Press.

Gallin, R. S., and B. Gallin. 1992. Hsin Hsing Village, Taiwan: From Farm to Factory. In *Chinese Landscapes: The Village as Place,* ed. Ronald G. Knapp. Honolulu: University of Hawaii Press.

Guldin, G. E. 1997. *Farewell to Peasant China: Rural Urbanization and Social Change in the Late Twentieth Century.* Armonk, N.Y.: M. E. Sharpe.

Guojia tongjiju nongcun shehui jingji tongjisi, ed. 1985-. *Zhongguo nongcun tongji nianjian* (Statistical yearbook of rural China). Beijing: China Statistics Press.

Huang, Shu-min. 1989. *The Spiral Road: Changes in a Chinese Village Through the Eyes of a Communist Party Leader.* Boulder: Westview Press.

Huang, Wen-ling. 1996. What a Waste. *Free China Review,* Vol. 46, No. 6:4–9.

Kirby, E. S. 1960. *Rural Progress in Taiwan.* Taipei: Chinese-American Joint Commission on Rural Reconstruction.

Knapp, R. G., and D. Shen. 1992. Changing Village Landscapes. In *Chinese Landscapes: The Village as Place,* ed. R. G. Knapp, pp. 47–72. Honolulu: University of Hawaii Press.

_____. 1991. Politics and Planning: Rural Settlements and Housing in Contemporary China. *Traditional Dwellings and Settlements.* Working Paper Series, No. 29, pp. 1–45. Berkeley: University of California.

Land Reform Law of the People's Republic of China. 1950. Beijing: Foreign Languages Press.

Lee, Yok-shiu F., and A. Y. So, eds. 1999. *Asia's Environmental Movements: Comparative Perspectives.* Armonk, N.Y : M. E. Sharpe.

Luo, H. 1984. Taiwan diqu nongcun zhuzhai ji shequ gengxian guihua zhi yanjiu (Research concerning planning for the renewal of housing and social areas in Taiwan). *Taiwan jingji* (The Taiwan Economy), Vol. 95.

Nee, V., and S. Su. 1990. Institutional Change and Economic Growth in China: The View from the Villages. *Journal of Asian Studies,* Vol. 49, No. 1.

Shen Tsung-han. 1970. *The Joint Sino-American Joint Commission on Rural Reconstruction.* Ithaca: Cornell University Press.

Thorbecke, E. 1992. The Process of Agricultural Development in Taiwan. In *Taiwan: From Developing to Mature Economy,* ed. G. Ranis. Boulder: Westview Press.

Williams, J. F. 1992. Environmentalism in Taiwan. In *Taiwan: Beyond the Economic Miracle,* ed. D. F. Simon and M.Y.M. Kau. Armonk, N.Y.: M. E. Sharpe.

Yan, Y. 1994. Better Life for Farmers. *Beijing Review,* Vol. 36, No. 52.

Yang, M.M.C. 1970. *Socio-Economic Results of Land Reform in Taiwan.* Honolulu: East-West Center Press.

Yuan, J., ed. 1987. *Dangdai Zhongguo de xiangcun jianshe* (Contemporary rural development in China). Beijing: Modern China Press (Dangdai Zhongguo chubanshe).

PART THREE

Changes Along China's Periphery

20

China's Changing Boundaries

Chiao-min Hsieh

China has been called the oldest nation. Founded by the Shang dynasty in the second millennium B.C., it has had an organized society for a longer period of time than any modern state. However, China's boundaries have varied greatly from one historical period to another.

The Chinese have never regarded China's boundaries as determined either by geography or history. As some scholars have noted, even in periods when China was weak and unable to resist invasion, the Chinese always have felt that territory once occupied for civilization must not return to barbarism; therefore, territory that was once Chinese must forever remain so, and if lost, must be recovered at all costs. The growth of the Chinese Empire had been built on this idea and principle; the barbarians were conquered, then absorbed and turned into Chinese by slow assimilation and cultural influence.

Since China's national boundaries have varied in different periods of Chinese history, which boundaries does China (the People's Republic of China, or PRC) accept as valid today? Upon close examination, it appears that China's territorial claims are based upon the boundaries of the Manchu Empire in 1840. The PRC renounced as inequitable the treaties concluded between the Chinese government and various foreign governments during the latter half of the nineteenth century, a period when the Manchu Empire had begun to decline. Chinese territorial claims are also closely related to China's struggle against foreign exploitation.

Although the PRC has not definitely committed itself to the restoration of the boundaries of 1840, it is determined to regain at least part of the Chinese territories occupied by the "imperialists" in the era of 1840–1919. One sign of this determination is the publication of maps in China, which have been referred to by neighboring countries as Chinese "cartographic aggression."

The Chinese regard territorial claims as political and ideological, rather than judicial, questions. The direct bargaining of territory for territory is accepted as a negotiating technique, but the belief that former Chinese territory must be restored to China still prevails. For example, China agreed in 1956 to recognize the McMahon Line as the boundary separating China from India and Burma because of the "friendly relations existing between both countries and China." Although the McMahon Line was honored in the case of Burma in 1960, it was repudiated in the case of India, with whom China was quarreling.

With respect to Burma (now called Myanmar) and Nepal, China considered both states as neutral countries and offered to conclude with them treaties of friendship and mutual nonaggression. While Burma accepted such a treaty, Nepal modified it by rejecting the nonaggression clause. In dealing with India, a much larger and more influential state, China adopted a very different course of bargaining. It is doubtful that India would accept any suggestion of exclusiveness in its relations with China. However, this is precisely what China demanded, namely, the realignment of Indian foreign policy to conform with that of China. It is not surprising, therefore, that India rejected this demand.

This chapter reviews past and present boundary conditions in China. The approach to different boundary situations varies, emphasizing historical, ethnic, physical, or political overtones that form the basis for conflict or change.

The Boundaries with Burma (Myanmar) and Laos

The Sino-Burmese border may be divided into two distinct sections. The northern section begins at the eastern end of the Sino-Indian boundary, extends southward along the watershed between the Irrawaddy and Salween Rivers, and terminates near Bhamo. This region is mountainous and somewhat remote. The southern section of the boundary begins at Bhamo and runs in a general southeasterly direction across hills, plateaus, and narrow valleys until it reaches the northwestern corner of Laos at the Mekong River. This region is relatively passable, and trade has flowed between Burma and Yunnan. The people of the region are mostly Shan or Tai.

Prior to the nineteenth century, no precise boundary existed, and the borderland included many cultures and political systems. The expansion of the British colonial empire into Southeast Asia led to a dispute concerning the exact location of the Sino-Burmese boundary. To solve this dispute, a conference was held in 1886 to delimit and demarcate the boundary.

The emigration of large numbers of Chinese from Yunnan to northern Burma after World War II and during the Chinese civil war created a border problem. This emigration was difficult to stop, since the Burmese side of the frontier was poorly patrolled. A border agreement would therefore enable Burma and China to control such movements.

Moreover, the armed border clashes between China and India in 1959 made it apparent to China and Burma that both would benefit from a boundary settlement. Burma, for example, wanted protection from border incursions, while China sought to isolate India from the other Asian states. In addition, China announced in 1949 that it intended to re-examine past boundary treaties and, at Bandung in 1955, expressed a desire to stabilize its boundaries. However, the only neighboring state with whom discussions had been conducted was Burma, and these were at Burma's request. Finally, in January 1960, the Burmese representative Ne Win traveled to Beijing to sign a preliminary border agreement, known as the Burma-China Boundary Treaty.

This treaty provided for the demarcation of the 1,358-mile Sino-Burmese boundary and resolved all territorial disputes between Burma and China. Moreover, according to the terms of this treaty of friendship, each party pledged neutrality, thereby renouncing participation in any military bloc or coalition. Thus, a "zone of peace" was established with Burma. In effect, however, this treaty gave China the power to influence, if not determine, Burmese defense and foreign policy.

The Sino-Lao boundary is 265 miles long and is demarcated by fifteen stone pillars (one pillar every 17.6 miles). The boundary stretches through rugged land inhabited by various minority groups. As a state, Laos lacks both ethnic and political cohesion. The peoples of the Sino-Lao borderland consider themselves members of their respective tribes and have no loyalty to the Laotian government. Thus, the central government ineffectively controls the border region, and China or Laos might claim either jurisdiction over the tribes living in this area. China, however, has not shown an interest in extending its political control into this borderland, and no boundary dispute has arisen between Laos and China.

The Boundaries with Pakistan and Afghanistan

There are questions not only about the location of the Sino-Pakistan boundary but also about its legality in view of the conflicting claims of India and Pakistan to Kashmir. Since India claims the whole of Kashmir, it naturally denies the legality of a Sino-Pakistan border as defined by Beijing. In marked contrast to the Soviet Union, which endorsed India's claim in 1955, the People's Republic of China supports neither India's nor Pakistan's claim to Kashmir.

The Sino-Pakistan border question was settled on March 2, 1963. The agreement was technically a provisional one, pending the final resolution of the status of Kashmir, and was not subject to ratification. India, however, objected to this agreement and refused to abide by it. Nevertheless, by mid-1964, the demarcation of the Sino-Pakistan "provisional" boundary was considered complete.

Landlocked Afghanistan shares in its northern region a common boundary with the People's Republic of China. From the north the boundary extends along the crest of the Little Pamir Mountains and then across the Upper Wakhjir River valley. The Sino-Afghan boundary is the result of Anglo-Russian diplomacy during the latter half of the nineteenth and early part of the twentieth centuries. The Anglo-Russian Convention, signed in St. Petersburg in 1907, established the Wakhan Corridor to prevent Russia and British India from sharing a common boundary and thus also produced the rather curious Sino-Afghan boundary of today. The people of the region have never accepted this boundary, and neither Afghanistan nor China has succeeded in extending its political control over the borderland. Moreover, the territorial rivalry between the British and former Tsarist empires no longer exists. Thus, the Wakhan Corridor is a remnant of the colonial period and today serves no function.

On March 2, 1963, the day on which the "provisional" Sino-Pakistan boundary agreement was signed, it was announced that China and Afghanistan had "agreed to conduct negotiations for the purpose of formally delimiting the boundary existing between the countries and signing a boundary treaty." At the first session of the negotiations, China said, "The settlement of the Chinese-Afghan boundary question will be another example to all neighboring countries for the peaceful settlement of questions between them through negotiations." Agreement on the twenty-mile boundary was reached and the final treaty signed. It was not subject to ratification and went into effect immediately. The main significance of the Sino-Afghan boundary settlement is that China formally recognized the Wakhan Corridor and took another step toward the diplomatic isolation of India. In a region abounding in mountain ranges and river valleys, the choice of physical features for the delineation of the border is not of great geographical significance, primarily because it does not represent a cultural, economic, or historical divide.

The Boundaries with the Small Himalayan States

There is general agreement between China and the small Himalayan states with respect to the location of the boundary that the latter share with China. Nevertheless, China considers the peaceful resolution of those border disputes that do exist with the Himalayan states as vital to its avowed objective of isolating India.

The Sino-Nepalese boundary extends 670 miles along the crest of the Himalayan Mountains and passes through Mount Everest and Mount Makalu. The present boundary is the result of an agreement signed by Nepal and China on March 21, 1960. This agreement specified that the boundary should run through the summit of Mount Everest, rather than south of it as shown on earlier Chinese maps. However, the Nepalese insisted that the entire mountain should be within Nepal, since Mount Everest's summit had never been reached except from the Nepalese side. The final boundary treaty, signed in October 1961, merely describes the boundary as "passing through" Mount Everest, without specifying the location of the boundary with respect to the summit.

Bhutan relies upon India for its military defense and the conduct of its foreign affairs, although the traditional route between Bhutan and India runs through Tibet. The location of the Sino-Bhutanese boundary is based upon tradition. The boundary follows physical features of the landscape but is not defined by treaty. Although such an ill-defined boundary could be a potential source of friction between Bhutan and China, the Chinese tend to minimize the boundary question. For example, on December 26, 1959, in a note to India, the Chinese stated: "Concerning the boundary between China and Bhutan there is only a certain discrepancy between the delineation on the maps of the two sides in the sector south of the so-called McMahon Line. But it has always been tranquil along the border between the two countries."

Sikkim, which is located between Nepal and Bhutan, is under even greater Indian control than Bhutan, since the strategic Chumbi Valley is located in Sikkim. China concedes that the Sino-Sikkimese boundary has been delimited since 1890 and demarcated since 1895. Moreover, China informed India in December 1959 that "the boundary between China and Sikkim has long been formally delimited and there is neither any discrepancy between the maps nor any dispute in practice."

The Boundaries with Other Communist States

The Sino-North Korean boundary is approximately 880 miles in length. It follows the Yalu River, the Tumen River, and for a distance of twenty miles, the watershed of the Paotou Shan. Although the Yalu has been accepted as the boundary since 1875 and the Tumen since 1909, apparently no detailed demarcation of the boundary in the two rivers has ever been made. Consequently, the islands of the two rivers have never been allocated to either country.

There are two possible sources of dispute between China and North Korea. One is the Supung Dam on the lower Yalu. This dam provides power for large areas of North Korea and China, but no convention has been signed that might clarify each country's jurisdiction over the dam. A serious dispute is not likely to arise between the two states in the foreseeable future because of their present close political association.

Another source of possible dispute is the Paotou Shan. From this mountain, the Yalu and the Tumen flow west and east, respectively, to form the boundary. Both China and North Korea have claimed this mountain and have issued statements and drafted maps to support their claims. It is interesting to note that the former Soviet Union has tended to support North Korea's claim rather than China's.

The origin of the present 796.4-mile boundary between China and Vietnam can be traced to the Treaty of Tientsin signed by France and China in 1885. The boundary conventions of 1887 and 1895 completed the detailed delimitation, which has remained undisputed through subsequent political changes. Unless the present close political relationship between the two countries deteriorates, it is unlikely that a serious border problem will arise.

The boundary between China and Mongolia is poorly defined and has never been a rigid frontier. This is a consequence of the geography and history of the borderland. Mongolia functions as a buffer state between Russia and China. Landlocked Mongolia is essentially a dry plateau extending 1,470 miles from east to west and 780 miles from north to south. In the southeast near Inner Mongolia, the elevation is about 4,000 feet and increases at the center of the plateau to 5,500 feet. The southern region of Mongolia is located within the Gobi Desert, a flat wasteland. Extensive grasslands cover the remainder of the country and are of considerable economic importance, since pastoralism is the dominant economic activity of the country. Because of the high aridity characteristic of Mongolia, crop cultivation is restricted primarily to the north, along the Selenga River.

The 2,700-mile Sino-Mongolian boundary separates Mongolia from the People's Republic of China, thus dividing Outer Mongolia from Inner Mongolia, which is an integral part of China. This separation expresses a fundamental difference between these two areas with respect to their geography and tribal history. Inner Mongolia receives more precipitation than does the Gobi Desert of Mongolia and, as a consequence, dry farming has been more successful in Inner than Outer Mongolia. In addition, the Mongols of Inner Mongolia, living nearer to and thus coming in more frequent contact with the intensive cultivators of the North China Plain, have developed a strikingly different character from that of the traditional nomads of the outer territory. There are about 1.2 million Mongols in Inner Mongolia, many more than in Outer Mongolia, but they are already outnumbered by Chinese, five to one. Despite this fact, the Chinese, following the Soviet example, have established the Inner Mongolian Autonomous Region.

Prior to the Japanese occupation of Manchuria, the Mongols of both Outer and Inner Mongolia had moved freely between the two regions. After the Japanese occupied Manchuria, the nomads were restricted by a definite boundary. On June 9, 1940, Japan and the Soviet Union agreed to demarcate the boundary separating their respective territories, Manchukuo and Mongolia. The present boundary is based upon this demarcation.

The boundary between China and Mongolia in the Altai Mountains in the west is also poorly defined. Since the major part of the Sino-Mongolian boundary runs through the driest part of the Gobi Desert, it would be difficult to demarcate it, and there appears to have been no urgency to do so.

The Russo-Chinese Boundary

Extending from the Pamir plateau in the west to the Pacific Ocean in the east, the Sino-Russo boundary is interrupted by the Mongolian People's Republic. With the latter in between, the 4,150-mile boundary is divided into two nearly equal sectors. The western, or Xinjiang-Turkestan, section extends for 1,850 miles through three major mountain ranges, while the eastern sector or Manchurian-Siberian section, runs primarily along the courses of the Argun, Amur, and Ussuri Rivers for 2,300 miles. Most of the boundary in both sectors was defined by a series of Sino-Russian treaties prior to the twentieth century.

The Western Sector

Historically, this area in Central Asia between Mongolia and Afghanistan has been known as Turkestan. Lying on the fringes of the Russian and Chinese Empires, it has often been an area of confrontation. The present-day boundary is the result of the Sino-Russian treaties of 1860, 1864, 1881, and 1895. However, in many respects these treaties are inconclusive, and several areas are still disputed.

Geographically, the border is in the midst of a complex of mountain systems, steppes, basins, and valleys extending across and along the frontier. The boundary cuts across the Pamir, the Tien Shan, and the Altai mountain ranges, and none of these are in alignment. Instead, there are many intervening valleys, which have served since ancient times as natural passageways through Central Asia.

In the southwest, the boundary starts in the Pamir—often called the "mountain core" of Central Asia—from which the major mountain ranges in Asia radiate. Strategically located, the Pamir lies within the border zones of Kashmir, Afghanistan, Pakistan, Russia, and China, contributing to the ambiguous boundary situation in the area.

Most of the boundary difficulties between Russia and China in Inner Asia probably stem from minority problems rather than from the location of the boundary per se. In other words, people rather than land are the source of tension in this border region.

Ethnically, the Sino-Russian frontier is inhabited by a variety of minority groups that include Kazakhs, Kirghiz, Uighurs, Tadzhiks, Mongols, Han Chinese, and Russians. This complicated mix of ethnic groups has resulted in a situation of conflicting loyalties among minority groups toward the Russian and Chinese governments.

After the dissolution of the Soviet Union in 1992, Russia retained only thirty-four miles of the border with China. The three new states of Kazakhstan, Kyrgyzstan, and Tajikistan—previously part of the Soviet Union—now have boundaries with China. They joined together as one delegation to negotiate on boundary problems with China, following suggestions by the Russians. However, it was agreed that all four of these members of the Commonwealth of Independent States (CIS) would sign separate agreements with China. The Russian-Chinese agreement that defined the western part of the border was signed in early 1994. The agreement between Kazakhstan and China was signed at the end of 1994, with the deferred resolution of two disputed sites. An agreement between Kyrgyzstan and China was signed in 1996, also with the deferred resolution of some disputed areas. Tajikistan and China were able to agree on a 112-mile portion of the boundary in the Pamir area but failed to define a longer section along the high mountain portion of the border. It seems likely that most of the Pamir border will remain undefined due to its rugged landforms.

The Eastern Sector

The complexities that characterize the western boundary are not present in the east. Here the boundary is a physical one, clearly marked by the Amur River and its tributaries and the Argun and Ussuri Rivers. However, discrepancies do exist, especially over the junction of the Zeya and Amur Rivers, the junction of the Amur and Ussuri Rivers, and several islands in the Ussuri River. The border has been defined in several treaties, namely, the Treaty of Nerchinsk (1689), the Treaty of Kiaklita (1727), the Treaty of Aigun (1858), the Treaty of Peking (1860), and the Treaty of Tsitsihar (1911). The Chinese government has denounced these treaties with the imperialist Tsarist government as the products of coercion.

The eastern part of the western border starts at the northeastern tip of Mongolia, which is sandwiched between Russia and China. The border extends eastward more than 2,480 miles. On the whole, the boundary stretches through a cold, semi-arid area handicapped by its isolation from both the Russian and Chinese centers of production, population, and communication.

The boundary begins in the middle course of the Argun River and follows the river from southwest to northeast along the western side of the Great Khingan Range. Although sparsely settled, the region does have two major cities: Manchouli, on the Chinese side of the boundary, and Zabaykalsk, on the Russian side. It is here that the Chinese Eastern Railway crosses from Siberia to Inner Mongolia.

As the Amur flows around the northern end of the Great Khingan Range, the plateau-like landform ends. The elevation decreases and the landscape begins to reflect the maritime influence of the Pacific Ocean.

Since the normalization of Sino-Russian relations in 1988, border trade has increased significantly in the Russian Far East (or RFE). About 2,000 people cross the border daily at the cities of Heihe and Suifenhe, and 10,000 Chinese citizens work and reside in the cities legally. The collapse of the Soviet Union led to a decline in the RFE's economy and population, and the region has become dependent upon foreign workers. This has resulted in security problems in the area. It has been reported that about 1 million Chinese are living along the Russian-Chinese border, and more than 200,000 are living in the RFE alone. The Chinese workers are engaged in farming, timber cutting, construction, and light industry.

The central government of Russia and the residents of the RFE feel uneasy about the region's growing dependence on China. The solution would be to develop the local economy by importing capital and technology from Japan and South Korea and labor from China and South Korea.

However, both Northeastern China and Far Eastern Russia are remote regions from the central government and do not draw much attention from Beijing or Moscow. Nonetheless, some attempts have been made to support cross-border harmonization, including the improvement of the transportation system that connects the two regions. In addition, the recently created Tumen River Economic Development Plan, which has the involvement of the Japanese, will benefit the RFE. This triangular relationship among China, Russia, and Japan will lead to the establishment of a new transportation, trade, and investment network in Northeast Asia.

In 1960, the Soviet Union took control of the border river, and in 1969, an armed conflict on Damanskii or Zhen-bao Island took place on the Chinese side of the Ussuri River. The two nations soon reached an agreement and returned to the status quo. Since then, effective border agreements were successfully negotiated in 1991 and 1997, and the eastern border was defined.

On the Russian side of the boundary, Russians dominate the entire area, with the exception of a Jewish population near Birobidzhan, the "Jewish autonomous region," and substantial Ukrainian elements along the Ussuri River and Lake Khanka. Thus, for the most part, the boundary functions as an ethnic line of separation.

At present, direct contact between Russians and Chinese—the traders, commercial companies, state enterprises, and government agencies—has benefited the border area. However, the social infrastructure of the Russian Far East has been overburdened by an influx of Chinese peasants and urban traders who have migrated there to seek a better life. The Russians were not prepared for this sudden development, and illegal Chinese immigration has become a serious problem.

Another problem has been the inadequacy of transportation infrastructure between Russia and China, though recent improvements have been made in connecting the two regions. However, some 25–30 percent of bilateral contacts are not implemented due to lack of efficient transportation. At the same time, however, regional leaders of the RFE have opposed new bridge projects citing the "threat" of the illegal Chinese.

The Sino-Indian Boundary

On April 29, 1954, diplomats from Asia's two leading countries signed the Sino-Indian Agreement of Trade and Intercourse between the Tibet region of China and India. The two countries expressed their friendship and optimism in what was later known as the Panch Shila or five principles of peaceful coexistence: mutual respect for each other's territorial integrity and sovereignty, mutual nonaggression, mutual noninterference in each other's internal affairs, equality, and mutual benefit.

Eight years later, with the lofty ideals of Panch Shila already destroyed by angry exchanges of notes, protests, and insults arising from boundary disputes, a massive army of the Chinese People's Republic crossed the Himalayas and delivered a swift setback to the frontier forces of the Indian government. Since then, the Indian and Chinese governments have published volumes of evidence reasserting their opposing claims to the disputed areas.

The Three Sectors

The present dispute involves more than 2,000 miles of boundary. For convenience of discussion, the disputed territory has been divided into three sectors: western, middle, and eastern.

In the western sector lies the boundary dividing Kashmir from Xinjiang and Tibet. The boundary is more than 1,000 miles long, and the contested area encompasses more than 15,000 square miles. The Aksai Chin area, in the northern part of this sector, is of greatest importance to the Chinese, since within this bulge of territory lies the great Xinjiang-Tibet motor highway.

The middle sector of the disputed boundary is about 400 miles long and extends along the crest of the Himalayas from the Sutlej River to the Nepalese border. In this sector, the total contested area is less than 200 square miles and includes several disputed points in Spati in the Nilang region, and near the Shipki Pass. The disputes here were the first to receive wide publicity but are of far less gravity than those in the other two sectors.

In the eastern sector, India claims that the McMahon Line forms the boundary between India and China. This line follows the crest of the Assam-Himalaya Range between Bhutan and Burma and is slightly more than 700 miles in length. China denies the validity of this line and claims a quite different boundary that runs along the base of the Himalayan Range. The territory between the two lines is now referred to in India as the North East Frontier Agency (NEFA) and encompasses about 32,000 square miles.

There are two distinct disputes in this sector. On the one hand, the Chinese and Indians contest possession of the entire Himalayan NEFA. On the other hand, the Chinese complain that the Indians have established posts at a number of points north of the "illegal" McMahon Line.

Historical Background

The boundary dispute between China and India stems from the semi-independent status of Tibet within China, British and Russian political ambitions and spheres of influence in the Himalayas, and the physical nature of a frontier that lacks any well-defined geographical or "natural" boundaries. These various elements are inextricably interwoven and complex. The eleven-volume "White Paper" published by the Indian government reflects the minute detail with which the problem can be studied. Here, however, only the high points of this long history will be considered.

Although China had maintained some form of influence over Tibet since the fourteenth century, it was not until 1720 that the Manchu emperor K'ang Hsi actually controlled the area's administrative affairs through the Buddhist leader, the Dalai Lama. From that time until the overthrow of the Manchus in 1911, Chinese officials, sent from Beijing and residing in the Tibetan capital of Lhasa, had in effect ruled Tibet.

At the turn of the twentieth century, the Chinese feared an invasion of Tibet from both British India and an expanding Russia. Alarmed by the Russian influence in Lhasa, Lord Curzon, then Viceroy of India, attempted to convince the Balfour government in London that the Russians were attempting to conquer all of Asia. In 1900 Curzon sent an armed mission to Lhasa under Francis Younghusband to establish by force a British trade outpost north of the Himalayas. In 1904 an Anglo-Tibetan convention was held. It was the first time that Tibet had been opened to the Western world.

Thus, as the British in India extended their political control northward, and the Russians in Central Asia pushed southward, the famous "Great Game" of Anglo-Russian diplomatic contention was initiated. Both powers tried to establish their influence in Tibet. After several diplomatic and military struggles, the Anglo-Russian Convention was signed in 1907, in which the two powers agreed not to station representatives in Lhasa and to refrain from direct relations with Tibet except through the Chinese minister.

With the protection of these two conventions, the Chinese initiated a program of complete dominance of Tibet. In 1910, the Chinese general Chao Erh-feng captured Lhasa and began the complete absorption of Tibet into the Manchu Empire. In 1911, the Manchu Dynasty was overthrown, and control over Lhasa disintegrated as the Tibetans rose in open revolt against the Chinese army.

In private bilateral talks between Britain and Tibet at the 1913 Simla Conference, their boundary was fixed along the crest of the Assam-Himalaya Range on a large-scale map of eight miles to the inch. This map, in two large sheets, was never shown to the Chinese delegate. Using a much smaller-scale map at Simla, and with considerable pressure, McMahon persuaded the Chinese delegate to sign the agreement. The Chinese Government continued to repudiate any border determined by British diplomatic manipulation of Tibet.

Thus, the present border between Tibet and India is the result of Britain's attempt to make Tibet an independent buffer state, as well as of the efforts of Britain and India to maintain administrative control over Assam.

In 1949, the exile of the Nationalist Chinese government and, in 1950, the assumption of complete control over Tibet by the new Chinese Communist government changed not only the status of the Tibetan-Indian border but the status of all of China's border areas. In 1954, under the Panch-Shila agreement, India admitted for the first time that Tibet was an integral part of China and, therefore, that anything that happened in Tibet was the domestic concern of China. No such admission had ever been made by the British government. The Indian government, however, refused to renounce its claims to the northern territories and continued to insist that the McMahon Line in Assam was its northeastern boundary and that this was not subject to renegotiation.

In March 1959, the inhabitants of Tibet revolted against their communist over-lords. In the months that followed, thousands of Tibetan refugees, including the Dalai Lama and his advisors, fled into India. India's granting of political asylum to these refugees was met by Chinese protests of Indian interference in Chinese affairs. Although the Indian-Tibetan boundary, per se, was never mentioned in the angry exchange of notes between New Delhi and Beijing, tension in the border areas mounted, and the Chinese officials in Tibet deprived India of the rights within Tibet that it had been guaranteed in the Panch-Shila Agreement of 1954. As a conse-quence, border intrusions and incidents became more frequent and violent.

The tension culminated in hostilities on October 20, 1962, when about 20,000 troops of the Chinese People's Liberation Army poured over the Thagla Ridge. Suddenly, on November 21, 1962, the Communists announced a unilateral cease-fire and withdrew to the previous line of control. The situation as it existed then remains the same today.

The Chinese Argument

The main argument of the Chinese is, of course, the "illegality" of the McMahon Line, and on this point, their argument seems valid. The line as established at Simla was not the result of a detailed study of the boundary problem but rather of politi-cal maneuvering designed to satisfy the needs of British policies in Central Asia.

The Chinese government has also pointed out some very obvious discrepancies in the Indian argument: First, maps published by the British surveyor-general prior to the Simla Conference show that the boundary in the western sector follows the Karakoram crest, which the Chinese claim is the traditional or customary line. Second, maps published by Britain and India prior to 1936 show this same tradi-tional or customary line in the eastern sector and not the McMahon Line. Third, prior to 1954, all Indian maps showed the boundaries of the western middle sector as undefined. On maps published in 1954 and after, these boundaries are precisely delimited.

In the western sector, the Indian government claims that its right to the Aksai Chin area is established by its jurisdiction over that area. Two events make this claim questionable. First, during the Chinese takeover of Tibet in 1950, 30,000 troops were moved through the Aksai Chin without the knowledge of the Indian govern-ment. Second, during the early 1950s Chinese construction crews, using local labor, were able to build the great Xinjiang-Tibet motor highway, again without the knowledge of the Indian government. It is evident, therefore, that India's claim of jurisdictional control over the Aksai Chin does not seem to be supported by actual control over the territory.

It should also be noted that several times during the 1962 crisis the Chinese foreign minister, Chou En-lai, offered to relinquish Chinese claims to the NEFA or eastern sector, an area of 32,000 square miles, if India would in turn abandon its claim to the 1,500 square miles of territory in the Aksai Chin area. India's continued insistence that China has no legitimate claim to Indian territory is rather difficult to understand, in light of the vague legal and historical case upon which this insistence is based and China's willingness to negotiate and to relinquish many of its territorial claims.

The Indian Argument

The government of India has put forward several arguments to back up its claims for the validity of its historical border in the Himalayans, notably the McMahon Line. Tibet, China, and representatives of the British Indian government at the Simla Conference in 1914 ratified this international boundary. The Indians claim that the fact that the Chinese representative signed the agreement, which has a map attached to it, is evidence that China accepted the agreement. In addition, the Chinese accepted the eastern part of the McMahon Line as the traditional boundary between Burma and China in their agreement with Burma of October 1960, thus indirectly accepting the validity of the McMahon Line as the traditional boundary between India and Tibet. A third point is that the McMahon Line corresponds to the watershed boundary along most of its length.

In the western sector, the Indian government is able to produce a large variety of maps and documents to support its claim. Since the tenth century, the Chinese have recognized important points on the present Indian alignment as the boundary between Tibet and Ladakh. Furthermore, there is evidence that the boundary of Xinjiang has never extended south of Kunlun and that the Kashmir government always maintained the trade routes.

In the middle sector, apart from the natural and geographical basis of the high Himalayan watershed range, the Indians believe that literary and religious tradition and ancient chronicles corroborate Indian border claims in a surprisingly precise way.

To sum up, India asserts that there was no boundary problem until 1959. It insists that the Kunlun arc and the great Himalayan ranges form the most impressive boundary in the world and that the boundary has been recognized in tradition and custom for centuries. It has determined the limits of administration on both sides and has been confirmed by valid international agreements in regard to different sectors at different times during the past 300 years. The Indian government believes that the Chinese government has inherited the treaty obligations of the previous Chinese governments and ought to respect this boundary, which is defined by nature, formed by history, and sanctified by the law of nations.

Summary

China's territorial claims seem to be predicated upon the boundaries of the Manchu Empire in 1840. After this date, the Manchu Empire started crumbling, and the Chinese feel that unfavorable and unjust treaties were forced upon them. Because the Chinese regard territorial claims as political and ideological rather than judicial questions, they have tried to attach political strings to border settlement treaties with its neighbors. The Chinese regarded China's international boundaries as "a problem left over from imperialism" and declared that they would reexamine them and "recognize, abrogate, revise, or renegotiate them." The two most critical boundary disputes are those with Russia and India.

The long Sino-Russian struggle over Central Asian borderlands is a one in which the Chinese have usually had to give way. The losses of territory in Xinjiang to the Russians resulted, in every case, from the ignorance of the Chinese in regard to the principles of boundary geography. In the early part of the twentieth century, the Russian side of the western border was heavily garrisoned. The Chinese side was poorly patrolled, and the Russians were accused of taking advantage of this fact and moving the portable boundary markers into Chinese territory. Most of the boundary difficulties between the Soviet Union and China in Inner Asia stem from minority problems rather than from boundary problems per se. The minorities have historically been outside the acumen of Russia or China. The Soviets feel that since they give the minorities more recognition, in any dispute, the minorities will side with them. The Soviets have launched a campaign to open up certain border areas to agriculture. The Chinese have responded by stationing hundreds of thousands of soldier-farmers along the frontier. The boundary in the east is a clearly marked physical one, and many of the problematic complexities of the western sector do not exist there. However, discrepancies do exist, especially over the junction of the Zeya and Amur Rivers, the junction of the Amur and Ussuri Rivers and several islands in the Ussuri River.

The 1954 Sino-Indian Agreement of Trade and Intercourse between the Tibet region of China and India contained five principles of peaceful co-existence. By 1962, these five principles had all been smashed, and the two countries have been engaged in bitter border disputes ever since. Much of the trouble in the area stems from the fact that many boundaries were created in this area by the Russians and British as they tried to set up buffer zones and states. The Chinese and Indians had little to say about the matter. Another problem is the geography of the area; it doesn't allow for easy delimitation along physical lines. Also, the people in the disputed areas are nomadic and characterized by their shifting allegiances. In the eastern section the boundaries are so complex, both physically and culturally, that there is no easy solution for delimiting them. The Chinese argument is based on early maps, the underhanded dealings that took place in defining the McMahon Line, their settlement of disputes with their other neighbors, and their willingness to relinquish claims to the eastern sector if India will drop its claims on the much smaller area in Aksai Chin.

India bases its argument on the fact that the McMahon Line corresponds to a watershed boundary along most of its length and China's acceptance of the eastern part of the McMahon Line as the traditional boundary between Burma and China. India also has maps and documents to support its claims. They also accuse the Chinese of displaying their historically aggressive nature in this dispute and feel that the Chinese should live with the boundaries they inherited from the Nationalists.

In dealing with the boundary problems with the small Himalayan states, China has followed a policy of trying to isolate India from the other Asian states. Furthermore, if the countries have political policies favorable to the Chinese, then China usually settles with them on a fair and equitable basis. However, if the prevailing ideology of the country is unfavorable to the Chinese, then the Chinese are cold and demanding in their negotiations.

There is some justification for disputing the Chinese-North Korean border. However, a serious dispute will probably not arise between them in the near future because of their close political ties. The situation with Vietnam is similar. Mongolia has been established as a buffer state between the Russians and the Chinese. However, the boundary between Outer Mongolia and China has never been a rigid one.

References

Bhutani, V. C. 1985. What Boundary Question? Sino-Indian or Indo-Tibetan? *China Report* 21(5):427–432.

Caroe, O. 1960. The India-Tibet-China Triangle. *Asian Review* 57(205).

_____. 1960. The Sino-Indian Question. *Journal of the Royal Central Asian Society*.

_____. 1960. Tibet, the End of an Era. *Journal of the Royal Central Asian Society* 47.

Clubb, O. E. 1971. *China and Russia*. New York: Columbia University Press.

Connell, J. 1960. The India-China Frontier Dispute. *Journal of the Royal Central Asian Society* 47.

Doolin, D. J. 1965. Territorial Claims in the Sino-Soviet Conflict. *Hoover Institute Studies* 7.

Ellegiers, D. 1970. *The Border Between Russia and China*. Brussels: Centre d'étude du Sud-Est Asiatique et de l'Extreme-Orient.

Fitzgerald, C. P. 1963. *The World Today*. Royal Institute of International Affairs.

Hsieh, Chiao-min. 1964. *Ageless Land and Countless People*. New York: Van Nostrand.

Jackson, W. A. D. 1968. The *Russo-China Borderlands*. New York: Van Nostrand.

Jing, H. 1989. The Truth About the Eastern Sector of the China-India Boundary. *China Report* 25(1):99–112.

Jones, S. B. 1943. *Boundary-Making: A Handbook for Statesmen, Treaty Editors, and Boundary Commissioners*. Washington, D.C.: Carnegie Endowment for International Peace.

Kim, W. B. 1994. Sino-Russian Relation and Chinese Workers in the Russian Far East: A Porous Border. *Asian Survey* 34 (12).

Kireev, G. 1997. Strategic Partnership and a Stable Border. *Far Eastern Affairs* (Moscow) 114.

Lamb, A. 1964. *The China-India Border: Origin of the Border Dispute.* London: Oxford University Press.

_____. 1966. *The McMahon Line: A Study in the Relations Between India, China, and Tibet.* London: Toronto University Press.

Lattimore, O. 1940. *Inner Asia Frontier of China.* New York: American Geographical Society.

Lazikin, A. 1994. Russia-China: Direct and Cross-border Trade and Economic Cooperation. *Far Eastern Affairs* (Moscow) 96–97.

Li, H. 1981. The Crux of the Sino-Soviet Boundary Question. *Beijing Review* 24(30):12–17, 31; (31):13–16.

Lukin, A. 1998. The Image of China in Russia Border Regions. *Asian Survey* 38(9).

Murty, T. S. 1987. *India-China Boundary: India's Options.* New Delhi: ABC Publishing.

Robinson, T. W. 1972. The Sino-Soviet Border Dispute: Background, Development, and the March 1969 Clashes. *American Political Science Review* 66(4).

Rose, A. 1912. Chinese Frontiers of India. *Geographical Journal* 39.

Rowland, J. 1967. *A History of Sino-Indian Relations.* New York: Van Nostrand.

Schwarz, H. G. 1963. Chinese Migration to Northwest China and Inner Mongolia. *China Quarterly.*

Syed, A. H. 1974. *China and Pakistan.* Boston: University of Massachusetts Press.

van Eckelen, F. W. 1967. *India Foreign Policy to Border Disputes with China.* The Hague: Martisum Nighoff.

Watson, F. 1966. *The Frontier of China.* New York: Frederick A. Praeger.

Wiens, H. J. 1963. China's North and Northwest Boundaries. *Contemporary China,* Vol. 5. Hong Kong: Hong Kong University Press.

21

The Return of Hong Kong:
Liabilities or Assets?

C. P. Lo

At midnight on July 1, 1997, soon after the speeches of Prince Charles of Britain and President Jiang Zemin of China, the Union Jack was lowered for the last time in Hong Kong. At the same time, the flag of the People's Republic of China and the flag of the Hong Kong Special Administrative Region (SAR) were raised on their masts, accompanied by the Chinese national anthem in the new convention and exhibition hall of Hong Kong. At this moment, Chinese national pride was at its highest: The ceremony marked the end of 155 years of British rule and the restoration of Hong Kong's sovereignty to China. The ceding of Hong Kong to Britain had been the result of a series of unequal treaties forced upon a weak China (under the Qing Dynasty) after her defeat in the Opium Wars (1840–1841). Now, the return of Hong Kong's sovereignty to China signified the defeat of imperialism and the victory of the new socialist China. Within one and a half hours of the changeover, the governing officials of the Hong Kong SAR, led by Mr. Tung Chee-hwa, the chief executive who replaced the governor, were sworn in, and the Chinese Liberation Army swiftly moved in to take over the British military installations in Hong Kong. By July 2, Chinese control of Hong Kong was complete.

   With the changeover, both China and Hong Kong have committed themselves to the daunting task of showing the world that Hong Kong under Chinese rule will continue to prosper and that the principle of "One Country, Two Systems," as proposed by the late Deng Xiaoping, actually works. Hong Kong's continued success is particularly important because China wants to entice Taiwan to reunify with China under the same principle. This chapter examines from a geographical point of view the impact of Hong Kong's sovereignty change on China as well as China's influence on Hong Kong.

Hong Kong's Economic Development
Under British Rule

At the time of the changeover, Hong Kong (with a population of 6,421,300 at the end of 1996 and a land area of 423 square miles) was recognized as a world city and financial center. As an "Asian tiger," it is the world's eighth-largest trading economy and has the world's largest container port. It has also been recognized as the world's freest economy. Within a short span of fifty years, Hong Kong attained a gross domestic product per capita of more than US$24,000, higher than some European Union nations. All of this was achieved under the free-enterprise system supported by the government, the so-called *laissez-faire* or positive nonintervention policy, which attracts foreign investment and favors an export-oriented economy. With the new affluence, the quality of life of the population greatly improved. In 1996, the average life expectancy for men was seventy-six years and that for women was eighty-two years. All of these statistics indicate that Hong Kong has reached an advanced stage of economic development (Howlett 1997).

The economic development of Hong Kong has benefited from a number of factors, the most important of which is Hong Kong's geographical location (Lo 1992). It is situated on the South China coast right at the entrance to the agriculturally rich hinterland of the Zhujiang (Pearl River) Delta (Figure 21.1). Hong Kong is open to the South China Sea and the Pacific Ocean, occupies a central position in Southeast Asia, and serves as a focal point for major shipping routes from Europe and North America. The advent of air transport further enhanced the centrality of Hong Kong (Yeung 1996). Hong Kong is therefore an excellent entrepôt port, a function that the British wanted Hong Kong to serve when the colony was founded in 1843 with the main objective of trading with China. The naturally deep and well-sheltered harbor of Hong Kong makes an ideal seaport to fulfill such a function (Chiu and So 1986).

All these geographical advantages would not have helped the development of Hong Kong had it not been for the fact that British rule provided a politically secure environment that shielded Hong Kong from much of the turmoil of Chinese politics. Without political interference, the people of Hong Kong devoted their full energy to working and making money. The political stability in Hong Kong has helped to attract skills, capital, and labor from China, particularly during the Chinese civil wars and immediately after the Chinese Communist victory in 1949.

As an entrepôt port, Hong Kong was vulnerable to world economic upheavals. Hong Kong's economy was stifled when the Korean War brought about a United Nations trade embargo on China in 1951. However, local entrepreneurs shifted their efforts to industrial development, which took off with the resources of capital, skill, and labor that were already in place. Early export-oriented light industries based on textiles produced at low cost were very successful. As capital was accumulated, Hong Kong manufacturing diversified, and more sophisticated goods were

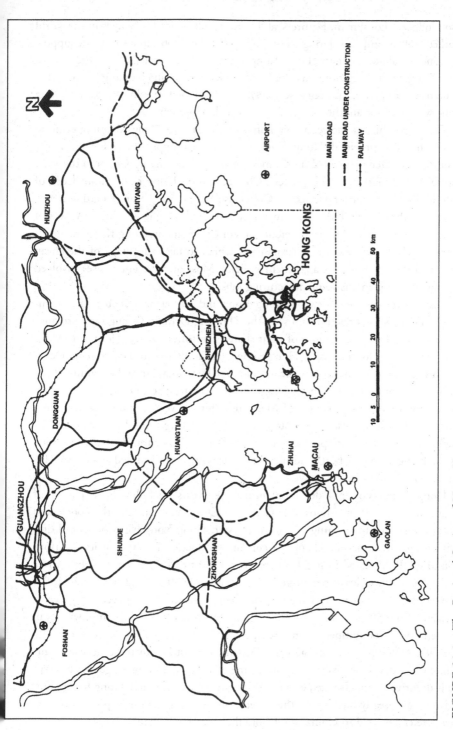

FIGURE 21.1   The Regional Setting of Hong Kong
SOURCE: Modified from Yeh 1996.

also produced. Local manufacturers' ability to rapidly adapt to changes in the world market explains much of Hong Kong's success in the 1960s and 1970s. In support of its industrial development, Hong Kong's service sector, particularly in the areas of finance, telecommunications, and business services (including banking, accounting, insurance, real estate, and legal services), has developed rapidly. By 1980, Hong Kong was ready for another stage of economic development.

The major shift in the economic policy of China after Deng Xiaoping came to power in 1978 provided Hong Kong with another opportunity for economic advancement. Deng reformed the Chinese economy, opening up China to the West and the world market. In this process, China relied on Hong Kong's knowledge of trade and its excellent service sector. China also found Hong Kong's hard currency (pegged to the U.S. dollar) to be a major asset in marketing its products. As special economic zones opened up in different parts of China, many Hong Kong manufacturers relocated their factories in Mainland China to take advantage of the cheap labor and the favorable tax abatement policies. Most of these enterprises involved low-skilled, labor-intensive manufacturing jobs.

Business transactions between Hong Kong and China have increased tremendously. Hong Kong's relationship with the Zhujiang Delta in Guangdong province is particularly intimate because of their shared Cantonese dialect (Johnson 1995). Many of the people who live in Hong Kong have family ties to the Zhujiang Delta, which was designated as an open economic zone by China. Since the signing of the Sino-British Joint Declaration on the Future of Hong Kong on December 19, 1984, the economies of Hong Kong and Mainland China have become more and more intertwined. It is interesting to note that by 1990, Hong Kong had recaptured its traditional role as an entrepôt port, with China as its major trading partner. China found Hong Kong to be the ideal port for exporting its finished goods and for importing products from the West.

Hong Kong's economic development in recent years has followed the pattern of what is known as "front shops, rear factories": China manufactures the goods, and Hong Kong sells them (Sung 1995). As a result, Hong Kong has experienced dramatic economic structural change. Secondary production in terms of its contribution to GDP decreased from 31.6 percent in 1980 to 14.7 percent in 1997, while tertiary production increased from 67.5 percent in 1980 to 85.2 percent in 1997. The consequence of this change is that employment in the secondary sector decreased from 50.1 percent in 1980 to 23.6 percent in 1997, while employment in the tertiary sector jumped from 48.4 percent in 1980 to 75.5 percent in 1997 (Howlett 1996; Census and Statistics Department 2002). All these developments demonstrate that Hong Kong has entered the de-industrialization stage that has typically occurred in an advanced economy. The presence of China as Hong Kong's hinterland has greatly facilitated this development. Hong Kong's prosperity was therefore closely tied to China well before the sovereignty change.

Hong Kong Under the British: Late Moves Toward Democratization

The British rule of Hong Kong can be described as one of benevolent despotism. Harris (1988) called it an "administrative no-party" system, because Hong Kong was not governed by representatives of the people through elections, at least not until the last governor, Mr. Christopher Patten, arrived in 1992. The governor of Hong Kong was appointed by the king or queen of Britain to be the head of Hong Kong and the commander-in-chief of the British forces stationed in Hong Kong. The British government issued two legal documents to Hong Kong: Letters Patent and Royal Instructions, which may be regarded as the constitution of Hong Kong (Miners 1986). These documents authorized the creation of the Executive Council and the Legislative Council as the central political organs of the government of Hong Kong. As head of government, the governor presided at meetings of the Executive and Legislative Councils. He possessed all powers except that of the judiciary, although he had to consult with the Executive Council when important government policy matters were to be decided. The governor therefore had the power of a despot, which is not unusual in Asia, where authoritarianism is a common form of government. The difference in the case of Hong Kong lies in the fact that as a British citizen, the governor espoused the beliefs of Western liberal constitutionalism. He allowed freedom of speech and listened to criticisms from the ordinary people.

The Executive Council was the policymaking body, whose members were appointed by the governor. The Executive Council considered all principal legislation before it was introduced into the Legislative Council. The Legislative Council's main function was to assist the governor in making laws, and it also controlled finance. Before 1985, the Legislative Council was composed entirely of government officials and unofficial members selected by the governor. Under the governor, there were three important branches of government headed by (1) the attorney general, (2) the chief secretary, and (3) the financial secretary, who controlled a large number of government officials (so-called "civil servants" under the British system). Together they formed a very efficient hierarchy of administrators responsible for the day-to-day running of Hong Kong.

After the signing of the Sino-British Agreement in 1984, the governor of Hong Kong decided to open up the Legislative Council membership for election, as a gesture to promote the confidence of Hong Kong's citizens. However, the form of election adopted was rather limited and, because of its limitation, was tolerated by China. Based on the recommendations of a white paper on representative government published in November 1984, election by functional constituencies and indirect election by district boards were adopted. Each functional constituency was to represent an economic, social, professional, or other sector of the community. In 1985, there were nine functional constituencies: commercial, industrial, financial, labor, social services, education, legal, medical, and engineers. Members in these functional constituencies were to elect twelve members to the Legislative Council. There were eighteen district boards covering the whole of Hong Kong, which is made up of three major geographical units: Hong Kong Island, Kowloon, and the New Territories (see Figure 21.1). The main function of the district boards is to advise government on matters affecting the interests or well-being of the people living and working in the districts. The district boards consisted of government officials, appointed unofficial members, elected members from the constituencies, and Urban Councillors. To implement indirect election to the Legislative Council, district boards were grouped into ten geographical constituencies, each based on one, two, or three district boards and representing 500,000 people. Hong Kong legal residents living in the appropriate district elected members of the district boards. All the elected members of the district boards in a geographical constituency elected a member to the Legislative Council to represent them. Thus, a total of ten members were elected based on geographical constituencies. In addition, one member each from two special constituencies, the Urban Council for the urban areas of Hong Kong and the Regional Council for the urban areas in the New Territories, was elected. In total, twenty-four members were elected to serve in the Legislative Council in 1985. In 1988, two additional functional constituency members were elected. However, the twenty-six elected members represented only a minority in the Legislative Council, because apart from the governor, three ex-officio members, and seven official members who served on the council, the governor appointed twenty additional members, making fifty-seven members in total. Clearly, the twenty-six elected members could not outvote the other thirty members who were obligated to support the governor.

On June 4, 1989, the world-shaking Tiananmen Square incident erupted in Beijing. China's actions were condemned by Western democracies, and China's image was greatly tarnished. The confidence of the people of Hong Kong was crushed. Well over a million people marched in the streets of Hong Kong to protest China's treatment of the pro-democracy student demonstrators in Beijing. The United Democrat Party, a political party founded by Martin Lee, a British-trained Hong Kong barrister, was particularly vocal in criticizing China's infringement of human rights, and it received much support from the citizens of Hong Kong. Brain drain, which had been taking place in Hong Kong since 1984, further accelerated. The British government believed that the people of Hong Kong should be given more say in their government as a means to protect them from China's interference after 1997. In 1991, major changes in the electoral system for the Legislative Council occurred. The ten district board–based geographical constituencies in the electoral college were abolished and replaced by a new system for direct elections to return eighteen members from nine double-seat geographical constituencies. New functional groups were also added, thus raising the number of functional constituency seats to twenty-one. For the first time, people in Hong Kong could vote for their geographical district representatives directly. As a result, the total number of elected members in the Legislative Council grew to thirty-nine out of a total of sixty members (including three ex-officio and eighteen appointed members). With this configuration, the United Democrat Party won twelve of the eighteen district seats. However, these moves toward democratization were still very limited.

When Christopher Patten, a politician and a former member of Parliament and chairman of the Conservative Party in Britain, became governor of Hong Kong in 1992, he moved vigorously toward a higher degree of democratization of the Legislative Council, which provoked a dispute with China. In preparation for the 1994 and 1995 elections, he announced in his October 1992 policy address a constitutional package that proposed several democratic reforms: to lower the voting age from 21 to 18, to replace corporate voting with individual voting in all of the Legislative Council functional constituencies, to make every member of the Hong Kong's 2.7 million working population eligible to vote in one of the nine additional functional constituencies, to adopt a "single seat, single vote" voting system for all geographical constituency elections, to abolish all appointed seats in the municipal councils and the district boards, and to establish an election committee with members drawn from the elected district boards. These proposals met strong opposition from China, which held that these proposed election arrangements contradicted the terms of the Sino-British Agreement of 1984. Despite the failure to obtain China's consent, Patten sent his proposals as the Electoral Provision Bill to the Legislative Council, which passed it in 1993. This resulted in a very strained relationship between Britain and China, which created a great deal of anxiety in Hong Kong about the future changeover. China announced that it would not honor the new Legislative Council arrangements. In other words, whoever was elected to the Legislative Council in 1995 for a four-year term would not be able to serve their full term. After July 1, 1997, the Legislative Council would be disbanded, and China would establish a provisional legislature to replace the Legislative Council after the 1997 changeover.

On September 17, 1995, for the first time, the sixty members of the Legislative Council were wholly elected: twenty members by geographical constituencies, thirty by functional constituencies, and ten by an election committee. Hong Kong was divided into twenty single-seat geographical constituencies, and there were twenty-nine functional constituencies, of which the labor functional constituency could elect two Legislative Council members. Under the "single seat, single vote" system of direct election, individual persons could vote in the geographical constituencies. The total number of registered voters was 2.57 million in 1995, compared with only 1.92 million in 1991.

However, the democratization of the Hong Kong government had come too late; none of the elected Legislative Council members would be allowed to serve after July 1, 1997, when China took over Hong Kong. In retrospect, allowing the partially elected Legislative Council to remain in place until 1997, as originally arranged under the Sino-British Agreement, would have better served the interests of Hong Kong's people. The provisional legislature that has replaced the Legislative Council is even less democratic, because its members were "elected" by a special Election Committee consisting of members approved by China. Although about half of its members had been Legislative Council members, their pledge of allegiance has shifted from the Hong Kong people (the electorate) to China. The most blatant exclusion from the provisional legislature is the United Democrat Party, which had won strong support from the people of Hong Kong in both the 1991 and 1995 elections.

Hong Kong Under Chinese Rule: Turning Back the Clock of Democratization?

Since 1997, the Hong Kong SAR has been governed by the Basic Law, which has been drafted by the Basic Law Drafting Committee and Basic Law Consultative Committee established in 1985. Members of these committees were citizens of Hong Kong from all walks of life, but rich merchants and entrepreneurs have had a higher degree of representation, and China has kept a close eye on the activities of these committees. The Basic Law underwent two drafts each, followed by a round of consultation to solicit public opinion on the drafts. After revision, the second draft of the Basic Law was formally promulgated in April 1990 by the National People's Congress of China. There are 160 articles in the Basic Law, which spell out specifically the relationship between the Central Authorities (China) and the Hong Kong SAR and the fundamental rights and duties of Hong Kong residents. It also establishes the structure of government.

The Basic Law states very clearly that the Hong Kong SAR is a local administrative region of the People's Republic of China and that it shall enjoy a high degree of autonomy (Article 2). The Hong Kong SAR shall be allowed to maintain its capitalist system and way of life for a period of fifty years (Article 5). However, it is also under the control of the Central People's Government in China (Article 12), which is responsible for Hong Kong's foreign affairs and defense (Article 13). In other words, the Hong Kong SAR is practicing the "One Country, Two Systems" principle.

The form of government has undergone only slight modification. The chief executive, who replaces the governor, "shall be selected by election or through consultations held locally and be appointed by the Central Government" (Article 45). The term of office is five years. The chief executive is to be selected by an 800-member election committee consisting of Legislative Council members, Hong Kong deputies to the National People's Congress, and Hong Kong representatives of the National Committee of the Chinese People's Political Consultative Conference (Annex I of the Basic Law). The first chief executive, Mr. Tung Chee-hwa, a Hong Kong shipping magnate and businessman who has connections with China, Taiwan, Britain, and the United States, was selected by an Election Committee of only 400 members.

The Executive Authorities replace the former Executive Council. These authorities comprise the chief executive, a Department of Administration, a Department of Finance, a Department of Justice, and various bureaus, divisions, and commissions that may be established by the SAR (Article 60). Advisory bodies can be established. In fact, the chief executive, Mr. Tung Chee-hwa, has freely appointed "experts" to advise him on social and economic problems in Hong Kong under the Executive Authorities structure.

The Legislative Council remains the same, at least in the number of members (sixty). These members are also to be elected. However, the method of election is different from that passed by the pre–1997 Legislative Council for the 1995 elections. The Basic Law stipulates that the method for forming the Legislative Council "shall be specified in the light of the actual situation in the Hong Kong Special Administrative Region and in accordance with the principle of gradual and orderly progress. The ultimate aim is the election of all the members of the Legislative Council by universal suffrage" (Article 68). However, election by the Election Committee, functional constituencies, and geographical constituencies through direct elections is maintained, albeit with significant changes.

Because a not-so-democratic provisional legislature replaced the popularly elected Legislative Council after the 1997 changeover, China was condemned by Western democracies, most vocally by Britain, Canada, and the United States, for the infringement of the Sino-British Agreement. Hong Kong Chief Executive Tung Chee-hwa declared his decision to hold elections for the Legislative Council, to replace the provisional legislature, as soon as possible, not later than May 1998. However, the specific method of election of Legislative Council members was to be decided by the provisional legislature. Clearly, there was a conflict of interest among the members of the provisional legislature, who obviously wanted to remain in office. By September 28, 1997, the provisional legislature passed the following method of electing sixty members to the Legislative Council in May 1998: (1) thirty members to be elected by functional constituencies, (2) twenty members to be elected by geographical constituencies through direct election, and (3) ten members to be elected by the 800-member Election Committee. Although the methods are similar to those for the 1995 elections, direct election in the geographical constituencies has changed from the "single seat, single vote" to a "proportional representative system," by which voters will elect a list of party-supported candidates rather than individual candidates. The boundaries of the geographical districts have also expanded so that the whole of Hong Kong will be divided into only five districts. All of these changes are viewed as a way for the provisional legislature to prevent the United Democrat Party from winning seats in the Legislative Council. For election of the functional constituencies, the method of voting has reverted to corporate voting, with the "one professional association, one vote" method. A large number of pro-Chinese associations have obtained voting rights in the functional constituency election (thus ensuring that members of the pro-China political parties will get elected in May 1998). As a result of all these changes, the total number of effective voters will be reduced drastically to about 150,000 from the 2.5 million voters in the 1995 elections. The provisional legislature, through the conservative pro-China action of its members, sent out a signal of great retreat from democracy, which does not help Hong Kong's (or China's) world image.

The third component of the Hong Kong SAR government is the judiciary, which consists of courts at all levels: the Court of Final Appeal, the High Court, district courts, magistrate's courts, and other special courts. The judicial system previously practiced in Hong Kong will be maintained. The Court of Final Appeal can invite judges from other common law jurisdictions to sit on it. It is very important for Hong Kong to maintain its legal system, which is inherited from the British. Such a fair and just legal system supports the economic development of Hong Kong. Most important of all, Hong Kong laws will guarantee the human rights of its citizens.

Finally, the Hong Kong SAR continues to maintain the civil service of the British system. These are the "public servants" who have been so efficient in running the day-to-day matters of Hong Kong. This system has contributed to the success of Hong Kong's economic development. Through the supervision of the Independent Commission Against Corruption, the Hong Kong civil service has been largely free of corruption, making Hong Kong the least corrupt country in Asia in which to do business. This is in great contrast to China, where doing business requires establishing a relationship that has monetary implications. Hong Kong merchants and industrialists who have dealings with China are well aware of this. One wonders whether the good name of Hong Kong will be tarnished as the Chinese influence increases.

The Importance of Hong Kong to China

Ever since the Communist victory in Mainland China in 1949, Hong Kong has always served as China's window to the outside world. Through Hong Kong, China has been able to obtain new ideas and technology from the West. Through investing in Hong Kong, China has been able to obtain hard currency to buy Western technologies and necessities to help China's economic advancement. This explains why China condoned the British control of Hong Kong, even during the period of Cultural Revolution, when water and food supplies were exported to Hong Kong without interruption. China's investment in Hong Kong has been quite substantial, especially since the economic reform and open-door policy launched by Deng Xiaoping in 1979. Hong Kong has played an important role in helping China's economic development. Hong Kong provides the expertise and an efficient port for China to trade with any country of the world. Hong Kong is also the stepping-stone for the West to invest in China. Investors from the West liked Hong Kong because Hong Kong was under British rule and practiced the law that they were familiar with. China has set up special economic zones modeled after Hong Kong, the best example being the Shenzhen Special Economic Zone, which looks like a small Hong Kong in its urban make-up. Because of Hong Kong, most of the early special economic zones are located in and around the Zhujiang Delta, of which Hong Kong is geographically a part. Therefore, Hong Kong has been instrumental to the rapid development of the Zhujiang Delta, now an important open economic area of China.

When Deng Xiaoping decided in 1984 to reclaim Hong Kong from Britain, China had already reached a stage of economic development that had brought China closer to Hong Kong. Although Deng Xiaoping still stressed the importance of Marxism and Maoism, he coined the phrase "socialism with Chinese characteristics" to describe the economic reform of China. In reality, China has adopted a capitalistic market-oriented development strategy, which seems to work very well. Deng sanctioned the economic changes, declaring that making money is glorious. In the years since the signing of the Sino-British Agreement, China has invested heavily in Hong Kong. As mentioned above, this promoted Hong Kong's further economic development. The "front shops, rear factories" relationship between Hong Kong and China emerged. China has already firmly established itself as an important exporter of manufactured goods in the world market (notably to the United States). China is so open now that foreign investors feel welcome and are comfortable investing directly in the country. Since the 1990s, China has actively sought membership in the World Trade Organization (WTO). After signing a crucial agreement with the United States in 1999, China succeeded in gaining its accession to the WTO. Other cities in China, notably Shanghai, have developed exporting facilities that rival the efficiency of those of Hong Kong. Hong Kong's role as "window to the West" and stepping-stone to China has diminished in recent years. It seems that Hong Kong now relies more on China for its economic growth than before. This was clearly borne out by the Asian financial crisis of 1998, during which Hong Kong's open economy suffered while China's GDP continued to grow at 8 percent. China assisted Hong Kong by supporting its currency during this period. This reversal of roles between Hong Kong and China is an important trend that explains the lessening importance of Hong Kong to China in recent years.

The return of Hong Kong to China is important because of its political implications. It marks the ascendancy of China as a world power and vindicates China's grudge against Britain for the humiliating treaties forced upon China in the nineteenth century. The event also conveys an important message to Taiwan of China's intent for a unified China in the near future, and it showcases the success of Hong Kong's re-integration with China under the "One Country, Two Systems" principle. In the past, Hong Kong served as a middleman between China and Taiwan. Traveling to China from Taiwan required one to go through Hong Kong (because of Taiwan's policy of no direct travel to China). Since the return of Hong Kong to China, it has become more difficult for the Taiwanese people to obtain a permit to visit Hong Kong. The Chinese government established this policy despite the fact that Taiwanese investment in China, particularly in the Fujian region, is very extensive. Taiwanese industrialists have relocated many of their factories in China.

However, Taiwan has had discussions with China about beginning formal reunification talks, and when the political climate in Taiwan is correct, direct flights between Taiwan and China will very likely be established. Hong Kong will easily lose its intermediary role between Taiwan and China.

The restoration of China's sovereignty over Hong Kong has brought financial benefits to China as well as the allegiance of the Hong Kong merchants and entrepreneurs that China could not have obtained during the period of British rule. China realizes that anti-communist sentiment is quite strong in Hong Kong. In a China-ruled Hong Kong, this opposition may be stifled or subdued.

On the other hand, China faces some risks in putting Hong Kong under its flag. Although Hong Kong has a limited democracy, the people of Hong Kong are accustomed to freedom of speech and a free press. There is a very popular Cantonese radio broadcast program in Hong Kong called, "A Tea Cup in the Turbulence," during which ordinary people call in to vent their anger about local issues, many of which are related to the Hong Kong and Chinese governments. The people of Hong Kong are generally well educated and, through the years of British rule, very experienced in political protests and demonstrations. The majority of Hong Kong citizens, especially university students, are very critical of China's human rights record, particularly the way China treats its dissidents. In the Basic Law, Article 4, it is clearly stipulated that the rights and freedoms of the residents of the Hong Kong SAR will be safeguarded in accordance with law. These rights given to the people of Hong Kong are not what China would like to give to all the Chinese in the mainland. However, the pressure on the central government of China to loosen up on its control over people's thoughts will be growing as China becomes economically more developed. Hong Kong provides a model for the people in Mainland China to emulate. This emulation of Hong Kong's development is already quite common in the Zhujiang Delta; the Hong Kong style of Cantonese speech, television programs, housing designs, popular music, and fashions are all being copied.

China is well aware of the risks. This explains why China has not adopted the electoral reforms for more democratic government instituted under Christopher Patten. This quest for a more democratic and open government is inevitable as a country's economy advances and as its nationals become more affluent and better educated. Without exception, economically advanced nations are also democracies with freedoms of speech and press. In this regard, the Hong Kong SAR, through its Basic Law, can provide a model of limited democratization that China can emulate in the future.

Development Potential of Hong Kong

Since July 1, 1997, the Hong Kong SAR has been fully under Chinese rule. The only factor that has changed as far as Hong Kong's economic development is the loss of the protective umbrella of British rule. That umbrella was required when China was in political turmoil, such as during the civil wars in the 1940s and the Cultural Revolution in the 1960s and 1970s. Now China is a much more open country and is practicing a market-oriented economy very similar to that of Hong Kong. The reintegration of Hong Kong with China could not have come at better time. It also means that the fate of Hong Kong is now closely tied to that of China, as it should be. What Hong Kong needs to do now is to project to the world an image of openness and adherence to democratic rule of law. Hong Kong's geographical advantages are still intact, and the completion of the new airport and an expanded container terminal, all oriented westward to the Zhujiang Delta hinterland, put Hong Kong ahead of other Asian countries in capturing the trade in the region. With its British-style administration-led government structure more or less unchanged, Hong Kong is still well equipped as a high-tech world city to provide efficient services to China and other parts of the world.

Events that occurred during the first five years after the changeover showed some good and bad signs. So far, China has kept its promise of noninterference in Hong Kong's government. In other words, China has observed its "One Country, Two Systems" principle. When the currency crisis swept across East and Southeast Asia in November 1997, which also impacted Hong Kong's exchange rates, the financial secretary of Hong Kong had a free hand to deal with it promptly and decisively. Obviously, Hong Kong has a strong foreign exchange base, but China would have had to intervene had Hong Kong exhausted its financial resources. Therefore, as noted earlier, China's economic strength helped Hong Kong to survive the crisis. The currency crisis also reflected one strong point of Hong Kong's economy: The Hong Kong government has excellent oversight of its banking system, and as a result, banks in Hong Kong have been prudent in lending money and granting credit, unlike other Asian countries. The mostly corruption-free business environment and civil service helped Hong Kong to survive this crisis, which bankrupted South Korea and Thailand and devalued the Korean, Thai, Malaysian, Indonesian, Singapore, Taiwanese, and Japanese currencies. The one weak point of Hong Kong's economy is its overconcentration of investment in real estate, which makes Hong Kong the most expensive place to live in the world. Any slack in the property market will hurt Hong Kong's economy and result in bank loan defaults. The currency crisis caused a sharp fall in Hong Kong's stock market because the Hong Kong government had to increase interest rates to protect the Hong Kong dollar. The crisis caused the collapse of a number of large Japanese-funded department stores in Hong Kong and even large corporations, such as Hong Kong Telecom and Hong Kong Bank. The real estate market lost more than 50 percent of its value. In view of the growing budget deficit, the Hong Kong government decided to cut salaries, raise taxes, and trim services. By March 1999, Hong Kong had entered an economic recession with a 6 percent unemployment rate, and no sign of economic revival was in sight as of early 2003.

In his first five-year term as chief executive, Mr. Tung Chee-hwa has done a reasonable job, which appeases China. He was not very well liked by the people of

Hong Kong, who have blamed him for all the economic woes that have befallen Hong Kong in recent years. He was criticized for his unwillingness to promote a greater degree of democracy in Hong Kong. Of course, as a businessman, his interests are synonymous with those of the rich merchants and entrepreneurs rather than the factory and office workers. As a caretaker of the transition period, Mr. Tung is a hard-working chief executive. He is patriotic and seems very eager to win China's approval. This is dangerous in the sense that he is inviting China's interference, in contradiction to the "One Country, Two Systems" principle. In his first administrative report delivered to the provisional legislature on October 8, 1997, he stated his concern about housing, education, and the tax system. He promised to increase public housing supply by 85,000 units each year. He also agreed to establish a plan to help Hong Kong residents to buy their own residential units through government-assisted loans. He believed in the need to emphasize moral education and improve the quality of teachers and instructional technology in schools. He promised to examine the tax system to ensure the competitiveness of Hong Kong in doing business in the world market. All these are commendable objectives. Obviously, Mr. Tung is interested in neither political reform nor environmental protection in Hong Kong. He has also given the impression that he does not work well with the British-trained civil service. Despite his low job approval rating, he was re-elected by the Election Committee to a second five-year term as the chief executive of Hong Kong in March 2002.

The most negative sign was that projected by the action of the provisional legislature. The provisional legislature's repeal of the electoral reform laws passed by the Legislative Council just before the changeover on July 1, 1997, discussed above, revealed an anti-labor attitude, which is not surprising since most members are rich merchants, industrialists, and business people. Another damaging impact was some members' attempt to change the method of election by functional constituency to give more votes to pro-China associations. The change in the method of direct election with the obvious objective of depriving the United Democrat Party of votes in the May 1998 election of the First Legislative Council was another bad move. All these indicate clashes in interest among the business people, the pro-Chinese camp, and a small number of independently minded members. Despite these changes, a surprising outcome of the May 1998 election of the First Legislative Council was that the United Democrat Party won the largest number of seats under the direct election category, an indication of the Hong Kong people's strong support for a democratic society.

A positive development for Hong Kong during this period was the opening of the new multibillion-dollar airport at Chek Lap Kok on July 6, 1998. Despite some initial problems, the huge modern airport has put Hong Kong into a leadership role as the air transportation hub of the Pacific region. In June 1998, President Clinton's official visit to China included Hong Kong as its last stop, thus symbolizing U.S. support for Hong Kong under China's sovereignty.

It is important for Hong Kong to maintain its competitiveness in the world market. The expensive property market and the rise in real wages have resulted in higher costs of Hong Kong goods and services. Because Hong Kong has relocated most of its industry to the Zhujiang Delta region, Hong Kong has to rely heavily on the service sector to sustain its economic growth. Hong Kong's investment has focused excessively on real estate, which is highly vulnerable to interest rate fluctuations, as borne out by the recent currency crisis. If manufacturing in the Zhujiang Delta area declines, the demand for services in Hong Kong will also decline. One should note that the economic development in the Zhujiang Delta has been very rapid. In due course, the economic structure in the Zhujiang Delta will change to one that will compete rather than complement that of Hong Kong. Hong Kong will have to develop high-value and high-skill industries to lessen the heavy reliance on the service sector in anticipation of the economic structural change in the Zhujiang Delta area. On the other hand, Hong Kong can benefit from a greatly enlarged hinterland in the form of a Greater Zhujiang Delta economic area. A number of ports in China are quite eager to capture Hong Kong's entrepôt function, the most notable competitor being Shanghai, which has been advancing rapidly. Even Hong Kong's neighbor, the Shenzhen Special Economic Zone, is competing for Hong Kong's container-port shipping business by offering lower fees to shippers. To remain the leader, Hong Kong needs to maintain a well-educated and talented quaternary workforce with strong international connections.

It is not clear how in-migrants from China will affect Hong Kong. There is still strict control by Hong Kong and China on the number of in-migrants who can permanently settle in Hong Kong. A future influx of Chinese in-migrants to Hong Kong will affect the quality of life of the Hong Kong people in relation to crime, income, housing, and education.

Finally, Hong Kong has to deal with its environmental pollution problems, which include deteriorating air and water quality. As the Zhujiang Delta region becomes industrialized and its market-oriented agriculture increases the use of chemical fertilizers and insecticides to boost productivity, the magnitude of the resulting air and water pollution also affects Hong Kong, which is located at the mouth of the Zhujiang River. The main culprits are the township-village enterprises. The Hong Kong government has long recognized the importance of environmental protection but has taken only limited action to alleviate the problem. For both the physical and economic health of the people of Hong Kong, decisive action should be taken by the Hong Kong SAR immediately. A case in point is the outbreak in December 1997 of the H5N1 virus (avian flu), believed to have spread from chickens imported from the Zhujiang Delta region, which has caused great economic damage to Hong Kong, affecting not only poultry farmers and retailers but also the tourist industry in Hong Kong.

Hong Kong's future development is closely tied to China's economic development. It is fortunate that China's future development potential is still very high, despite the recent currency crisis, which seems to have had a lesser effect on China than other countries in the region. China has not devalued its currency *(Renminbei)*. To support its development, China has to rely on Hong Kong's stock market to attract capital. Over 100 China-backed and state-owned enterprises (the so-called "red-chip stocks") are listed on the Hong Kong Stock Exchange, raising a total equity capital of US$92 billion directly or indirectly through Hong Kong in 2002. China's direct investment in Hong Kong exceeded US$7.4 billion by the end of 2001, making China the second most important investor in Hong Kong, after Britain. China has planned for large-scale engineering, construction, and investment in its central and western regions in the near future. This provides opportunities to Hong Kong for direct investment in China. Hong Kong and China will work closely together in the areas of finance, services, and industry for a long time in the future. The people of Hong Kong will have to be able to adapt quickly to changes in the world economy and in its relationship with China if it wants to maintain its status as an Asian tiger. Human ingenuity has always been Hong Kong's strong point. Currently, Hong Kong's main problem lies in the deteriorating quality of its education. Remedies are being sought by the Hong Kong government. There is now increasing cooperation between Hong Kong and China in research at the university level. Because of its high remunerative structure, Hong Kong is in a strong position to attract talented youths from China to strengthen its high-tech workforce.

Conclusion

The return of Hong Kong to China on July 1, 1997, was a most important event for both China and Hong Kong. Politically, it marks the end of the colonial status of Hong Kong and the re-integration of Hong Kong with its motherland, China. This re-integration would have been most joyous to the Chinese in Hong Kong had it not been for the fact that China is a communist country with a poor record on human rights, which the Tiananmen Square Incident of June 4, 1989, seemed to have confirmed. All the quarrels between China and Britain over the election reforms stemmed from the distrust between the two countries over each other's motives. Political reforms in Hong Kong for a more open government and a directly elected Legislative Council are seen by the British as the means to ensure Hong Kong's independence from China's interference after the 1997 sovereignty change. Unfortunately, the British government did not have the foresight to give Hong Kong a more democratic form of government earlier.

To China, Hong Kong is an important asset politically and economically. Politically, the re-integration of Hong Kong in 1997 and Macao in 1999 presages Taiwan's final unification with China. These events give pride to the Chinese government and the Chinese people. The sovereignty changeover lifts China into world power status and vindicates the humiliation that China experienced during the period of British imperialism. A prosperous Hong Kong will certainly add to China's world prestige. Economically, Hong Kong, as the world's eighth-largest trading economy, will help accelerate China's economic development in the twenty-first century. Hong Kong is the place for China to gather capital in support of all the construction projects going on in China. As a world city, Hong Kong has the efficient telecommunications and financial expertise to support China's trade with the West.

The major liability to China is the strong demand by the people of Hong Kong for a more open and democratically elected government, which is denied to the Chinese people in the Mainland. The "One Country, Two Systems" principle is a means to pacify the people of Hong Kong, but it also serves to remind the Chinese people in the Mainland that Hong Kong's decadent capitalistic system cannot be practiced in China. Even after Hong Kong's re-integration with China, Hong Kong remains different. But Hong Kong's ideas have found ways to infiltrate China, as travelers in the Zhujiang Delta region can attest.

To Hong Kong, re-integration with China presents challenges and opportunities. Without the British umbrella that used to shield Hong Kong from China's interference, Hong Kong has to project a more open and democratic image to the world to show that the government is free from Chinese interference.

Hong Kong's future development will be closely tied to that of China. Hong Kong has to be vigilant of changes in the world market and adapt quickly in order to maintain its leading role as an entrepôt port and financial center in Asia. The new airport has provided the first step for Hong Kong to achieve such a goal. Hong Kong's investment in education and technology, supported by an open and fair society in a pollution-free environment, will ensure Hong Kong's success in the next millennium. If China can keep its "hands-off" promise, Hong Kong will be able to advance economically on its designated path, and because of that, will be a greater asset to China. Hong Kong also provides a model for China to emulate should China undertake future political reforms to support its economic development.

References

Census and Statistics Department. 2002. *Hong Kong Annual Digest of Statistics*. Hong Kong: Government Printing Department.

Chiu, T. N., and C. L. So. 1986. *A Geography of Hong Kong*, Hong Kong: Oxford University Press.

Harris, P. 1988. *Hong Kong: A Study in Bureaucracy and Politics*. Hong Kong: Macmillan.

Howlett, B. 1996. *Hong Kong 1996: A Review of 1995 and a Pictorial Review of the Past Fifty Years*. Hong Kong: Government Printing Department.

Howlett, B. 1997. *Hong Kong 1997: A Review of 1996*. Hong Kong: Government Printing Department.

Johnson, G. E. 1995. Continuity and Transformation in the Pearl River Delta: Hong Kong's Impact on Its Hinterland. In *The Hong Kong-Guangdong Link: Partnership in Flux*, ed. R. Y.-W. Kwok and A. Y. So, pp. 64–86. Armonk, N.Y.: M. E. Sharpe.

Lo, C. P. 1992. *Hong Kong*. London: Belhaven Press.

Miners, N. J. 1986. *The Government and Politics of Hong Kong*, 4th ed. Hong Kong: Oxford University Press.

_____. 1990. Constitution and Administration. In *The Other Hong Kong Report, 1990,* ed. R.Y.C. Wong and J.Y.S. Cheng, pp. 1–28. Hong Kong: Chinese University of Hong Kong Press.

Sung, Y.-W. 1995. Economic Integration of Hong Kong and Guangdong in the 1990s. In *The Hong Kong-Guangdong Link: Partnership in Flux,* eds. R. Y.-W. Kwok and A. Y. So, pp. 224–250. Armonk, N.Y.: M. E. Sharpe.

Yeh, A. G.-O., ed. 1996. *Planning Hong Kong for the 21st Century.* Hong Kong: Centre of Urban Planning and Environmental Management, University of Hong Kong.

Yeung, Y.-M. 1996. Hong Kong's Hub Functions. In *Planning Hong Kong for the 21st Century,* ed. A. G.-O. Yeh, pp. 143–161. Hong Kong: Centre of Urban Planning and Environmental Management, University of Hong Kong.

22

Taiwan and Mainland China: Divided or United?

Chiao-min Hsieh

The current division of China between the Mainland and the island province of Taiwan (encompassing Penghu, or the Pescadores, and several inshore islands) is but the latest chapter in a 5,000-year history of alternating periods of unity and division. This is best encapsulated by the Chinese proverb: "After long unification, there will be divisions, and after long divisions, there will be unity" *(He Jiu Bi Fen, Fen Jiu Bi He)*. During the 2,681 years from 770 B.C. to A.D. 1911, China was divided into separate kingdoms for a total of 1,519 years, or 57 percent of the time, and united in a single empire for 1,162 years, or 43 percent of this long period. Some historians have observed that advanced technologies and prosperous economies emerged during China's years of division. Some even claim that mega-states usually develop during the period of a civilization's decline. The modern history of China—from the overthrow of the Qing (Manchu) Dynasty and the establishment of the Republic to the completion of the communist takeover of the mainland in 1949 and the resultant flight of the Nationalist government to Taiwan—seems to lend some credence to this thesis.

Not long after the 1911 revolution, the two major Republican factions—the Kuo-min-tang (KMT) or Nationalists, and the Chinese Communist Party (CCP)—split and began a lengthy and bitter civil war for total control of the country. They cooperated for only two brief periods, in 1924 and 1937, for tactical reasons, and not even the Sino-Japanese War (1937–1945) could bring the factions together against a common enemy. Since the communist victory in 1949, the central government has worked steadily to reunify the country, with considerable success. Tibet was brought under firm Chinese control in stages during the 1950s, Hong Kong reverted to China in 1997, and Portugal formally surrendered Macao in 1999. This leaves only Taiwan, which was annexed by Japan from 1895 to 1945 and separated politically from the Mainland in 1949, outside the umbrella of Mainland China.

Four countries were divided shortly after the Second World War: East and West Germany, North and South Korea, North and South Vietnam, and Taiwan and Mainland China. Although these four pairs of "separated regimes" have similar characteristics, their different historical backgrounds have led to different reunification solutions. For example, the military forces of the Soviet Union and America partitioned Germany and Korea after the Second World War, and France and the Viet Minh partitioned Vietnam in 1954. China was separated by a civil war. Each faction in the four divided countries claimed legitimacy and established separate political entities. In general, East Germany, North Korea, North Vietnam, and Mainland China followed communist ideology, while West Germany, South Korea, South Vietnam, and Taiwan developed along the lines of democracy and capitalism. Because of the ideological split, reunification of each of these four regimes was transformed from an internal political issue into a complicated international one.

Each reunification has taken a different form. North Vietnam used military power to seize South Vietnam (the military model). The Germanys were reunited through political and economic dialogue (the peace model). The Koreas are still divided. The relationship between Mainland China and Taiwan remains uncertain.

The unification problems faced by Korea and China are similar because both have taken military actions in the past but have more recently made moves toward peaceful solutions. Even though North Korea has threatened world peace by possessing nuclear weapons, the threat is aimed mainly at the United States rather than South Korea. The unification of Taiwan and Mainland China, however, will be much more complicated than any reunification of Korea. North and South Korea are about the same size in land area and population. Both regimes are recognized internationally by about the same number of countries. In 1973, both North and South Korea developed "dual recognition" in international affairs. Furthermore, Korea's model of "one country, one nationality, two governments and two foreign policy positions" is similar to Germany's model of "one nationality, two countries, two governments and two foreign policy positions," a policy that eventually enabled reunification. In contrast, only 28 countries, mostly in Africa and Latin America, recognize Taiwan, compared with 160 nations that officially recognize Mainland China. Their territories and populations are unequal, and perhaps most importantly, they are separated both physically and psychologically by a body of water about one hundred miles wide. These factors ensure that the continuing stalemate that stretches across the Taiwan Strait affects not only the security of the western Pacific but also the security of the entire world.

Background of the Taiwan Strait Conflict

When the Communists defeated the Nationalists (KMT) in October of 1949 and established the People's Republic of China (PRC), the defeated KMT relocated the government of the Republic of China (ROC) to Taipei on Taiwan. President Chiang Kai-shek vowed to retake the Mainland by force and challenged the legitimacy of the "bandit" regime in Beijing. Chairman Mao Zedong was determined to capture Taiwan, and the "liberation of the island" became his most important mission. The "civil war" had ended on the Mainland but had shifted to the Taiwan Strait.

In 1950, the PRC decided to invade Taiwan but put its invasion plans on hold after the outbreak of the Korean War. American President Harry Truman, afraid that the Communists would conquer Taiwan, dispatched the Seventh Fleet to "neutralize" the Taiwan Strait. When the United States recognized the Republic of China in Taiwan as the legitimate government of China, the PRC's liberation of Taiwan was postponed indefinitely.

The Chinese Communists considered the American action an "occupation" of Taiwan. The 1955 Chinese invasion of the Tachen and Chiangshan Islands along China's east coast encouraged the United States to sign a mutual defense treaty with the ROC and commit itself to the defense of Taiwan against Communist invasion.

In the summer of 1956, the PRC shelled the islands of Quemoy and Matsu, two Nationalist outposts off the shore of Fujian province. President Dwight Eisenhower, taking a strong stand against this Communist "aggression," threatened to use nuclear weapons, if necessary. Beijing continued to focus the world's attention on the offshore islands until the shelling ended in October of 1958. This was the last time Beijing would threaten the offshore islands or Taiwan with invasion, possibly due to internal problems such as the Cultural Revolution of 1966–1976.

Though the military struggle along the Taiwan Strait ceased, a propaganda war took its place. Many a battle was fought on the diplomatic front as China and Taiwan sought international recognition.

With the signing of the "Shanghai Communiqué" between President Richard Nixon and Premier Zhou En-lai in 1972 and the establishment of U.S.-China diplomatic relations in 1979, Beijing announced a cease-fire policy along the Taiwan Strait. It also launched a series of peace initiatives aimed at reunifying Taiwan with Mainland China. The Communists abandoned the slogan "Liberate Taiwan!" and now proclaimed a policy of "peaceful unification." In response, the Nationalists shifted from a call of "Recover the Mainland and defeat the bandit invasion!" to the "Three Principles Doctrine to Unify China."

Beijing's reunification drive reached a high point in 1981 when Marshal Ye Jianying made a nine-point proposal and for the first time invited the KMT authorities in Taipei to the negotiation table. His plan would permit Taiwan to "enjoy a high degree of autonomy as a special administrative region." Deng Xiaoping suggested that the Hong Kong formula, namely, "one country, two systems," be applied to Taiwan and also promised to be more "generous" toward Taiwan.

Premier Zhao Ziyang, while visiting the United States in 1984, offered this peaceful scenario for unification: "After the country is reunified, Taiwan, as a special administrative region of China, can retain much of its own character and keep its social system and lifestyle unchanged. The existing party, government, and military setup in Taiwan can also remain unchanged."

However, two important factors caused Taiwan to reject these peaceful overtures from the Mainland. First, the KMT government in Taipei would have been reduced to the status of a local provincial government of China. Second, economic prosperity provided the Taiwanese workers with incomes ten times higher than those of the Mainland. Taiwan's response to the new Beijing initiative was summarized in the statement: "No contact, no negotiation, and no compromise." The islanders also flatly turned down an offer for "three exchanges"—mail service, trade in goods, and air and shipping services across the Strait.

The Thaw of the Frozen Relations

In 1987 President Chiang Ching-kuo of the ROC unexpectedly lifted the ban on travel across the Strait, allowing retired soldiers to travel to the Mainland to visit their relatives. With this move, a momentous change in relations between the two adversaries began. Later, the visit of millions of people from Taiwan to the Mainland had enormous consequences for Taiwan-Mainland China relations.

The motivation for Taipei to expand access to Beijing was threefold. First, there were humanitarian considerations. Many retired soldiers, who followed Chiang Kai-shek to Taiwan in 1949 and also helped Chiang Ching-kuo's effort in building the cross-island highways, had served the KMT loyally.

Second, there were political reasons. Veterans had helped the KMT to defend Taiwan and also supported the KMT candidates in political elections. To keep veterans' votes from straying to the Democratic Progressive Party (DPP) camp, the KMT leaders granted their requests.

Third, there were Nationalist considerations. The KMT leaders and Chiang Ching-kuo favored neither the Communists nor the Taiwanese independence movement. They wanted "one China," instead of "one China and one Taiwan." Permitting the people in Taiwan to visit the Mainland and encouraging cultural exchanges would arrest the independence movement and reverse the trend toward Taiwanization.

The relaxed conditions between Taiwan and Mainland China generated two notable changes. The first was an economy favorable to both the Mainland and Taiwan, and the second was diplomatic moves that might not necessarily be good for Taiwan. Two-way trade expanded from $1.5 billion in 1988, to $2 billion in 1989, to over $4 billion in 1990. At the same time, companies from Taiwan were investing in Mainland China, especially in Fujian province, which was not only the nearest province to Taiwan but also the ancestral home of many of the Taiwanese. Taiwan changed its foreign policy to one of "flexible diplomacy," abandoning Chiang Kai-shek's doctrine of *"Han Z'e pei lian li"* (never co-exist with the dead enemy).

For example, in 1989, Taiwan established full diplomatic relations with Grenada, Liberia, and Belize, even though they had diplomatic ties with Beijing. The PRC immediately terminated relations with these countries, since diplomatic relations might be construed as support for Taiwan's position of "two Chinas" or "one China and one Taiwan." Beijing successfully established formal ties with Saudi Arabia, Indonesia, and Singapore in 1990 and South Africa in 1998, while preventing Taiwan from expanding its diplomatic space in the international community.

The most effective way of reducing the level of hostility between the governments of Taiwan and the Mainland has been cultural exchange fueled by the flow of athletes, musicians, journalists, and others from Taiwan to the Mainland. This exchange, however, has not been reciprocated. Taipei still refuses to permit many Mainlanders to visit Taiwan, even for humanitarian reasons, and only a few thousand have done so.

In the 1990s, the pace of change in cultural and economic exchanges accelerated. An economic relationship developed across the Taiwan Strait. Tourism, trade, and investment expanded greatly. China's minerals and other raw materials, superabundant manpower, and huge and rapidly growing markets, combined with its political stability and welcoming policies, encouraged Taiwanese capitalists to invest enormous sums in Mainland enterprises, including some being privatized as the government relaxed controls. This certainly contributed to a better life for Chinese both on the Mainland and in Taiwan.

The Context of the Taiwan-Mainland Dispute

The area of politics and diplomacy is by far the most problematic. Theoretically, there are three possible outcomes to the dispute: (1) the status quo, or continuation of the now half-century-old situation of China divided but with each party claiming to be the legitimate government of the entire country; (2) reunification of Taiwan with China as a single political entity; and (3) formal independence for Taiwan alongside China. The majority of the inhabitants in Taiwan generally have favored the status quo, but more people support unification than support independence on the condition that the Mainland become a more free, prosperous, and democratic country. The emergence of one outcome over the others depends, in large part, on internal developments within Taiwan and only marginally on U.S. policy, which has consistently favored a "one China" solution with the terms of reunification to be worked out by the parties themselves.

Mainland China has no plans to deviate from its objective of bringing Taiwan back under the rule of Beijing. In 1980, for example, Deng Xiaoping enunciated the three top national priorities as containment of hegemonies, economic modernization, and reunification of China. Until very recently, even the terms of this reunification were unchanging. Taiwan is considered a renegade province, a subordinate part of the one and only China. Any attempt to split Taiwan from China in any kind of "two Chinas" solution would be met with force.

The longer the delay in the unification process, however, the less the chance of success. A new generation of leaders in the CCP and the KMT is less familiar with the history of the region and, therefore, less equipped or inclined to resolve the problem of unification.

Accordingly, Beijing has abandoned its former insistence on complete ROC surrender to the absolute sovereignty of the PRC and is now willing to apply the "Hong Kong formula" of "one country, two systems" to Taiwan. There is still, however, considerable disagreement over how a formula would be applied, since Taiwan is so very different from Hong Kong.

In the 1990s, Taiwan established the Strait Exchange Foundation, or SEF, and Mainland China set up a counterpart agency called the Association of the Relations Across the Taiwan Strait, or ARATS, as intermediary bodies to handle cross-strait exchanges. Gu Chen-Fu from Taiwan and Wang Dao-Han from the Mainland were assigned as the heads of their respective organizations.

In April 1993, Gu and Wang held a conference in Singapore that started a fresh dialogue between the two sides. This conference was a breakthrough for Taipei-Beijing negotiations and was the first Taiwan-Mainland contact since 1949. The hostility and confrontation of past years had slowly given way to reconciliation and negotiation.

The meeting was so successful that a second talk was planned for 1995, but it was suddenly suspended in retaliation for President Lee Teng-hui's visit to the United States. China stopped cross-strait talks, as well as other relevant discussions with Taiwan.

In 1996, a series of missile tests conducted by Beijing in waters off Taiwan's coast led the United States to send warships to the Taiwan Strait. Beijing recalled its ambassador, and Sino-American relations plunged to a new low.

Since the Gu-Wang conference in 1993 in Singapore was very successful and Gu's visit to Beijing in 1998 proceeded smoothly, hopes were high that Wang's first visit to Taipei in October 1999 would prove beneficial to both sides of the Strait. At this critical moment, President Lee Teng-hui, in an interview with a German broadcasting company in July 1999, suddenly announced that the situation between Taiwan and the Mainland was a "special state-to-state relationship." In the future, Taiwan's talks with the Mainland would be viewed as "bilateral talks between two states."

Lee's statement immediately and predictably created tension between the two sides of the Strait, received a vitriolic response from Beijing, made Washington very uncomfortable, and drew global attention in the international media. His position deviated from the "one China" principle that was the basis for the normal development of cross-strait relations.

For President Lee, "one China" did not exist, and the unification of "one China" could take place only in the future when China became democratic. Ever since the People's Republic of China was founded in 1949, he maintained, two different political entities, each being of equal status and neither currently holding complete sovereignty, have ruled the two sides of the Strait of Taiwan. And he rejected totally the notion that cross-strait relations were but an internal relationship between a legitimate government and a renegade group, or between a central government and a local government. Beijing's characterization of Taiwan as a renegade province of the PRC was historically and legally incorrect.

For its part, Beijing reiterated its familiar position that there was no theoretical or historical basis for Lee Teng-hui's "two states" theory. The existing pattern of cross-strait relations was at bottom a historical legacy of the 1946–1949 civil war in China and therefore was an internal affair of China.

Before Chiang Kai-shek was defeated and driven to Taiwan, the ROC, administering both the Mainland and Taiwan, was undoubtedly a state with independent sovereignty. But since it was replaced by the PRC in 1949, never has the so-called ROC, in its original sense, existed.

From Beijing's point of view, the political forces in Taiwan were merely the remnants of a former government on the island, which has yet to be reunited with the PRC. They asserted that nearly all members of the international community acknowledged there was only "one China" and that Taiwan has never held sovereign status in any real sense of international law. Lee's naked separatist remarks were an attempt to destroy cross-strait relations and were suggestive of Taiwan's independence movement.

According to Beijing, the divisive act of President Lee seriously damaged the basis of cross-strait relations, affected the stability of the Taiwan Strait situation, and harmed the peaceful reunification process. During this period of rising tensions, the United States tried to steer a middle course by reassuring Beijing of its commitment to a peaceful resolution of the "one China" question and pledging to Taiwan its continued support.

Taiwan's Political Landscape and Its Presidential Election

In Taiwan, the leadership of Chiang Ching-kuo and Lee Teng-hui played a defining role in politics. In spite of "young Chiang's" early authoritarian training in the Soviet Union, he devoted the last years of his life to the liberalization and democratization of Taiwan. He twice selected a Taiwanese, Lee Teng-hui, as vice president, paving the way for Lee to succeed him after his death in early 1988. He tolerated the formation of the Democratic Progressive Party (DPP) in December 1986, lifted the ban on travel to the mainland in 1987, and revoked the state of martial law in place since 1949. The lifting of martial law started a period of radical reconstruction of Nationalist rule, culminating in 1996 with the first direct presidential election.

Chiang hated communism, but he did not favor the DPP's advocacy of Taiwan as an independent state. He supported unification across the Taiwan Strait because he believed that Taiwan's economic growth and political democratization could serve as a model of development for the Mainland and force the regime there to give up its communist dictatorship.

Lee Teng-hui served as president from 1988 to 2000. His background and ideas differed markedly from those of Chiang. A native Taiwanese, Lee's formative years were spent in Japan. He worshipped Japan's culture and had little regard for China. He spoke better Japanese than Mandarin and occasionally interjected Japanese words when speaking Chinese. Lee was a Presbyterian and received a Ph.D. degree from Cornell University. Not knowing Lee's political intention, but impressed with his academic achievement and religious beliefs, Chiang decided to groom him as a political leader. Lee served as mayor of Taipei, provincial governor, and vice president. As vice president, Lee adhered to the KMT position of his benefactor, Chiang. His sympathies, however, were more with the DPP and the independence movement, making him inclined to act favorably in that direction.

The leadership and policies of these two men ultimately led to free elections and a range of views on the future of Taiwan. Taiwan has about eighty-two political parties, even though some parties have never been active in elections. The three major parties, namely, the KMT, the DPP, and the People First Party (PFP), founded in 1993 and derived from the KMT, hold quite different views on the question of cross-strait relations. The KMT advocates, "No haste, be patient." The DPP maintains, "Taiwan has independent sovereignty. It is not a part of the People's Republic of China." The PFP has a "one China" policy that advocates, "Reunification, but not yet."

From 1951 to 1993, the KMT was the dominant party in elections. Since 1993, however, the DPP has gradually won more elections. In 1997, the DPP defeated the KMT for the first time, a trend that may continue into the future. In the long run, the early-leading KMT may eventually become less important in Taiwan's political life.

Beijing is concerned that the KMT's power to control the island is in decline. The rapid rise of the Democratic Progress Party (DPP), which urges self-determination for Taiwan's independence, also poses a threat to China's hopes for reunification.

In 1996, in the first direct presidential election on the island, the KMT nominee, Lee Teng-hui, won with 54 percent of the vote. The 2000 election was much more competitive among three candidates: the ruling KMT party nominee, Lien Chan (then vice president); the People First Party (PFP) candidate, James Soong; and the DPP candidate, Chen Shui-bian.

All of these candidates had impressive educational backgrounds and extensive experience in public life. Near the end of the campaign, about one-third of the voters were undecided, and the election was so close that it was difficult to predict.

At the beginning of the campaign, the most popular candidate was Soong, the independent breakaway candidate. Born on Mainland China, he received his Ph.D. from Georgetown University. James Soong had served the KMT-led government as Taiwan's first directly elected provincial governor in 1994. Soong and Lee Teng-hui had helped each other politically to such an extent that the relationship was characterized as "father and son."

Two years later, President Lee announced plans to abolish the provincial government (the seat of the provincial government is in Taichung and the central government is in Taipei) to avoid the confusion over the name "province," as used in Beijing, and to clarify Taiwan's status as a state and not a province. This announcement severed President Lee's relationship with his right-hand man, Governor Soong. Soong openly opposed Lee's announcement and tried to preserve the provincial status. Their relationship became so strained that in 1998, when President Lee had to choose a presidential candidate for 2000, Lee decided to give the nomination to Lien Chan, his vice president, instead of Soong.

Soong furiously announced that whether or not he received support from the KMT, he would run for president himself. Since Soong's decision would threaten Lien's chances, the KMT expelled him, and Soong organized the PFP.

Vice President Lien Chan came from a wealthy family. His grandfather, Lien Wang, was a noted historian. His father served as Minister of Internal Affairs under Chiang Kai-shek. He received a Ph.D. from the University of Chicago. His father was born in Taiwan, his mother in northeastern China, and he himself was born in Sian. He had long served the KMT in various positions, including premier and vice president in the 1990s. With such a prominent political career, Lien had a good chance to be elected.

The third candidate was Chen Shui-bian, the leader of the DPP opposition party, a native of Taiwan, and a graduate of National Taiwan University's law school. He first served as an attorney and became one of the leaders of the DPP. In 1994, Chen was elected mayor of Taiwan's capital, Taipei, the highest office ever attained by a DPP member. A split in the KMT vote gave Chen his narrow lead to victory. At the end of his term, however, Chen lost his bid for re-election to a popular, brilliant KMT politician, Ma Ying-jeou, who has a Harvard law degree and is a descendant of a Mainlander. Two years later, Chen's race for the presidency would be a repeat of his 1994 Taipei mayoral victory: a three-way contest that split the KMT votes to Chen's benefit.

History indeed repeated itself in the 2000 presidential election. The KMT had been in power in Taiwan for more than a half-century and believed that if Lien and Soong could be united in the campaign, there would be no chance for DPP candidate Chen Shui-bian to win. Lien and Soong, however, could not agree on who would be the presidential candidate. Once Soong became an independent candidate, their continued attacks on each other ensured that Chen would be the benefactor. The campaign fit the Chinese slogan, "The fight between snipe and clam will end with both as captives of the fisherman." Chen was the fisherman and won the election.

Although the backgrounds, qualifications, and abilities of these three candidates for president were important considerations, the most important factor in the race was their stand on the issues important to Taiwan. The state of cross-strait relations was a given element, and the perceived impression of Taiwan politicians was that the Taiwan populace preferred the status quo to either immediate independence or immediate unification.

Soong's policy sought to establish "quasi-international relations between relative sovereignties," a middle way between the PRC-endorsed "one country, two systems" model and Taiwan's independence policy. He also advocated a thirty-year nonaggression pact between Taiwan and the PRC, to be followed by a European Union–style relationship.

Lien was the official candidate of the dominant KMT, with huge financial support and a vast network of party organizations. Political stability in Taiwan would be the ultimate reward for his "three no's" and "three musts" policies. The three no's referred to no Taiwan independence, no hasty unification, and no confrontation. The three musts were maintenance of peace, the continuation of exchange, and the search for a win-win situation that benefits both sides of the Taiwan Strait.

Of the three leading presidential candidates, Chen Shui-bian was the advocate for Taiwan independence, even though he supported economic ties and day-to-day interactions with the Mainland. Chen was a proponent of the DPP's preferred slogan, "Strengthen the base and go west."

Beijing authorities claimed that any of the candidates in Taiwan were acceptable except the one who supported Taiwan independence. The PRC's dislike of Chen was obvious.

In Taiwan, the presidential race of the three closely matched candidates would depend on the two faces of the population: one being the descendants of the Mainland who favored the KMT, and the other being the native Taiwanese who supported the DPP.

The central issue of the election was the economic prosperity and stability of the island. In other words, domestic concerns would play the leading role. Political corruption, or "black gold," is loathed by Taiwan's voters. At a critical moment, the former president, Lee Teng-hui, spread word of a financial scandal in which huge sums of government money had turned up in the pocket of Soong's son in America. Soong immediately denied the charge. Although there was little truth in the matter, the incident damaged Soong's reputation tremendously. Once the rumor was spread, Soong's untarnished image was lost, and his lead in the polls declined rapidly.

One factor in Chen's victory was the support of Lee Yuan-tseh, a Nobel laureate and the president of the Academic Sinica. His academic accomplishments had earned him great respect and influence in Taiwan. Lee, a native Taiwanese, has indicated his support for the idea of Taiwan independence. Besides his personal support, he also united many Taiwanese industrial leaders for Chen's campaign. Lee's actions gave Chen a chance to win the presidential campaign in 2000.

The Economic Factors Affecting Taiwan-Mainland Relations

Chen Shui-bian won the election to become the president of the ROC in Taiwan in March 2000 and ended the fifty-five-year rule of the Nationalist Party. Chen's presidency will not be easy: Taiwan's economy is in the deepest recession on record, relations with China are paralyzed, and Taiwan's society is restless.

Critics of Taiwan's situation have said, "We have a ruling party that cannot rule and an opposition party that does not know how to effectively oppose the government." President Chen indeed faces an uncertain future; however, his party did win the recent legislative election. As a result, the legislature may be less of an obstacle to Chen's executive power than in the past.

The biggest problem for Taiwan will be its economic situation. Chen's administrative strategy is to separate politics from economics, but the main cause of the economic problem is political in nature. Tens of thousands of Taiwanese companies, especially hundreds of garment and footwear factories, have already shifted their operations to the Mainland, pouring in an estimated $60 billion.

More serious is the role of technology companies, such as computer companies producing basic components like keyboards and disk drives. Some people fear that Taiwan is jeopardizing its security by transferring technology and money to China and contributing to the support of military forces there. Because of this fear, former president Lee Teng-hui formulated the "no haste, be patient" policy, designed to prevent China from gaining access to Taiwan's technologically sophisticated industries, including those involving computers and semiconductors. Lee tried to curb the increasing flow of investment into Mainland China and asked Taiwanese investors to remember their roots on the island. But more than 30,000 Taiwanese businessmen, such as plastics tycoon Wang Yung-ching, blasted the government's business policies toward China. Some accused Mr. Wang of undermining the Taiwanese economy, but Wang issued a statement: "There is only one road for Taiwan to walk, and that is the road of breaking the restriction on individual development."

After much debate on the issue, Taiwan recently lifted longstanding restrictions on direct investment in China. It is a milestone in the island's relations with Mainland China. The old government policy of "no haste, be patient" has been replaced by a new policy of "aggressive opening, effective management." The new policy is the result of a realization that the capital flows have been a one-way street from Taiwan to China, and Taiwan needs to increase the return flow. They also hope that companies will send home more of their profits from China.

Economic issues may be the most important factor in establishing a stable and peaceful relationship between Taiwan and China. Now that both Beijing and Taipei have entered the World Trade Organization, or WTO, this commercial tie may have greater potential to transform relations between Taiwan and China than the frequently discussed diplomatic and military topics. Trade between Taiwan and China now totals more than $25 billion annually and continues to grow at a pace of 7 percent per year. In many ways, the commercial policy situation in Taiwan with regard to China is very much like the U.S. policy toward China. Since 1990, the U.S. Congress has supported expanded commerce with China despite friction over human rights abuses, a rising U.S. trade deficit with China, and many other bilateral tensions. Important factors contributing to this policy change were the decline in faith in economic sanctions as a tool of national policy and the U.S. business community's strong lobbying for improved economic relations with China. Similar lobbying by the Taiwanese business community has turned the Taiwanese administration, once openly hostile to Beijing, into an advocate for expanding direct commerce. These business connections with China have changed the political landscape in Taiwan regarding cross-strait relations. Political tensions are still high, but growing commercial ties have created an incentive for peaceful cooperation on both sides of the Taiwan Strait. China, too, would likely welcome increased trade, direct shipping, and investment opportunities as a way to strengthen its economy. Therefore, China willingly supported Taiwan's WTO membership application.

Taiwan's admission to the WTO allows it access to a forum where it can deal with China directly as an equal through a mediated process. In addition, other interested members can join in WTO disputes between two members. Thus, in a commercial dispute between Taiwan and Beijing, members such as the United States, the European Union, and Japan could formally enter into the discussion. WTO membership will have an important diplomatic value in the Taiwan Strait dispute.

The parties to the Taiwan Strait problem have tried military threats, political maneuvers, and diplomatic negotiations, but all have failed. Commercial ties, however, have proven a powerful force for improved cross-strait relations. The more China is able to advance political and economic links, the less likely it will resort to a military option.

Deep pessimism about Taiwan's future is widespread, paralleling Japan's experience, with more and more people faced with debts. President Chen's approval rating dropped from 80 percent in May 2000 to 34 percent in March 2001.

Taiwan has no experience with a coalition government. Given the public perception of the KMT as a "sore loser" and the DPP as a "dilettante," it will take years before a mature, pluralistic culture will be formed. It is increasingly unlikely that the future of Taiwan will be settled by military force. The future of Taiwan will be settled by economic globalization: China will continue to develop and gain more expertise, and Taiwan will continue to be attracted to China's cheap labor, raw materials, ample power supply, immense market, and familiar cultural and linguistic environment.

The Prospect of Reunification

During the second half of the twentieth century, the U.S.-Soviet relationship dominated world affairs. At the beginning of the twenty-first century, the U.S.-China relationship is the most important one. For nearly thirty years, from Henry Kissinger's groundbreaking trip to China in 1971 to the events of September 11, 2001, the two countries have endured difficult periods of friction over trade, human rights, religious freedom, Tibet, and Taiwan. With President Bush's administration, the relationship between the two countries will not be easy. The tragedy of

September 11, 2001, however, may shift tensions and ease the two countries toward common goals and friendlier relations.

The most difficult part of U.S.-China relations will continue to be the Taiwan problem. If Taiwan and China can solve their issues through negotiation and to their mutual benefit, that will tremendously improve the relations between the United States and China.

At present, deep pessimism about Taiwan's future is widespread, and President Chen's approval rating is falling. The people of Taiwan want new internal policies and a new relationship with Mainland China. Taiwan's future in no small way depends on the health of its economy and how its relationship with Mainland China progresses. A sound economic relationship with Mainland China is good for Taiwan's economic development. As someone suggested, "We Taiwan Chinese and you Mainland Chinese are two separate entities, yet we are both under one super-entity." As economic integration progresses, all Chinese people will gradually see each side's viewpoint and will eventually achieve unity.

In the political future of Taiwan, five people will play prominent roles: Lee Teng-hui, Chen Shui-bian, James Soong, Lien Chan, and Ma Ying-jeou. Lee and Chen are both natives of Taiwan and will form an alliance to promote Taiwan's independence; Soong and Lien will unite over Taiwan's links to Mainland China. Ma is an influential political star, having defeated Chen in the mayoral election.

Recently, Chinese Vice Premier Qian Qichen invited DPP members to visit China and called for renewed dialogue and strong economic ties across the Taiwan Strait. This new initiative of wooing Taiwan is in sharp contrast with China's former policy prohibiting contacts with the DPP. President Chen Shui-bian also has said that in dealing with the Mainland, Taiwan should have "three more and three less," meaning more economics and less politics, more contact and less misunderstanding, and more trust and less threats.

The different states of political, economic, and social development between Taiwan and the Mainland ensure that the unification process will be a long one. If both sides can maintain friendly relations and continue the trend toward economic interdependence, there is a good chance for a synergy to emerge, which will lead to ultimate success. With the civil war generation long gone, the leadership in both Taipei and Beijing will be more pragmatic, more rational, and more willing to co-exist peacefully. Whether the outcome of unification is "two nations, two systems" or a loose confederation or some kind of commonwealth under the name of China is an open-ended question. If Chinese history provides any guidance, Taiwan will eventually unite with the Mainland. From a world perspective, unified China will eventually reduce the likelihood of war in the Pacific region. The land and human resources of the Mainland and the economic and technological assets of Taiwan, acting together, could quickly contribute to world prosperity and would be a great contribution to world peace.

References

Chang, Parris. The Evolution of the Chinese Communist Party Since 1949, ed. M. Shaw.

_____. China's Relations with Hong Kong and Taiwan. Unpublished paper.

Chang, Parris H., and Martin Lacaster, eds. 1993. *If China Crosses the Taiwan Strait: The International Response.* University Press of America.

Chao, Linda, and Ramon H. Myers. 1998. The First Chinese Democracy: Political Life. In *The Republic of China on Taiwan.* Baltimore: Johns Hopkins University Press.

Cheng. T. J., et al., eds. 1995. *Inherited Rivalry: Conflict Across the Taiwan Straits.* Boulder: Lynne Rienner.

Chiu, Hungdah. 1979. The Question of Taiwan in Sino-American Relations. In *China and Taiwan Issue.* New York: Praeger.

Chou, Y. S., and A. J. Nathan. 1987. Democratizing Transition in Taiwan. *Asia Survey* 27.

Chough, R. N. 1993. *Reaching Across the Taiwan Strait: People-to-People Diplomacy.* Boulder: Westview Press.

Copper, J. F. 1996. *Taiwan: Nation, State, or Province?* 2nd ed. Boulder: Westview Press.

_____. 1995. *Words Across the Taiwan Strait.* University Press of America.

_____. 1988. *Quiet Revolution: Political Development in the Republic of China.* University Press of America.

Downen, R. L. 1984. *To Bridge the Taiwan Strait.* Washington, D.C.: Council for Social, Political, and Economic Studies.

Fong, Peter. 1989. *The Unification of China* (in Chinese). Taipei.

Free China, July 3, 1998; Oct. 3, 1997; Dec. 27, 1997; Nov. 21, 1997; Nov. 28, 1997; Aug. 23, 1997; July 11, 1997; Dec. 5, 1997; May 20, 1994; Aug. 26, 1994.

Goldstein, Carl. 1992. Strait Ahead. *Far Eastern Economic Review* 54.

Hsieh, Chiao-min. 1964. *Taiwan—Ilha Formosa: A Geography in Perspective.* London, Washington, D.C.: Butterworth.

Huang, Chi, Woosang Kim, and Samuel S. G. Wu. 1992. "Conflict Across the Taiwan Strait," 1951–1978. *Issues and Studies* 28.

Jia, Qingguo. 1991. Beijing's Perspective of the Changing Relations: An Analysis. Unpublished paper, Cornell University.

Kao, L. 1991. "A New Relationship Across the Taiwan Strait." *Issues and Studies* 27.

Kuo, Li-min, ed. 1992. *Mainland China's Policy Toward Taiwan: Selected Documents, 1941–1991* (in Chinese), 2 vols. Taipei: Yung-yeh.

Lacaster, M. L. 1984. *The Taiwan Issue in Sino-American Strategic Relations.* Boulder: Westview Press.

_____. 1989. *Policy in Evolution: The U.S. Role in China's Unification.* Boulder: Westview Press.

Lee, Lai-to. 1991. *The Reunification of China: PRC-Taiwan Relations in Influx.* New York: Praeger.

Leng, Tse-Kang. 1996. *The Taiwan-China Connection: Democracy and Development Across the Taiwan Strait.* Boulder: Westview Press.

Ma, Ying-jeou. 1992. *The Retrospect and Prospect of the Relations Between the Two Sides of the Taiwan Strait* (in Chinese). Taipei: Mainland Affairs Council.

Moody, Jr., P. R. 1991. "The Democratization of Taiwan and the Reunification of China." *Journal of East Asian Affairs* 5.

Myers, Ramon H. 1991. *Two Societies in Opposition: The Republic of China and the People's Republic of China After Forty Years.* Palo Alto: Hoover Institute.

Senese, D. J., and D. D. Pilecunas. 1989. *Can the Two Chinas Become One?* Washington, D.C.: Council for Social, Political, and Economic Studies.

Stolper, Thomas E. 1982. *China: Taiwan to the Offshore Islands.* Armonk, N.Y.: M. E. Sharpe.

Su, Chi. 1995. *The Future of Taiwan-Mainland Relations.* Taipei: Mainland Affairs Council.

Tsai, George W. 1991. An Analysis of Current Relations Between Taiwan and Mainland China: A Political Perspective. *Issues and Studies* 27.

Wang, Y. S. 1990. The Mainland-Taiwan Issue: A Challenge for Taiwan's Reunification. In *The Foreign Policy of the Republic of China on Taiwan: An Unorthodox Approach.* New York: Praeger.

Yang Shangkun. 1990. Yang Shangkun on China's Unification. *Beijing Review* 33 (48): 11–17.

Zhao, Quansheng, and Robert Slitter, eds. 1991. *Politics of Divided Nations: China, Korea, Germany, and Vietnam.* Baltimore: School of Law, University of Maryland.

23

The Geography and Political Economics of Inner Mongolia Beyond 2000

Robert W. McColl

Few areas of the world are undergoing fundamental economic and social changes as dramatic as those in the People's Republic of China. Because of its immense geographic size, these changes are uneven within China and result in increasingly distinctive regional geographies. The dramatic economic growth and social changes currently characteristic of China's coastal areas are not shared by interior provinces, most of which remain economically retarded. In the next century, China will need to address these geographic discrepancies or it will face serious political problems. Because of its geographic proximity and economic diversity, as well as its small population relative to the national core area of North China and Beijing, the Inner Mongolian Autonomous Region (IMAR) will be a key factor in China's new geography.

History

Inner Mongolia and all of Inner Asia historically played a function similar to that of current coastal China. It was a zone of trade, the location of major capital cities, and an area of dynamic social and political change. Inner Asia was the zone through which the ancient Silk Road passed. It was the zone from which Central Asian nomadic and pastoral tribes attacked, dominated, traded, and influenced developments in sedentary and agrarian China. It was along this frontier that the Chinese built the Great Wall, which some, such as Owen Lattimore, argue was as much to keep the Chinese in as the barbarians out.

The Mongols and their predecessors were tribal nomads. They did not organize politically by territory. They left no buildings or archaeological sites. They organized by groups ("banners" is the modern term). The Chinese organized by territory and then social structure. The Chinese built permanent cities and left abundant archaeological footprints. The result is that once the Chinese occupied these Inner Asian tribal lands, they plowed the land, built cities, and created territorial administrative boundaries that served the objectives of an agrarian population.

What is today the Inner Mongolian Autonomous Region was once composed of several distinct provinces such as Chahar, Rehol, Suiyuan, and large portions of other provinces such as Gansu and Heilongjiang (Figure 23.1). The idea of a Mongol autonomous region was created in 1947, two years before the actual founding of the People's Republic of China.

Geography

The present east-west extent (over 3,000 km) of the IMAR reflects Maoist-era military considerations—defense against possible threats of invasion from a Soviet-backed Mongolia. And, reflecting its remote position in the mind of the Chinese, and its original population composition, the IMAR ended up becoming the third largest political unit in China. The first and second largest areas (Tibet and Xinjiang) share similar geo-political characteristics. Not until after 1950 did the IMAR have any meaning economically for China proper. In fact, even presently the IMAR has little functional meaning to the national economy, but this may well change.

As it evolved after 1949, the IMAR developed distinct links with agrarian and industrial China proper. Although many geographers stress the latitudinal (east-west) physical zones of the IMAR, its functional economic development has been distinctly longitudinal (north-south corridors).

Virtually all of the cities in the IMAR are dominated by Han Chinese and act as gateways linking the IMAR to the agrarian and industrial core areas of eastern China. They have little identity or association with the Mongol way of life or culture.

Just as coastal China's cities provide access to distinct geographic regions (Manchuria to Korea and Japan, Shanghai to the Yangtze Valley and the entire Pacific Basin, Guangzhou to Southeastern China and Southeast Asia), the cities of the IMAR focus upon distinct regions. The eastern IMAR is most closely tied to developments in Manchuria. The central region is tied to Mongolia, North China, and Beijing. The western region is linked to western Mongolia, Gansu province, and Xinjiang. Each of these IMAR regions has distinctive physical as well as economic geographies and potentials.

The IMAR's long east-west extent means that there are numerous and major topographic and climatic differences among its subregions (see Figure 23.2). Chinese geographers have identified six distinct topographic zones: (1) the northeast and Hinggan Mountains, (2) the Hulunbeir Plateau, (3) Northern Inner Mongolian Plateau, (4) the Songliao Plain east of the Hinggan Mountains and adjoining

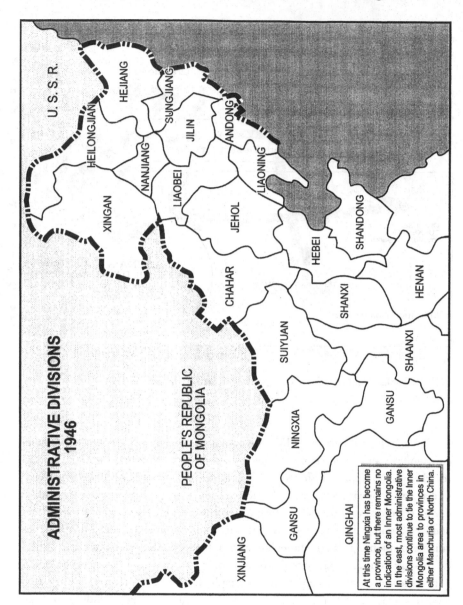

FIGURE 23.1   Administrative Divisions in 1946

Manchuria, (5) the Hetao Plain, and (6) the Ordos Plateau. However, a modern economic geography based upon maximum homogeneity but still recognizing distinct physical geographies can reduce this to three regions: (1) eastern areas that abut Manchuria, (2) the central region adjoining the North China Plain and Beijing areas, and (3) the westernmost areas of greatest barrenness and lowest population, linked to Ningxia and Gansu provinces.

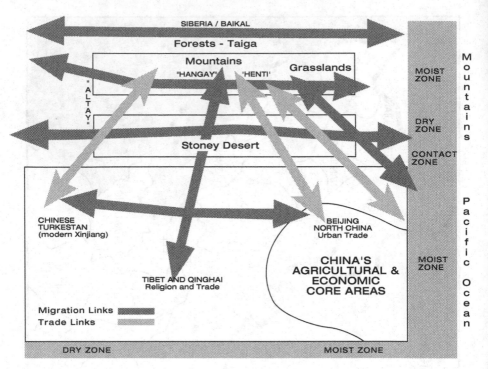

FIGURE 23.2   Schematic of Geographic Relations: Inner Asia—Inner Mongolia and China

Climatically, the easternmost IMAR is continental and moist (300 mm average precipitation per year). It also is an extensive grassland steppe, with major forest reserves in the Hinggan Mountains. It shares borders, and thus links, with Russia and Mongolia as well as several provinces in China's Northeast (Manchuria). It also is a major center of energy resources, with deposits of coal, oil, and natural gas. Combined, these features and its location make it a prime candidate for rapid economic development and population expansion—primarily through migration from crowded China proper.

In stark contrast, the westernmost area of the IMAR is primarily stony desert with little vegetation to support even nomadic herders. Still, it shares the western loop of the Yellow River and has significant natural resources, including coal and rare earths, of which the IMAR is the world's leading source. As with Montana and Wyoming in the United States, or Siberia in Russia, the distance from major consuming areas is a hindrance to economic development. However, its remoteness is perceived as a significant military advantage.

Between these extremes is the central area, which is focused upon the great loop of the Yellow River (Huang He). This is the most populated and developed of the IMAR's regions. Here are the major cities of Baotou (a center of iron and steel production) and Jining (a major transport hub).

These geographic and cultural features obviously affect the future of the IMAR and its various regions. However, to predict future developments, one must understand the past as well as the basic human and geographic circumstances that underlie change.

The geographic area that now constitutes the Inner Mongolian Autonomous Region historically has been a transition zone for Chinese expansion. In fact, one can think of the entire region as a kind of land-based interior "coastline" or zone of transition between the fields of the sedentary agrarian Han Chinese and the vast "desert ocean" of nomadic peoples. This transition zone was where the early Chinese emperors built the Great Wall, defining the boundary between the Mongol steppe horsemen and the agricultural Han Chinese. There never was enough water to permit Chinese-style farming, and the Chinese were never a culture of herders. However, the nomads of the steppes also were the traders of early Chinese history. They brought the goods from the West and Central Asia to China and carried Chinese goods (especially silk and tea) to the West. One consequence was that China's first capitals and major cities were located along this inner Asian margin. Cities such as the famous Xi'an (Changan), Datong, and others unfamiliar to Western ears, marked key trading centers and political foci. Inner Asia's deserts were China's first "ocean" realm, and Inner Mongolia was its "coast." As noted above, just as today China's major cities and economies concentrate along the Pacific Rim, so historically did its cities concentrate along the transitional grasslands of Inner Asia.

And then things changed, dramatically. For several centuries, Inner Asia became wholly unimportant economically and militarily. From the eighteenth century on, China focused almost exclusively on its coastal margins and away from Inner Asia. This geographic shift created an economic desert parallel to the IMAR's physical desert. Trade and related socioeconomic developments once associated with the overland Silk Road now shifted to the east and to sea routes. Coastal ports replaced inland emporia. The land and population of the IMAR reverted to nomadic herding and occasional banditry. What had been a front door was now a back door, and it was ignored.

China proper had become self-sufficient in food. It had little to fear militarily from Inner Asian tribes or people, and luxury goods for trade and the court now came and went by sea.

Traditional land use in IMAR was pastoral nomadism. Mongols raised sheep, goats, and horses along with some cattle. They used the land sparingly. They did not settle permanently or plow the fragile soils. They were organized by social groups (not genetic families as with "clans") or by *neg usnihan,* common water/pasture areas.

The political organization by banners reflects their history of nomadism, which was focused on groups and their individual leaders more than upon territory. Today, China pays respect to this tradition, but the modern "banners" are in fact the equivalent of counties. They are territorially defined, and they are controlled by Chinese, not Mongols. Table 23.1 presents the major cultural differences between the Mongols and Han Chinese.

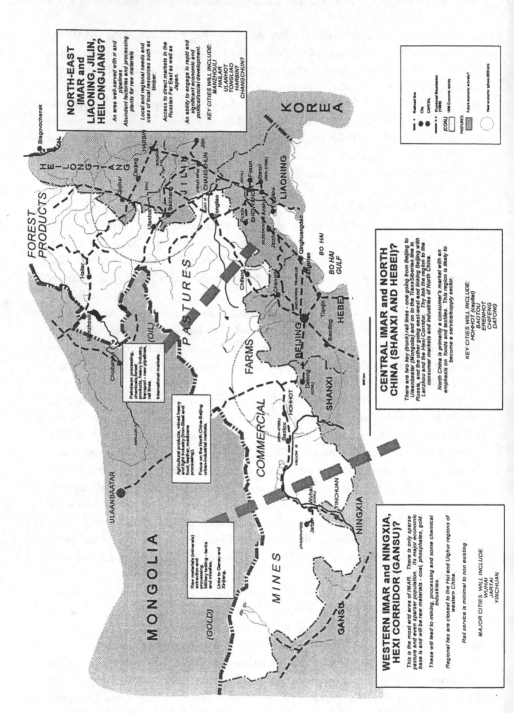

FIGURE 23.3   Inner Mongolian Autonomous Region Economic Divisions 2050??

TABLE 23.1   The IMAR and China: Some Cultural Comparisons

Mongols	Transition	Han Chinese
Nomadic herders	Sedentary herding	Settled farmers
Mobil "Ger"	Cement block houses	Fixed houses, tiled roofs
No cities, only temples	Communal centers	Industrial cities and hamlets
Seasonal markets		Permanent markets and infrastructure
Belief in spirits and the supernatural		Emphasis on social order and the family
Special colors are blue for the sky/heaven white for purity		Special colors are yellow for the earth and emperor red for luck
Political organization is of large areas with moving populations		Political organization is of people in fixed locations
Personal loyalty is to a strong personality/leader		Personal loyalty is first to the family or clan, then to the most powerful local person

Basic Ingredients for the Future:
The IMAR in 2050

As China modernizes and the internal as well as export demand for industrial and manufactured products burgeons, the need for basic materials and energy supplies has become a driving factor in economic development. Because the IMAR has such materials and because of its proximity to consumers, these factors will be crucial in defining the nature of the IMAR in the coming decades (see Table 23.2 and Figure 23.3).

Coal

Coal is one of the IMAR's major industrial resources. With close to two hundred coal mines, four in excess of ten billion tons, coal in the IMAR is important for local development, and it is increasingly being exported as electricity to neighboring industrial districts in both North China and Northeast China.

TABLE 23.2 Economic Activities in the Major Regions of the IMAR

| | Where | | |
What	City	Region	Who
Iron and Steel	Baotou	Center	Han Chinese
Manufacturing			
heavy industry	Baotou	Center	Han Chinese
Food Processing	Qiqihar	Northeast	Mongol
	Tongliao	Northeast	Mongol
Coal Mining	Wuhao	Center (west)	Han Chinese
	Chifeng	Northeast	Han Chinese
	Manzhouli	Northeast	Han Chinese
	Huolingol	Northeast	Han Chinese
Timber	Orgohan	Northeast	
Traditional Medicines		Northeast	Han & Mongol
Textiles	Baotou	Center	Han Chinese
	Hohhot	Center	Han Chinese
	Dongsheng	Center	Han Chinese
Leather			
Livestock Raising	grasslands	Center and East	Mongol
Processing	Hohhot	Center	Han Chinese
Wool and Cashmere			
Mutton and Cattle	grasslands	Center and East	Mongols
Processing	Hohhot	Center and East	Han

SOURCE: National Atlas Compiling Committee, *Economic Atlas of China* (Oxford University Press and the Cartographic Press, 1993).

The earliest coal mines were developed at Datong in what now is Shanxi province. Datong, a capital of the Liao dynasty (A.D. 916 to 1125), translates as "great gathering" and reflects the city's early function as a major market center for exchange between Mongols and Han Chinese. However, in addition to the mines at Datong, coalfields are found along the entire eastern border between the IMAR and China proper. The IMAR is the second-largest coal-producing area in all of China, a factor that impels it to important economic status for the foreseeable future. For example, the Junggar coalfield located in western IMAR has reserves estimated at almost forty billion tons and will become the largest open-pit mine in

China. Furthermore, as a natural co-relation, this also is the major area for dinosaur fossils.

Geologically (and thus geographically) related to coal are oil and natural gas. Thus far, oil has been the dominant discovery, and it has been concentrated in eastern IMAR. Of special interest is the major oilfield and development at Daqing in neighboring Heilongjiang province. In obvious geologic as well as geographic proximity are the new oil and gas fields being developed in eastern Mongolia—the Tamtsag Basin. These regions are environmentally similar to the oil and gas basins of eastern Wyoming and western Canada in North America.

Iron Ore

Deposits of iron ore are concentrated in the mountains north of modern Baotou—a city created to produce iron and steel for use in military equipment, especially in the production of tanks. However, this city and its industry were created by a national defense need (the safety of an interior location and an ability to develop and test hardware in secret). It is not an economic operation of value in competition with similar facilities in East China or in Manchuria. Thus, it is unlikely that the Baotou works will see much in the way of future expansion. However, they may be modified and may be of major importance to consumers in the interior, as transportation costs would be lower from this region than from Manchuria or the coast.

Herbs and Traditional Medicines

There is an increasing demand by the masses of eastern China for herbs and traditional medicines, and they are another of the important economic products of the IMAR. More than four hundred medicinal herbs, as well as deer antlers, bear's gall, and musk, are supplied by the IMAR. In the future, many of these will be produced on herbal and animal farms, since demand continues to vastly exceed supply. Han Chinese entrepreneurs will most likely develop these industries, and they will be concentrated in the eastern mountains and plains. They will have great economic value, use little land, and provide distinct opportunities for both Mongol and Han Chinese.

Agriculture

In addition to eastern China's virtually insatiable demand for energy and cheap labor, the increasingly affluent urban populations of eastern China have created new diet demands, especially for meat other than pork. Inner Mongolia is ideally sited to meet these needs, in addition to being an optimal location (by distance and low population density) for increases in crop agriculture.

The IMAR currently has about 5.3 million hectares of land under cultivation with less than 60 percent of this being irrigated. This farmland represents only 5 percent of the total area of the current IMAR. The most intensive agriculturally developed area is the Hetao Plain, located between the mountains and the western loop of the Yellow River. The IMAR also has an estimated 87 million hectares of steppe grassland, an ideal base for expanding China's meat-producing industry. The cultivated areas are almost exclusively highly mechanized state farms and former state farms. Their size and operations are reminiscent of the large farms of Russia or the U.S. Midwest.

The Past as Prologue

Virtually all new heavy industrial development affecting or being affected by IMAR is concentrated in the east, along the border with Manchuria. This is a symbiotic relationship and has every indication not only of continuing but of expanding at a geometric rate. In addition, there will be an increase in feedlot-style operations to meet the meat demands of the much more densely settled and urban regions of eastern China. Table 23.3 lists the comparative advantages of the IMAR's various regions.

This brief overview of each of the IMAR's major regions and their economic bases reveals which areas will be the most important as China enters the twenty-first century. In addition, I discuss various natural economic links that may lead to a more rational geopolitical division of the region. After all, there are realities that simply cannot be changed, short of a catastrophic event (natural or man-made). This could result in a provincial structure that differs significantly from that of the present.

Eastern Inner Mongolia

A land of forests, grasslands, and mineral resources dominated by oil and coal, eastern IMAR abuts the industrial and energy-hungry region of Northeast China. And like Northeast China, the eastern IMAR has moderate population densities. It also shares a cultural history and language base (the Manchus were more closely related to Mongols than to Han Chinese). Its key cities and economic regions will be dominated by the following.

Tongliao.   Tongliao is a major gateway city with excellent rail and road infrastructure. Its immediate economic function will continue to be dominated by food processing, especially orchard fruits for export to the rest of China. Its future will see increased processing of traditional medicines and increased development of its coal mines, with the coal exported as electricity to energy-hungry industrial plants in Northeast China.

Ulanhot.   Ulanhot's economy is dominated by oil seed processing, iron smelting, chemical processing, tobacco, and woolen and cashmere textiles. These products are generally lower in quality and design than similar products produced in East China. Thus, Han Chinese entrepreneurs and factories will continue to increase their investments and use cheap Mongol and Han Chinese labor to increase their profits.

TABLE 23.3 Comparative Advantages of the IMAR's Regions

Activity	Character	Location
Agriculture	Mechanized large farms	East and Center
Settlements	Nucleated, sedentary	East and Center
	Dispersed, pastoral	Center and West
Raw Materials	Mines	East and West
	Timber	East
	Wool and hides	Center and West
	Oil	East
	Phosphates	West
	Gold	West

Manzhouli. As a border city or "port" linking the steppes and farmlands, Manzhouli exports its local coal supplies in the form of electricity (the most cost-effective method). And, because of its local moisture and climate advantages, it, too, is an important food-processing center for orchard fruits for export to other regions.

Erenhot. Also a "port," Erenhot, located on the border with Mongolia proper, is active in the export of machine products produced locally as well as in Northeast China to Russia and Mongolia. It is likely to continue this function, with some hiatus due to domestic Russian financial and political problems. However, as the Mongol economy expands, Erenhot could become the Inner Asian equivalent of coastal Shanghai.

The demographic profile will reflect increased in-migration of Han Chinese and a clear division between the industrially employed Han and the agrarian and pastoral Mongols.

The Central Eastern Region

Horoqin Sands. Despite the name "sands," this region is an extensive grassland prairie abutting the northeastern province of Liaoning. Historically, this region has been one of the IMAR's most productive animal-raising and mixed-crop areas. Later the Han Chinese introduced massive mechanized state farms in the area. However, given the increased demand and profit to be gained from meat production, it seems likely that this region may well become China's feedlot belt in the near future.

Chifeng. Chifeng is another border ("port") city that, like Manzhouli, exports its local coal resources as electricity. However, its proximity to the city of Jinzhou in Liaoning may make it analogous to Shenzen outside Hong Kong. Jinzhou is a center for specialized electronics—computers, meters, and so on—and, as its local labor costs soar, the proximity of Chifeng and its cheaper labor will make the city a natural "colony" for Jinzhou.

The Central IMAR

This region is dominated by the capital, Hohhot, and the heavy industrial city of Baotou. It also is the region that controls north-south and east-west rail traffic. It is the natural gateway to the region inside the great loop of the Yellow River and to Beijing and the North China Plain.

Hohhot.   Hohhot is the capital of Inner Mongolia, and it serves as a kind of modern "Datong" or great gathering center. There are important agricultural specialty areas both east and west of the city proper. It has a large dispersed population and an agricultural base on a fertile plain. The city itself is the site of diversified industries, including light manufacturing and food processing. It is a natural first stop for many Han Chinese migrating from the North China Plain and eastern China.

Baotou.   With both coal and iron ore available nearby, Baotou initially was developed as an interior, and thus militarily secure, site for one of Communist China's first modern integrated iron and steel complexes. Since the transport of its output was too expensive to compete with the iron and steel of Northeast China, it has focused on military and transport equipment, especially battle tanks, which are tested in the barren wastes of the nearby Ordos and Gobi Deserts and moved by rail to their final destinations. The establishment of this industrial complex led to a dramatic infusion of Han Chinese into the area.

The Western IMAR

This is the most desolate and remote area of all Inner Mongolia. Physiographically, it is part of the stony Gobi Desert that stretches from Mongolia to Gansu. There are no significant agricultural or pastoral resources here—except for wild camels. There are no roads, but none are needed. However, there are important mineral resources. Wuhai is an industrial city and the site of a major coal mine. The nearest diversified city is Yinchuan in the Ningxia Hui Autonomous Region.

The Ordos Loop and Hetao Plain

The Hetao Plain is the most fertile and potentially expandable agricultural region in the IMAR. An ancient irrigated area that became a wetland-swamp when it was abandoned due to Mongol depredations, it has been reclaimed and agriculturally developed by Han Chinese since 1947. Its current population is predominantly Han Chinese, and its primary urban centers are the rail cities of Wuyuan and Linhe.

However, even with its access to the major rail line of the IMAR, the Hetao area's crops cannot compete with similar crops grown closer to major areas of consumption; transportation costs are simply too high. To prosper economically, the area will have to invest in specialty crops, especially orchards, for which it has some comparative advantage. These fruits can be canned locally to stand the long transport distances and yet remain economically competitive.

Demographics

Chinese population statistics are notoriously flexible. They often reflect desires or political agendas set outside the local region. They can be used to gain development or reduce attention and taxes. Consequently, and especially in the frontier provinces such as Inner Mongolia, it is necessary to rely on personal observations and resources in addition to official statistics.

According to statistics from the State Statistical Bureau, at the advent of the new millennium, the IMAR's population exceeds 21 million. The vast majority (over 85 percent) are Han Chinese, not Mongols. Mongols constitute only about 3 million (about 11 percent) of the total population. Yet this Mongol population is larger than that of the independent state of Mongolia to the north and represents over 72 percent of all Mongols living in China. The majority of the Mongols are concentrated in low-paying factory jobs or in agriculture (crops and animals).

Conclusion

When combined, the above geographic, demographic, and economic factors lead to some inevitable conclusions regarding the IMAR as it most likely will appear by 2050. First, the IMAR's natural economic and political geographies will likely result in de facto if not de jure regional divisions. East-west distances are simply too great and the links too tenuous. The only thing that would override this logic would be a desire to appear sensitive to the Mongol minority and thus avoid an ethnic-based political confrontation. However, even the Mongols are not homogeneous from east to west, not to mention how few Mongols there really are.

Second, the demographic profile will reflect a vastly increased number of Han Chinese who have migrated to the region for both agricultural and industrial jobs. Mongols will decrease in both total numbers and as a percentage of the overall regional population. To meet the appearance of continued political recognition of its minorities, the central government will continue to create Mongol autonomous banners, but it is likely that the overall area of an Inner Mongolian Autonomous Region will be substantially reduced from what we see today.

Finally, the resulting political geography will reflect the IMAR's increasingly specialized economic geographies—resources, manufacturing, and mixed agriculture.

References

Heilig, G. K. 1999. *China Food: Can China Feed Itself?* (CD-ROM). IIASA LUC Project, Laxenburg, Austria.

Liang, L. 1983. *Geography.* China Handbook Series. Beijing: Foreign Languages Press.

National Atlas Compiling Committee. 1993. *The Economic Atlas of China.* Beijing: Oxford University Press, Cartographic Press.

Sun, J., ed. 1988. *The Economic Geography of China.* London: Oxford University Press.

Zhou, S. 1992. *China Provincial Geography.* Beijing: Foreign Languages Press.

Zhung guo ren kou: Neimenggu (China's Population: Inner Mongolia). 1987. Beijing: Xinhua.

24

Xinjiang (Eastern Turkistan): Names, Regions, Landscapes, and Futures

Stanley W. Toops

What is now the Xinjiang Uyghur Autonomous Region (XUAR) has been labeled a variety of names. This Inner Asian area is composed of the Tarim Basin, the Turpan Basin, the Dzungarian Basin, and the Ili Valley. This area is known as Eastern Turkistan to distinguish it from Western Turkistan, the former Soviet Central Asia.

The Xinjiang Uyghur Autonomous Region, the largest of China's political units, covers an area of 637,066 square miles (1,650,000 square kilometers), one-sixth of China's total area, three times the size of France. Located in the northwest of China, Xinjiang is bounded on the northeast by Mongolia, on the west by Russia, Kazakhstan, Kyrgyzstan, and Tajikistan, and on the south by Afghanistan, Pakistan, and India. Xinjiang's eastern borders front the Chinese provinces of Gansu, Qinghai, and Tibet (Figure 24.1).

How to conceptualize this place and its peoples is the fundamental issue in organizing the geography of Xinjiang. Three themes are useful in considering the geography of this place: names, regions, and landscapes.

Names

What is in a name? The map of Xinjiang reveals a number of names—place names as well as the names of people. These names structure the meanings of our perspectives of Xinjiang.

411

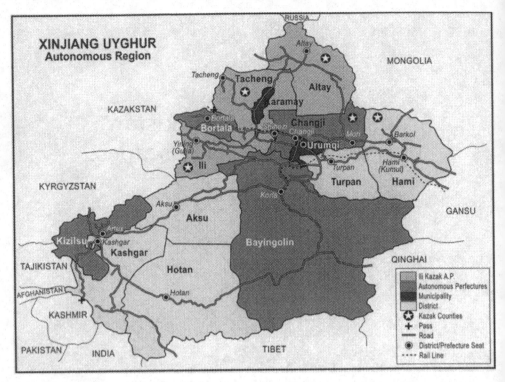

FIGURE 24.1   Xinjiang Uyghur Autonomous Region

Places

Today, the name of the region is the Xinjiang Uyghur Autonomous Region. This is the name used by the People's Republic of China. Prior to the founding of the autonomous region in 1956, the region was a province known simply as Xinjiang. Xinjiang, in Chinese, means new dominion, new territory, or new frontier. In the character of *jiang*, one sees the elements for bow, earth, and field. The Manchu gave the region this name in 1888 when the region became a province of the Qing Empire. Prior to the name of Xinjiang, the region was known in Chinese as Xiyu or "Western Region(s)." So the Chinese names of this region indicate (1) a region located in the west (in the pre-modern period) and (2) a territory or frontier new to China (in modern times).

Many people in the region are Uyghur. The Uyghur are a Turkic ethnic group who are Sunni Muslim and write with an Arabic-style script. To the Uyghur, the region has a different name, Sharqi Turkistan, which means "Eastern Land of the Turks." However, this name cannot be used in Xinjiang, so in Uyghur the name of the region is Shinjang, a Uyghurization of the Chinese word Xinjiang.

Other place names give some indication of the physical and human geography of the region. The Tian Shan, or "Heavenly Mountains," divides the region. In Uyghur they are called the Tengri Tagh ("Mountains of God"). To the south is the Taklamakan Desert; no one knows exactly what the meaning is. Some have speculated that it means, "If you go in, you don't come out." This is not a likely meaning. Another thought is that it has something to do with fields and thus concerns prior agricultural activity. To the north is the Dzungarian Basin, named after the Mongols who were destroyed by the Manchu in the late 1700s.

The main city is Ürümqi, deriving from Mongol for "Beautiful Pastures." The Chinese call it Wulumuqi. Earlier, the city was known in Chinese as Dihua, "Land That Has Been Sinicized." The southern center of the region is Kashgar; this name in Uyghur refers to "green," perhaps the green of the oasis, or the green of jade, or the green of the roof tiles. In Chinese it is called Kashigaer, shortened to Kashi. Turpan, the site of the old Uyghur capital, means "lowland"—this is the second-lowest place on earth. In Chinese it is called Tulufan.

So the place names refer to God, to fields, to lowland, to Mongols, to pastures. History and geography are both told through these names. The current name of the whole region has no meaning in Uyghur; in Chinese it means a frontier that is new. To the Uyghur the region is not a frontier but a center, not new but old.

Peoples

The names of the Turkic peoples of Xinjiang also provide insight. Turk means "strong" in the Turkic languages. The more populous Turkic peoples of Xinjiang are the Uyghur, the Kazak, the Kyrgyz, and the Uzbek.

The Uyghur (variants: Uighur, Uygur, and Uigur) are the titular ethnicity of the Xinjiang Uyghur Autonomous Region. The name Uyghur, in Turkic, means "united"; a similar root is shared with word yoghurt. In Chinese, they are called Weiwuer. Up until the 1930s, Uyghurs in Xinjiang spoke of themselves as *yerlik*, locals. Thus one from Kashgar was known as a *Kashgarlik*, from Hotan, a *Hotanlik*. In the 1920s the name Uyghur was proposed for this group at a conference in the Soviet Union; this connected present-day ethnic groups with those of the past. By using these various Turkic names for groups such as the Uyghur and Uzbek, differences rather than similarities are highlighted among these groups.

The Kazak (variants: Kazakh, Qazaq) live in the northern portion of Xinjiang, where they have been nomads for centuries. They are also Turkic and Muslim. Today many Kazaks also speak Uyghur, but the reverse is not true. Most of the world's Kazaks live in the newly independent Kazakhstan. Kazak means "free" in their language. In Chinese they are called Hasake.

The Kyrgyz (variants: Kirgiz, Kirghiz, Kyrghyz, Qirghiz) are another group of nomadic Moslems. They practice alpine herding techniques in the Pamir of southwest Xinjiang. The Kyrgyz are a Turkic-speaking group; their language is a close cousin to Kazak. Most of the world's Kyrgyz live in Kyrgyzstan. Kyrgyz may refer to "forty daughters," *kirik kiz,* some of the group's ancestors. In Chinese they are called Ke'erkezi.

The Uzbek are a sedentary group of Turkic-speaking Moslems. The Uzbek language is quite similar to Uyghur. Only a very few Uzbeks live in Xinjiang, mostly in the large cities. They are traders, craftspeople, and farmers. The culture is very similar as well. Uzbekistan is where most Uzbeks live. Uzbek comes from two words, *uz* meaning "self" and *bek* meaning "rule" or "leader"; the Uzbeks rule themselves. In Chinese they are called Wuzibieke.

The Turkic names of the peoples of Xinjiang speak of strength, of unity, or freedom, of their ancestors, and of self-rule. The names of the people provide clues to their self-identity. Their names of their places also show the meaning of this land.

Regions

The Taklamakan Desert and the Tian Shan Range have proven to be effective barriers to administrative control and cultural influence. Three regions (South, Central, and North) of historical association have been formed in what is today called Xinjiang. This regional construction clarifies Xinjiang's overall geography.

South: Kashgaria (Alte Sheher)

The southernmost region is Kashgaria, that area centered around Kashgar, also known as Alte Sheher, Turkic for "Six Cities," referring to the oases around Hotan, Yarkand, Yengi Hissar, Kashgar, Uch Turfan, and Aksu. Being in the southwest, Kashgaria was always the last region to be invaded by the powers to the north and east.

Kashgaria, being in the southwest, received the waves of the great religions first. Buddhism arrived in second century, coming overland from India via the Kushan Empire. The Hotan area was a center of Buddhist culture. Similarly, Alte Sheher was the first segment of the area to come under the influence of Islam, during the ninth and tenth centuries. The imprint of Islam was increased with the Timurid Renaissance in the fourteenth and fifteenth centuries, centered on Kashgar.

Central: Turpan (Uyghurstan)

The second region is that centered in Turpan. The area has also been known as Uyghurstan, after the Uyghur Empire (A.D. 840–1200). Before the Han control of the area, Turpan was under the control of the Xiongnu people. After a period of dissolution, the Tang controlled the region, to be followed by the Uyghur. Turpan was the capital of the Uyghur Empire. The Turpan Uyghurs were the first Turkic group to acknowledge Mongol power in 1209. During the rule of the Chagatai Khanate followed by Mogholistan, Turpan continued to be a protectorate, maintaining its Buddhist and Nestorian heritage against the Islamicization of the Moghuls until the onslaught of Timur.

This area was the first to be controlled by the Qing incursion (in response to the rising power of the Oirat Mongols). Turpan and its environs always remained at the forefront of any possible incursion of Chinese forces or the steppe peoples to the north of Tian Shan. Under the People's Republic, the ethnic makeup of this central region has seen a dramatic rise in the number of Hans, making the area stand out in sharp contrast to the more Turkic Alte Sheher.

North: Dzungaria and Ili

The third region, bounded by the Tarbagatai, the Altai, and the Tian Shan, lies in the Dzungarian Basin and the Ili Valley. In the regional, historical geography of Xinjiang, Dzungaria and Ili have always been the realms of the horse and the nomad—Wusun, Xiongnu, White Hun, Ruran, Turk, Karluk, Khitan, Chagataites, Moghol, Oirat (who left one of their names to the basin), and Kazak. Coming from the north and east, these nomadic groups have in succession tamed the vastness of the Dzungarian Basin, producing a cultural force that occupied the steppe. Often the nomadic peoples of Dzungaria had under their protection the less-mobile peoples of Turpan and Kashgaria.

Slightly better watered than the Dzungarian Basin, Ili has been the headquarters for the nomads of the north and thus the focal point of their power. The Wusun sought by Zhang Qian were seated in the Ili Valley. The Western Turks sought by the Byzantines ruled in Ili. Later, the Kara Khitai and then the Chagatai Khanate focused their rule in Ili. The relative safety of Ili allowed the Oirat Mongol to foray into the south. Ili was the focus of the Qing-mandated migrations of Manchu, Han, Hui, Uyghur, and Xibo after the Oirat (Dzungar) were decimated. Kazak migrants and Russian control over the area in the latter part of the 1800s served to differentiate Ili from other portions of Xinjiang.

These three regions of Kashgaria (Alte Sheher), Turpan (Uyghurstan), and Dzungaria (with the subregion of Ili) provide a way to organize Xinjiang into regions of historical associations, political control, and cultural identity. Dzungaria has had a dominant nomad ecology. Kashgaria has remained an oasis culture. Turpan represents a transitional area. Their locations have also oriented the regions. Kashgaria has looked more toward Mecca, India, and Central Asia. Turpan has been the closest to China. Dzungaria has had closer linkages to the north, the Mongols, or the Russians. Today, the South is mostly Uyghur, the Central is more Chinese than Uyghur, while the North has Kazak, Uyghur, and Han populations.

Landscapes

Xinjiang has many layers of landscapes. The first layer is the natural landscape of mountains and desert, oases and steppe. The second layer is the ethnicity of modern-day inhabitants of Xinjiang—Han, Uyghur, and Kazak. The third layer is the economic landscape of production and distribution, roads and rails, oil and agriculture. The fourth layer is the territorial landscape of administration.

Natural Landscape

Xinjiang lies in the middle of the Eurasian landmass, comprised mostly of intermontane basins and mountain ranges. The northern perimeter is bounded by the Altai Mountains, approaching 13,120 feet (4,000 meters) in altitude. On the southern perimeter, the Pamir reach over 19,680 feet (6,000 meters) and the Kunlun Range has several peaks over 22,900 feet (7,000 meters), including the world's second highest peak, K2 (Godwin Austen) at 28,244 feet (8,611 meters). The Tian Shan, or the Heavenly Mountains, divides the region in the middle. North of the Tian Shan is the Dzungarian Basin, and south of the Tian Shan is the Tarim Basin. The forking of the Tian Shan forms some intermontane troughs, the Ili Valley, and the Turpan Depression (at its lowest point, 505 feet, or 154 meters, below sea level). The diversity of this landscape ranges from the second lowest point to the second highest point on the earth.

Ethnic Landscape

Xinjiang ethnic diversity forms a basis for regionalization. With a variety of ethnic groups living in the area, all of their experiences and traditions can be brought to bear on any issue. An understanding of the distribution of the ethnic groups provides clues to the cultural landscape of the area.

Of the thirty different ethnic groups in Xinjiang, thirteen have made Xinjiang their home. The thirteen are Uyghur, Kazak, Kyrgyz, Uzbek, Tatar, Xibo, Manchu, Mongol, Daur, Han, Hui, Tajik, and Russian. They represent different language groups, religions, and customs (see Table 24.1). According to China's 2000 census, Xinjiang's population of 18.46 million is composed of 45.2 percent Uyghur, 40.6 percent Han, 6.7 percent Kazak, 4.5 percent Hui, and 3 percent other. Within the cities one may see a great variety of ethnic groups, but most of the minority groups live on the periphery. A definite Central Asian component is the population base in the region, particularly outside of the capital Ürümqi.

Economic Landscape

The traditional economic landscapes of this Silk Road region were herding, oasis agriculture, and trade. On top of that, the state has added the modern economic elements, including distribution (road, rail, air), as well as production (oil, textiles, agribusiness) and consumption (urban and rural).

The state's project of developing Xinjiang restructured the economic landscape. Transportation linkages lead to Ürümqi and thence to Beijing in a hierarchical, centralized fashion. Traditional economic centers such as Kashgar, Turpan, and Gulja are superseded by Ürümqi's industries based on petrochemicals and textiles. Oil

TABLE 24.1   Xinjiang's Ethnic Groups

Group	Language	Religion	2000 Census	Location
Uyghur	Turkic	Islam	8,345,622	All over
Kazak	Turkic	Islam	1,245,023	North
Kyrgyz	Turkic	Islam	158,775	South
Uzbek	Turkic	Islam	12,096	North
Tatar	Turkic	Islam	4,501	Central
Han	Sinic	Syncretic	7,489,919	Central
Hui	Sinic	Islam	839,837	Central
Xibo	Tungusic	Shaman	34,566	North
Manchu	Tungusic	Syncretic	22,329	Central
Mongol	Mongolic	Buddhist	149,857	South, North
Daur	Mongolic	Buddhist	6,405	North
Tajik	Indo-European	Islam	39,493	South
Russian	Indo-European	Christian	8,935	North
Others			102,153	
Total			18,459,511	

SOURCE: *Tabulation on the 2000 Population Census of Xinjiang Uyghur Autonomous Region* 2002.

found in the north at Karamay (*kara* means black and *may* means oil in Uyghur) and the current oil exploration in the Tarim have added to Xinjiang's economic value to China. Agriculturally, the hallmark of Xinjiang's development has been the Production and Construction Corps (PCC) formed from demobilized elements of the People's Liberation Army in the 1950s. Large amounts of central investments and subsidies were directed to rebuilding the land. At the same time, these central funds and demobilized troops contributed to the consolidation of central control. In terms of consumption, Ürümqi has been the focus of the economy with people paying high prices, earning not so high wages, and living in high rises. In rural areas in the south, farther from the markets, people still live in poverty. Border trade was nonexistent in the 1960s, limited in the 1970s, and grew in the 1980s and 1990s.

Territorial Landscape

The state delineates the path that administrative structures take. The province was designated the Xinjiang Uyghur Autonomous Region on September 20, 1955. Most of the ethnic groups of Xinjiang are represented by some autonomous area. The Ili Kazak Autonomous Prefecture (A. P.) recalls the Eastern Turkistan Republic. The other autonomous prefectures are Changji Hui A. P., Bortala Mongol A. P., Bayingolin Mongol A. P., and Kizilsu Kyrgyz A. P. There are autonomous counties for Tajik, Xibo, Hui, Mongol, and Kazak clusters (see Figure 23.1).

The other features of the administrative divisions include the municipalities. Regional-level municipalities are administered directly by the XUAR authorities; these include Ürümqi, Karamay, and Shihezi. Ürümqi, as the capital, rates municipality status. Karamay and Shihezi are essentially new cities created in the desert,

Shihezi for agricultural production and Karamay for oil production. A hierarchy of authority has been established, centered on Ürümqi.

The natural, ethnic, economic, and territorial landscapes combine in layers to make up the complex and diverse landscape of Xinjiang. Government policies from 1949 through the early 1980s sought to confirm the focus on Ürümqi and reoriented the region to Beijing. The border with Central Asia was sealed. Through the 1980s and 1990s, government policy has allowed an opening to the outside world. In Xinjiang's case, this meant trade with bordering countries, particularly Pakistan, Kazakhstan, and Russia, re-establishing Central Asian connections. For Xinjiang to improve economically, border trade and foreign investment has been critical.

The break-up of the Soviet Union and the arrival of the newly independent Central Asian states have created a new dynamic on Xinjiang's doorstep that presents alternatives to the desires of Beijing. From Kashgar, Beijing seems far away, and Mecca doesn't seem that far away at all. Xinjiang's cultural and economic similarities with its neighbors Kazakhstan, Kyrgyzstan, Uzbekistan, and even Pakistan weigh in the balance against the political and territorial similarities with Gansu, Qinghai, and Tibet as part of China.

Futures

Some guide to the future of Xinjiang is present in the restructuring project embarked upon by the PRC. The names of territorial units reflect an administrative geography shaped by the government. Once, a Uyghur drew my attention to the map, pointing out that although Xinjiang was called a Uyghur Autonomous Region, after including the "autonomous" areas of the Kazak, Kirgiz, Mongol, and Hui, where does a Uyghur "autonomy" exist? The differences in the names and identities of places and peoples are subsumed under the greater identity of China.

The traditional regions are also changing. The historical division of three regions— Kashgaria (Alte Sheher), Turpan (Uyghurstan), and Dzungaria (with the subregion of Ili)—has begun to fade. Certainly the centralized hierarchical structure of the government, the party, and the territory has worked to focus Xinjiang upon Ürümqi, a modern center of power, away from the traditional power centers of Kashgar, Turpan, and the Ili Valley. Are all of the Uyghurs and Kazaks now *Junggoluk,* that is, *Zhongguo ren?*

The cultural landscape reflects enormous changes that are continuing. Han migrants come to the region in the 1950s and 1960s during the period of Mao's leadership with political goals foremost. During the period of Deng's leadership, Han migrants have again arrived but with economic goals foremost. What does the post-Deng leadership bring? The railroad has been extended across the central and northern portions of Xinjiang, where Han concentrations are the densest in the region. Ürümqi is essentially a Chinese city.

In 1999 the southern railroad reached Kashgar. In 1997 I was riding a bus from Kashgar to Ürümqi. When we reached the head of the rail construction near Aksu, the riders, both Uyghur and Han, were marveling at the railroad. A Uyghur turned to me and said, "Look at the progress of the railroad. It will bring many things." He was right.

In July 1998 President Jiang Zemin went to a Central Asian summit. China signed treaties and agreements with Kazakstan to solidify their mutual border and an oil pipeline across the border. At the summit, China, Russia, Kazakstan, Kyrgyzstan, and Tajikistan signed agreements to boost economic ties and to crack down on organized crime and political separatism. According to Jiang, "These evils include national separatism, religious extremism, terrorism, illegal arms trafficking, smuggling, and drug trade." Jiang followed these meetings with a speaking tour in Xinjiang. His stated goals for Xinjiang include opening its doors wider to the outside world and exploring the international market in Central and Western Asia. It was also reported that bombs were exploded in Hotan during Jiang's trip (BBC News 1998). This echoes the tensions that surfaced in 1996 and 1997 with bombs in Ürümqi and Gulja (Ining) and attacks on Muslim clerics who are perceived as pro-Beijing (Hutzler 1997). Government authorities cracked down hard on these actions, which further alienated the Uyghurs. Most Uyghurs are not in favor of separatism, given the experiences of the former Soviet Central Asian States.

After the events of September 11, 2001, China intensified its "Strike Hard" campaign in Xinjiang. In the summer of 2002, the United States and the United Nations both supported China in identifying the so-called East Turkistan Islamic Movement as a terrorist organization. Yet there is no such organization. There are a number of expatriate Uyghur organizations variously identified as supporters of "East Turkistan." These groups are political, emphasizing human rights, but are not terrorist in nature. Local people are very wary of the current political situation in Xinjiang. The combination of economic pressures, political antagonism, and cultural tension is a potent one indeed. Ethnic issues in border areas can cause problems for locals as well as the state. On top of this already complex mix, the new furor over current and potential oil reserves in Xinjiang is like adding fuel to a smoldering fire. Locals and leaders, Han and Uyghur, do not know what will happen to this land in the heart of Asia. The future is uncertain.

Summary

Three geographic themes serve to organize the geography of Xinjiang. Names of people and places show the identity of Xinjiang as a Central Asian area. The region can be subdivided into three historically distinct subregions—South, Central, and North. The landscapes of Xinjiang reveal the restructuring of the region as the bridge between China and its neighbors.

The main external forces that affect Xinjiang are Beijing's reform program and Central Asian cultural ties. A reawakening of the relationship with Central Asia has linked the region to the outside. China's economic reform program has transformed the region. External forces are powerful; equally strong *in the region* are the local forces of culture and history, environment and settlement. The local is still Central Asian, while being under the control of China. The resultant map of Xinjiang is the local response to the regional implementation of national policies in the face of international forces.

A Bibliographical Note

The pre-modern background is in Barfield, Fletcher, Lattimore (1940), Polo, and Yu. Pre–1949 Xinjiang is examined in Benson, Jarring, Lattimore (1950), and Wu. Xinjiang, 1949–1979, is examined in Barnett, Benson et al., Dreyer (1976), and McMillen (1979, 1981). Xinjiang after 1979 is examined in Benson et al., Cannon, Christofferson, Dreyer (1986), Gladney (1991, 1997, 2002), Harris, Hoppe, Pannell and Ma, Rudelson, and Toops (1992, 1995, 1999, 2000). XBS (2001) provides recent statistics. BBC, Hutzler, and Eckholm provide accounts of recent events. The entire volume of *Inner Asia,* volume 2, number 2, 2000, profiles Xinjiang history, economy, demography, colonization, ethnicity, and nationalism. The recent book *Xinjiang: Chinese Central Asia,* edited by Starr, examines the region in depth.

References

Barfield, T. 1989. *The Perilous Frontier.* Cambridge, Mass.: Blackwell.

Barnett, A. D. 1993. *China's Far West: Four Decades of Change.* Boulder: Westview Press.

BBC News. 1998. Central Asia Tightens Co-Operation. July 3.

_____. 1998. China Ends Kazakh Border Dispute. July 4.

_____. 1998. President Jiang On Xinjiang. July 13.

Benson, L. 1990. *The Ili Rebellion: The Moslem Challenge to Chinese Authority in Xinjiang, 1944–1949.* Armonk, N.Y.: M. E. Sharpe.

Benson, L., J. Rudelson, and S. Toops. 1994. *Xinjiang in the Twentieth Century: Historical, Anthropological, and Geographical Perspectives.* The Woodrow Wilson Center, Asia Program, Occasional Paper No. 65.

Cannon, T. 1989. National Minorities and the Internal Frontier. In *China's Regional Development,* ed. D. Goodman, pp. 57–76. New York: Routledge.

Christoffersen, G. 1993. Xinjiang and the Great Islamic Circle: The Impact of Transnational Forces on Chinese Regional Economic Planning. *China Quarterly* 133: 130–151.

Dreyer, J. T. 1976. *China's Forty Millions.* Cambridge: Harvard University Press.

_____. 1986. The Xinjiang Uygur Autonomous Region at Thirty. *Asian Survey* 26: 721–744.

Eckholm, E. 2002. "U.S. Labeling of Group in China as Terrorist is Criticized." *New York Times,* September 13, 2002, p. 1.

Fletcher, J. 1970. China and Central Asia, 1368–1844. In *The Chinese World Order,* ed. J. K. Fairbank, pp. 206–225. Cambridge: Harvard University Press.

_____. 1978. Ch'ing Inner Asia, 1860–1862, and Mongolia, Sinkiang, and Tibet. In *The Cambridge History of China, X, Late Ch'ing, 1800–1911.* Part 1., ed. J. K. Fairbank, pp. 35–91 and 351–408, Cambridge: University Press.

Gladney, D. 1991. *Muslim Chinese.* Cambridge: Harvard University Press.

_____. 1997. Rumblings from the Uyghur. *Current History* 98: 287–290.

_____. 2002. Xinjiang: China's Future West Bank. *Current History* (September): 267–270.

Harris, L. C. 1993. Xinjiang, Central Asia, and the Implications for China's Policy in the Islamic World. *China Quarterly* 133: 111–129.

Hoppe, T. 1987. An Essay on Reproduction: The Example of Xinjiang Uighur Autonomous Region. In *Learning from China?* ed. B. Glaeser, pp. 56–84. London: Allen and Unwin.

Hutzler, C. 1997. Chinese Border Shaken By Religious and Economic Discontent. *Associated Press*, November 24.

Jarring, G. 1986. *Return to Kashgar*. Durham, N.C.: Duke University Press.

Lattimore, O. 1940. *Inner Asian Frontiers of China*. New York: American Geographical Society, No. 21.

_____. 1950. *Pivot of Asia*. Boston: Little, Brown and Co.

McMillen, D. H. 1979. *Chinese Communist Power and Policy in Xinjiang, 1949–1977*. Boulder: Westview Press.

_____. 1981. Xinjiang and the Production and Construction Corps: A Han Organisation in a Non-Han Region. *Australian Journal of Chinese Affairs* 6: 65–96.

Millward, J. 1998. *Beyond the Pass: Economy, Ethnicity, and Empire in Qing Central Asia, 1759–1864*. Stanford: Stanford University Press.

Pannell, C., and L. Ma. 1997. Urban Transition and Interstate Relations in a Dynamic Post-Soviet Borderland: The Xinjiang Uygur Autonomous Region of China. *Post-Soviet Geography and Economics* 38: 206–229.

Polo, M. 1968. *The Travels of Marco Polo*, trans. R. E. Lathan. New York: Penguin Books.

Rudelson, J. 1997. *Oasis Identities: Uyghur Nationalism Along China's Silk Road*. New York: Columbia University Press.

Starr, F., ed. 2003. *Xinjiang: Chinese Central Asia*. Armonk, N.Y.: M. E. Sharpe.

Stein, M. A. 1921. 3 vols. *Serindia*. London: Oxford University Press.

Toops, S. 1992. Recent Uygur Leaders of Xinjiang. *Central Asian Survey* 11: 77–99.

_____. 1995. Tourism in Xinjiang: Practice and Place. In *Tourism in China*, ed. A. Lew and L. Yu, pp. 179–202. Boulder: Westview Press.

_____. 1999. "Tourism and Turpan: The Power of Place in Inner Asia/Outer China." *Central Asian Survey* 18 (3): 303–318.

_____. 2000. "The Population Landscape of Xinjiang/East Turkistan," *Inner Asia* 2 (2): 155–170.

Xinjiang Uyghur Autonomous Region Bureau of Statistics (XBS). 2001. *Xinjiang Tongji Nianjian* (Xinjiang Statistical Yearbook) *2001*. Beijing: China Statistics Press.

Xinjiang Uyghur Autonomous Region Population Census Office (XPCO). 2002. *Xinjiang Weiwuer Zizhhiqu 2000 Nian Renkou Pucha Ziliao* (Tabulation on the 2000 Population Census of Xinjiang Uyghur Autonomous Region). Ürümqi: Xinjiang People's Press.

Yu, Y. S. 1967. *Trade and Expansion in Han China*. Berkeley: University of California Press.

25

Province, Nation, and the Chinese Mega-State

Charles Greer

The return of Hong Kong to China in 1997 brought world attention to China's changing map. Much of the focus was on Britain's declining power, the resurgence of "Greater China," and widespread criticism of China's system of domestic governance. But behind this lay the important fact that China had created a unique entity, the Special Administrative Region, to integrate the former colonial territory. In the same year, a new national municipality was created at Chongqing, taking 30 million people and nearly 15 percent of the territory from Sichuan province, for the specific purposes of fostering economic development and facilitating management of the social, economic, and environmental aspects of the nearby Three Gorges Project.

These examples of administrative innovation are only the latest in five decades of communist rule. They follow the early creation of municipal units at several levels of government to extend urban control over proximate rural land and the implementation of autonomous regions, districts, and counties for special administrative conditions in areas of concentrated minority population. In the 1950s and 1960s, when development policy called for self-reliant economic bootstrapping, rural collectives were implemented to mobilize social resources in substitution for unwanted international capital resources. Since 1978, the special economic zones and open ports, the special development and special technology zones within cities, as well as the creation of Hainan province, all have addressed geographically specific elements of the country's development program.

These contemporary units are reminiscent of administrative innovations from China's past, when special purpose jurisdictions were used for frontier military colonies and *tu-si* (local authority) agencies in non-Han areas, for the Yellow River and Grand Canal management administrations, and for special status accorded to certain cities. Recognition of China's success at preserving and creatively adapting this feature of the country's pre-modern administrative heritage, especially during the decades of intensely doctrinaire Communist Party activism, is important to understanding the true nature of contemporary China.

And there is another aspect of this heritage that receives less attention than the special areas, but which is more important to understanding China as a country in the modern world. This is the system of basic administrative units—the Chinese provinces. The present chapter focuses on the provinces, with the goal of showing how they help define what China is today and hint at the role China will have in the global society of the twenty-first century. The system of provinces inherited from the pre-modern past was maintained through the difficult decades after 1949 and now is facilitating the industrialization of China's vast society. It may well provide a major strength in that country's future competition with other regions of the world.

The Provinces and Decentralization

A certain amount of scholarly attention has been given to the provinces in recent years, relating to the increase in their autonomy from central government control. A process of decentralization is recognizable, widely attributed to the effects of government policies implemented since 1978, by which provincial governments have taken on some economic decisionmaking responsibilities formerly held by the central government. A significant literature has emerged, examining what this decentralization process reflects about operation of the country's economic and political systems.

Indeed, the startling idea has been put forward that China's provinces might assert themselves to the point of threatening the cohesion of the People's Republic itself; but little research on the topic suggests this is likely to happen. The strongest statements in this direction are from Zheng (1994), who argues that the seeds for a federal system have already been sown in China, and from Ohmae (1995), who foresees global market forces emerging to eclipse the present system of international state governments. In Ohmae's scheme, certain regions will gain the power to supersede provincial and even national prerogatives, including the politico-military forces that have shaped states in the past. He postulates such "region states" in China centered on Guangzhou–Hong Kong, Yunnan-Laos-Vietnam, Taiwan-Fujian, the Tumen Delta, Dalian, and other areas.

Far more common, however, is the viewpoint (e.g., Womack and Zhao 1994) that in spite of regional economic specialization fostered by some provincial independence in foreign trade, the provinces of China will remain dependent on the country's internal markets and on the structure provided by central government.

Even in coastal southeastern China, where the strongest regionalism exists, breakaway provinces do not seem likely. Here the autonomy goes beyond economics to include social and cultural factors, even significant regional military establishments, but the cases of Guangdong (Goodman and Feng 1994), Fujian (Long 1994), and Shanghai (Jacobs and Hong 1994) all are convincingly shown to contribute more to the cohesion and augmentation of China's power than to the operation of centrifugal forces.

Some of the most interesting contributions to this discussion are provided by historical perspectives (e.g., Fitzgerald 1994; Lary 1997), which show that the relationship between provinces and the central government has fluctuated for hundreds of years, often reflecting regular shifts in the economic and administrative fortunes of the imperial system as a whole. Indeed, an interesting question to pursue is whether current conditions represent an evolution of China's historical center-province dynamic, continuing trends established in early modern times (through the eighteenth century), but which were interrupted by a century of turmoil after 1840 (see Naquin and Rawski 1987). Such a perspective is important for going beyond narrowly focused issues of contemporary political economy in seeking to understand the present and future roles of China's provinces.

The lack of attention in contemporary geographic literature to China's system of provinces is obvious as well. The provinces are geographic as well as economic and political entities. They are features of the humanized landscape that can be mapped and studied spatially, and which are produced by the processes of human society in its resource habitat. They are building blocks of the Chinese state, evolving away from their agrarian origins, but not diminishing in their potential for shaping the future of society.

One problem in launching a study of the provincial system is finding a context for comparative analysis of these units as geographic entities. However, a useful beginning point is suggested, if not pursued, in the observation of R. Bin Wong (1997:206): "As European states agree to delegate certain powers to supranational bodies while the central Chinese state finds its powers and responsibilities reduced, a new kind of parallel between Chinese and European state making is emerging: both China and Europe must redefine relationships among different levels of political power."

Comparison of Chinese provinces and European states is the subject of this chapter. Detailed consideration of the topic would require more space than is available, so the focus here is on three preliminary steps of the comparison: a survey of basic similarities and differences between the two systems of geographic entities, identification of factors underlying these similarities and differences, and discussion of possible implications for the future in the two different world regions.

Provinces and Countries: Similarities and Differences

As units of territory and population in the contemporary world, the components of China and Europe show marked similarities, although as units of industrial economy their differences are great. In their respective evolutions to organize society in the two different culture realms, they display certain historical similarities, although their disparate functions within those realms reflect the contrasting traditions for integrating society at the highest level.

Geographic Similarities

Europe's countries and China's provinces are juxtaposed graphically in Figure 25.1 and numerically in Table 25.1. The map shows outlines of these regions at the same scale and at comparable latitudes, for the impact of visual comparison. The table places the countries and the province-level units together in one list ranked by population size. An examination of the thirty-four countries in Europe and the thirty-two province-level units in China reveals that they vary dramatically in population and territorial size but within generally comparable ranges at both ends of the continent. Hong Kong is China's smallest territorial unit at 1,000 square kilometers, and Xinjiang its largest at more than 1.5 million square kilometers. But aside from China's special-purpose national municipality jurisdictions and its huge frontier autonomous regions, the main provinces range in size between 100,000 and 600,000 square kilometers, comparable to the range for the main countries of Europe.

In population size there is even less difference between the components of the two systems. In Europe, population ranges from more than 80 million to less than 2 million, and in China, with the new Chongqing municipality subtracted from Sichuan province, from more than 90 million to less than 3 million. Comparative densities of population confirm what is widely understood about China's tradition of intensive horticultural use of agricultural land, contrasted to Europe's more extensive cultivation practices (Table 25.2, page 430). A large portion of China's territory still has not achieved the demographic transition that accompanies the advent of industrial society, and thus even greater densities can be expected in much of China.

Economic Differences

The dramatic difference between Chinese provinces and the countries of Europe as industrially productive economic entities is illustrated in Table 25.3 (pages 431-432). This ranking from largest economy to smallest shows the degree of superiority in economic productivity that Europe had retained over China through the middle 1990s. The sixteen most advanced countries of western and northern Europe all rank above the strongest of the provincial economies, several of them by more than an order of magnitude. The Chinese entities are comparable to the countries of the former Soviet sphere in size of economy and in the role that foreign trade plays, though their generally larger populations result in lower per capita GDP figures. All of this confirms what is generally understood about the relative levels of industrialization in the two regions.

Nevertheless, certain features of the economic system make the prognosis for China's provinces less disadvantageous than might seem to be the case. First, to the

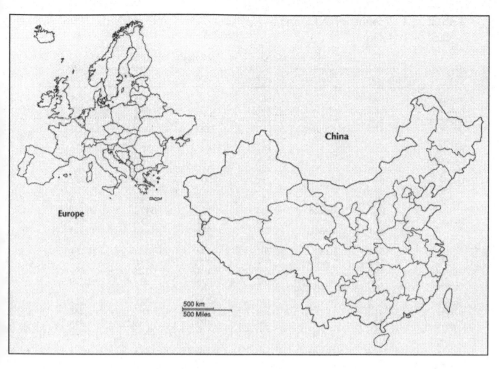

FIGURE 25.1   China and Europe: A Geographic Comparison

degree that larger populations represent greater potential for growth in productive capacity and markets, the role of the provinces in the global economy can be expected to improve more rapidly than that of the European countries. Growth rates may well remain higher in China than in Europe for some time to come. And growth will be particularly fast in certain units; national municipalities and coastal provinces will continue to take the lead in global competitiveness as long as the central government maintains its willingness to accept the disparity with poorer provinces that this creates. Even with these caveats, however, the Chinese units, with possible exceptions for cases such as Hong Kong and Shanghai, are not likely to match the economic levels of their European counterparts in per capita terms in the foreseeable future.

Historical Parallels

Longevity is a feature that the component units of Europe and China have in common, though in China there has been less shifting of boundaries in recent centuries. A historical atlas (Shepherd 1964) shows that kingdoms with identifiable approximations of some of today's national territories had emerged in parts of western and northern Europe by 1100 C.E., though the shape of other republics and empires is

TABLE 25.1   Countries of Europe and Province-Level Units of China, 1996,
Ranked by Population

Chinese Unit	Population (millions)[a]	European Country	Area (1,000 sq km)	Density (per sq km)	GDP[b] (1996 Bn US$)	GDP per capita)
Sichuan	114.3		567	201	50.8	$ 453
Henan	91.7		167	549	44.4	486
Shandong	87.4		153	571	71.8	823
	81.9	Germany	357	230	2,353.2	28,733
Jiangsu	71.1		103	690	72.3	1,018
Guangdong	69.6		178	391	78.5	1,146
Hebei	64.8		188	345	41.6	644
Hunan	64.3		210	306	31.9	498
Anhui	60.7		139	437	28.2	468
	58.3	France	544	106	1,540.1	26,417
Hubei	58.3		186	313	35.8	617
	58.1	UK	228	238	1,145.8	19,721
	57.2	Italy	301	190	1,207.7	21,114
	51.6	Ukraine	607	85	44.0	853
Guangxi	45.9		236	194	22.5	492
Zhejiang	43.4		102	425	50.0	1,139
Liaoning	41.2		146	282	38.0	931
Jiangxi	41.1		169	243	18.3	448
Yunnan	40.4		394	103	18.0	448
	39.7	Spain	505	79	581.6	14,650
	38.6	Poland	313	119	134.5	3,484
Heilongjiang	37.3		469	80	28.9	779
Guizhou	35.6		176	202	8.7	252
Shaanxi	35.4		206	172	14.2	399
Fujian	32.6		121	269	31.4	980
Shanxi	31.1		156	199	15.7	508
Jilin	26.1		187	140	16.1	622
Gansu	24.7		454	54	8.6	350
Inner Mongolia	23.1		1,183	20	11.9	513
	22.7	Romania	238	95	35.5	1,564
Xinjiang	16.9		1,600	11	11.0	623
	15.6	Netherlands	41	381	392.4	25,154
Shanghai	14.2		6	2,367	35.0	2,684
Beijing	12.6		17	741	19.5	1,813
	10.5	Greece	132	79	122.9	11,705
	10.3	Belarus	207	50	19.3	1,874
	10.3	Yugoslavia	102	101	–	–
	10.3	Czech Rep.	79	130	54.9	5,330
	10.2	Belgium	31	333	264.4	25,922
	10.0	Hungary	93	108	44.8	4,480
	9.8	Portugal	92	106	104.0	10,612
Tianjin	9.5		11	864	13.3	1,478
	8.8	Sweden	450	20	250.2	28,432
	8.5	Bulgaria	111	76	9.5	1,118
	8.1	Austria	84	97	226.1	27,914
Hainan	7.3		34	215	4.7	663
	7.2	Switzerland	41	175	293.4	40,750
	5.3	Slovak Rep.	49	109	19.0	3,585
	5.2	Denmark	43	122	174.2	33,500
Ningxia	5.2		66	79	2.3	450
	5.1	Finland	305	15	124.0	24,314

(continued on next page)

TABLE 25.1   Countries of Europe and Province-Level Units of China, 1996, Ranked by Population *(continued)*

Chinese Unit	Population (millions)[a]	European Country	Area (1,000 sq km)	Density (per sq km)	GDP[b] (1996 Bn US$)	GDP per capita)
Qinghai	4.9		721	8	2.2	452
	4.5	Croatia	57	80	19.1	4,244
	4.4	Moldova	34	129	1.8	409
	4.3	Norway	324	13	157.8	36,698
	3.7	Lithuania	65	57	7.8	2,108
	3.6	Ireland	69	51	69.6	19,333
	3.6	Bosnia-Herz.	51	71	–	–
	3.4	Albania	29	118	2.7	794
	2.5	Latvia	65	39	5.0	2,000
Tibet	2.4		1,228	2	0.8	329
	2.2	Macedonia	26	85	2.0	909
	1.9	Slovenia	20	95	18.6	9,789
	1.5	Estonia	45	33	4.4	2,933
Other Countries/Units						
	148.1	Russia	17,075	9	440.6	2,975
Chongqing[c]	30		82	366	–	–
Taiwan	21.5		360	60	272.2	12,659
Hong Kong[d]	6.2		1	5,924	154.8	24,968

[a] Chinese population at end of 1996, from *China Statistical Yearbook 1997*, China Statistical Press. European population at mid-year, 1996, from *World Population 1996*, United Nations Department for Economic and Social Information and Policy Analysis, Population Division.
[b] China Statistical Yearbook 1997; and World Bank, *World Development Indicators, 1998*.
[c] Created from territory of Sichuan Province in 1997.
[d] Became a Special Administrative Region of China in 1997.

unrecognizable. The outlines of the modern map solidified gradually through subsequent centuries, albeit with great plasticity of units, as empires were contested and nation-states emerged in the present century and even the recent decade.

In China, administrative units the approximate size of today's main eastern provinces have been utilized, albeit using a variety of nomenclature, to implement the rule of unified imperial dynasties since the fourth century of the Common Era (Whitney 1970). The configuration was considerably altered during periods of disunity, and the boundaries of even the most established provinces were adjusted through history as territory was traded among them by successive dynastic governments. Nonetheless, several of the coastal and southern provinces had emerged as early as the twelfth century with locations and shapes (if different names) recognizable on today's map (Ginsburg 1966). By Ming times (fourteenth through seventeenth centuries), they had re-emerged from the Mongol redrawing of the map with even more recognizable outlines. That system of primary provinces, solidified during Qing times (seventeenth through twentieth centuries) and inherited by the

TABLE 25.2   Population Density of Province-Level Administrative Units in China (per square kilometer)

Country	Population Millions	Rank	Area 1,000 sq km	Rank	Average pop density/ sq km
China	1,219	1	9,573	3	127
India	953	2	3,166	7	301
United States	265	3	9,529	4	28
Indonesia	98	4	1,919	16	103
Brazil	158	5	8,547	5	48
Russia	148	6	17,075	1	9
Pakistan	34	7	880	34	152
Japan	126	8	378	61	333
Bangladesh	123	9	148	93	834
Nigeria	104	10	924	32	113
Mexico	93	11	1,958	15	47
Germany	82	12	367	62	229
Vietnam	76	13	331	65	230
Philippines	72	14	300	72	239
Turkey	63	15	779	37	80
Iran	62	16	1,648	18	38
Egypt	61	17	998	30	61
Thailand	60	18	513	50	117
United Kingdom	59	19	244	79	241
France	58	20	544	48	107
Italy	58	21	301	71	191
Ethiopia	57	22	1,134	27	50
Ukraine	51	23	604	44	85
Myanmar (Burma)	46	24	677	40	176

People's Republic, has been adjusted since 1949 mainly in areas of minority concentration such as Sichuan and Inner Mongolia.

Functional Differences

Functional differences between the provinces and countries are apparent from the basic data included in Table 25.1. The use of national municipalities, autonomous regions, and a special administrative region for Hong Kong as province-level units in China contrasts with the European situation, where all units, regardless of size, are, in concept, nation-states. The literature is voluminous on both the evolution of Europe's nation-state system and the development of China's imperial state, but important studies continue to be produced (Armstrong 1982; Tilley 1975; Wong 1997).

The fundamental difference to recognize between the two systems is the national essence of European countries versus the administrative essence of Chinese provinces.

TABLE 25.3   Countries of Europe and Province-Level Units of China, 1996, Ranked by GDP

Chinese Unit	Population (million)^a	European Country	GDP^b 1996 Bn US$	GDP per Capita	Trade^c Volume Bn US$	Trade as % of GDP
	81.9	Germany	$2,353.2	$28,733	983	42
	58.3	France	1,540.1	26,417	566	37
	57.2	Italy	1,207.7	21,114	460	38
	58.1	UK	1,145.8	19,721	547	48
	39.7	Spain	581.6	14,650	224	39
	15.6	Netherlands	392.4	25,154	378	96
	7.2	Switzerland	293.4	40,750	151	51
	10.2	Belgium	264.4	25,922	332	126
	8.8	Sweden	250.2	28,432	152	61
	8.1	Austria	226.1	27,914	125	55
	5.2	Denmark	174.2	33,500	95	55
	4.3	Norway	157.8	36,698	85	54
	38.6	Poland	134.5	3,484	62	46
	5.1	Finland	124.0	24,314	68	55
	10.5	Greece	122.9	11,705	–	–
	9.8	Portugal	104.0	10,612	58	56
Guangdong	69.6		78.5	1,146	112	143
Jiangsu	71.1		72.3	1,018	22	31
Shandong	87.4		71.8	823	18	25
	3.6	Ireland	69.6	19,333	84	121
	10.3	Czech Republic	54.9	5,330	55	100
Sichuan	114.3		50.8	453	4	8
Zhejiang	43.4		50.0	1,139	14	29
	10.0	Hungary	44.8	4,480	29	65
Henan	91.7		44.4	486	3	6
	51.6	Ukraine	44.0	853	33	75
Hebei	64.8		41.6	644	4	10
Liaoning	41.2		38.0	931	14	36
Hubei	58.3		35.8	617	3	9
	22.7	Romania	35.5	1,564	20	56
Shanghai	14.2		35.0	2,684	28	80
Hunan	64.3		31.9	498	2	6
Fujian	32.6		31.4	980	16	51
Heilongjiang	37.3		28.9	779	4	15
Anhui	60.7		28.2	468	2	9
Guangxi	45.9		22.5	492	2	11
Beijing	12.6		19.5	1,813	15	76
	10.3	Belarus	19.3	1,874	13	68
	4.5	Croatia	19.1	4,244	12	63
	5.3	Slovak Republic	19.0	3,585	20	105
	1.9	Slovenia	18.6	9,789	18	97
Jiangxi	41.1		18.3	448	1	7
Yunnan	40.4		18.0	448	2	12
Jilin	26.1		16.1	622	2	16
Shanxi	31.1		15.7	508	2	14
Shaanxi	35.4		14.2	399	1	13
Tianjin	9.5		13.3	1,478	10	77
Inner Mongolia	23.1		11.9	513	1	8

(continued on next page)

TABLE 25.3 Countries of Europe and Province-Level Units of China, 1996, Ranked by GDP *(continued)*

Chinese Unit	Population (million)[a]	European Country	GDP[b] 1996 Bn US$	GDP per Capita	Trade[c] Volume Bn US$	Trade as % of GDP
Xinjiang	16.9		11.0	623	1	11
	8.5	Bulgaria	9.5	1,118	–	–
Guizhou	35.6		8.7	252	1	7
Gansu	24.7		8.6	350	1	7
	3.7	Lithuania	7.8	2,108	8	103
	2.5	Latvia	5.0	2,000	4	80
Hainan	7.3		4.7	663	2	43
	1.5	Estonia	4.4	2,933	6	136
	3.4	Albania	2.7	794	–	–
Ningxia	5.2		2.3	450	0.2	9
Qinghai	4.9		2.2	452	0.2	9
	2.2	Macedonia	2.0	909	–	–
	4.4	Moldova	1.8	409	2	111
Tibet	2.4		0.8	329	0.2	25
Other Countries/Units						
	148.1	Russia	440.6	2,975	151	68
Chongqing[d]	30		–	–	–	–
Taiwan	21.5		272.2	12,659	217	80
Hong Kong[e]	6.2		154.8	24,968	379	245

[a] Chinese population at end of 1996, from *China Statistical Yearbook 1997*, China Statistical Press. European population at mid-year, 1996, from *World Population 1996*, United Nations Department for Economic and Social Information and Policy Analysis, Population Division.

[b] *China Statistical Yearbook 1997* and World Bank, *World Development Indicators, 1998*.

[c] Sum of import and export values, 1996. *International Financial Statistics*, April 1998, International Monetary Fund, and *China Statistical Yearbook 1997*.

[d] Created from territory of Sichuan Province in 1997.

[e] Became a Special Administrative Region of China in 1997.

European countries evolved with focus on national character—the ethno-linguistic and cultural heritage of the dominant group around which a kingdom and eventually a nation-state was formed within the larger European culture realm. This realm was never, in spite of numerous attempts, integrated into one political system. Portions were integrated into various empires, from that of Rome to that of the Third Reich, but this function never overcame the more powerful force of national identity.

In China, of course, it was precisely the resurgence, in the sixth century C.E., of imperial integration of the entire (self-identified) civilized world that set a different historical stage for units of regional identity to evolve. The essence of a province grew around its region's role as an administrative subdivision of the grand empire. Since the bureaucratic state oversaw all of society's functions—economic, civil, and cultural—so too did the provinces within their own jurisdictions. In fact, the provinces managed much the same mix of societal processes as did the European states, but with an official administrative emphasis rather than a national one.

It is important to note that the feature of nationality, so heavily emphasized in Europe, has not been absent in China. Moser (1985) points out that differences in regional culture that are easily identified among various provincial groups of Sinitic peoples can be as great as cultural differences among the different nationalities in Europe. This comparison hinges on the root meaning of the word "nation." Although "nation" is presently used as an approximate equivalent for "country," in the root sense (based on the Latin *natio* for "birth") a nation is (from *Webster's New World Dictionary*) a "stable, historically developed community of people with a territory, economic life, distinctive culture, and language in common." Certain national population groups emerged in Europe centuries ago as the basis of kingdoms and then eventually of nation-states; some have done so within the past decade. In China, nation-statehood has not been a possibility for perhaps 1,500 years, because the forces of nationality were effectively subordinated to those of a more universal societal superstructure. Nevertheless, the equivalent groups have provided population bases for many of the provinces. This important historical similarity underlies the functional differences between provinces and countries.

The imperial heritage also helps to explain the greater geographic variety among China's provinces than that among Europe's countries. The empire's boundaries surrounded a much greater range of physical eco-regions, which supported significantly different human population densities and various land-use systems. The historical process of sinification transformed a given territory from pre-agricultural or nonintensively cultivated land to the intensively cultivated agri-scape of the Sinitic cultural realm. This process operated within the empire's political boundaries over the course of two millennia. It resulted in various degrees of sinification, depending largely on the nature of the areas's resource base and the length of time it was subjected to the sinification process. The effects of this process at the end of the twentieth century are reflected in the population densities of China's province-level units (Table 25.2).

Setting China's urban-based administrative units aside, the highest densities (400 persons per square kilometer and above) appear in areas of the eastern plains, which have relatively little nonagricultural land and reflect the longest duration of the Chinese cultural system. Provinces there have remained the centers of population concentration, production, construction, and commercial activity as society industrializes. New forces have begun modification of the agrarian landscape for mineral, energy, urban, and transportation development and for China's increasing participation in the global economy, but they have not launched a fundamental restructuring or redistribution of that landscape.

Intermediate population density values, between 100 and 300 persons per square kilometer, appear in areas to the west, south, and northeast of the core provinces.

Han Chinese settlement has a shorter history in these areas, and a larger percentage of less-intensively cultivated land remains. The regions of least dense contemporary settlement were of little or no economic utility to pre-industrial Chinese society, although they were of strategic importance. These frontier areas retain their historical roles of secondary economic regions and primary strategic zones. They have additional importance as natural resource bases for the industrializing economy and for heightened control in the globalizing security environment.

Post-Imperial Mega-State

The geographic weight of China's imperial heritage is reflected in Table 25.4, showing size and rank in both population and territory of the world's twenty-five most populous countries. Six of these countries can be called mega-states, in the sense that they rank very high in both land area and total population. Large territory in conjunction with large population suggests that the country must manage the sizable, and usually complex, societal systems generated by use of a variety of physical eco-regions. The six mega-states reflect the very different results of the period that shaped them—that of colonial/industrial expansion by the European nation-states. The United States, Brazil, and to a lesser degree Russia represent replacement of indigenous population and culture systems by those expanding from Europe. The Asian mega-states represent survivors of the colonial period, Indonesia being unique in that it was not a single entity before European contact.

Not only are China and India an order of magnitude larger in population than all but a few other countries of the world, but they represent the only two remnants of pre-modern, agrarian-based empires to have survived reasonably intact the period of European colonial expansion and industrial transformation. China's indigenous society survived less scathed than did India's; although had Pakistan and Bangladesh not been created in the wake of Indian independence, the two post-imperial mega-states would be even more similar in their size and impact on global affairs. It remains to be seen which features of each of these mega-states will be able to adapt for survival in the globalizing industrial world.

The significance of China's remnant imperial essence has been overlooked in much analysis of Chinese society in the period since 1949. So much focus was placed on the role of communism during its doctrinaire heyday through the 1970s that the power of pre-communist societal forms and processes was not given its due. This has begun to change with the flagging of radical influence and the improved clarity with which both Chinese and foreigners are able to view Chinese society.

Lucian Pye illustrated the imperial effect in the area of foreign affairs with his observation (1990: 58) that "China is a civilization pretending to be a state." His point was that some behaviors of members of China's highest leadership are very puzzling when compared with behaviors of heads of other modern states. However, these behaviors make sense when viewed within the historical context of China's

TABLE 25.4   The Twenty-Five Most Populous Countries

	Population (millions)	Rank	Area (1,000 sq km)	Rank	Average Density (sq km)
China	1,219	1	9,573	3	127
India	953	2	3,166	7	301
United States	265	3	9,529	4	28
Indonesia	198	4	1,919	16	103
Brazil	158	5	8,547	5	48
Russia	148	6	17,075	1	9
Pakistan	134	7	880	34	152
Japan	126	8	378	61	333
Bangladesh	123	9	148	93	834
Nigeria	104	10	924	32	113
Mexico	93	11	1,958	15	47
Germany	82	12	367	62	229
Vietnam	76	13	331	65	230
Philippines	72	14	300	72	239
Turkey	63	15	779	37	80
Iran	62	16	1,648	18	38
Egypt	61	17	998	30	61
Thailand	60	18	513	50	117
United Kingdom	59	19	244	79	241
France	58	20	544	48	107
Italy	58	21	301	71	191
Ethiopia	57	22	1,134	27	50
Ukraine	51	23	604	44	85
Myan Mar (Burma)	46	24	677	40	176
Zaire	45	25	2,345	12	19

SOURCE: *1997 Britannica Book of the Year*

imperial bureaucracy, when the primary obligation of China's leaders was to use power and authority for preservation of the moral order and the cultural attributes of their known world's central civilization rather than of a political entity operating among an assemblage of peers.

Conclusion

The administrative geography of empire—the system of provincial building blocks that together make the Chinese state—is a fundamental feature that survived China's semi-colonial experience. This important structure conditioned the country's transition to industrial society and the incumbent changes in the economic, social, and cultural framework. It helps to explain why China displayed a high degree of resistance to external forces for change, and why the country could control, to a greater degree than many others, the type and pace of industrialization that occurred. However, the price for this careful approach was a very slow process. China took more than a century (1842 to 1949) to accomplish what Japan essentially accomplished between 1853 and 1868. Additionally, the fact that this system addressed institutional and ideological change before fully pursuing technological change was puzzling to those whose political and economic systems would have begun with material issues.

Today the provinces as basic entities reflect this experience. Held back by the retarded pace of change until conditions in the country as a whole became favorable, they remain an order of magnitude behind their European counterparts in measures of economic development. Yet what they have preserved is a framework for integration of the provincial economies, a feature that Europe's countries have been striving for decades to achieve. The sovereignty of nation-states provided the impetus for Europe's colonial/industrial expansion, but now this sovereignty must be subdued for the European Union to become successful and effective in our contemporary world.

R. Bin Wong is correct in his observation that "both China and Europe must redefine relationships among different levels of political power." But the two redefinitions must proceed in fundamentally different directions: Europe to continue its historical search for a mechanism of integrating society at a level above the nation-state, and China to adapt its inherited structure for more efficient institutions and processes of industrial society.

The structural heritage of China's provinces, their endurance, even resurgence in recent decades, rests on their efficacy as units of political economy, nationality, and administration. Their importance as geographic entities in the configuration of the country is worthy of much more attention than it receives.

References

Armstrong, J. A. 1982. *Nations Before Nationalism*. Chapel Hill, N.C.: University of North Carolina Press.

Fitzgerald, J. 1994. 'Reports of My Death Have Been Greatly Exaggerated': The History of the Death of China. In *China Deconstructs*, ed. D. S. G. Goodman and G. Segal, pp. 21–58. London: Routledge.

Ginsburg, N., ed. 1966. *An Historical Atlas of China*. Chicago: Aldine.

Goodman, D. S. G., and Chongyi Feng. 1994. Guangdong: Greater Hong Kong and the New Regionalist Future. In *China Deconstructs*, ed. D. S. G. Goodman and G. Segal, pp. 177–201. London: Routledge.

Jacobs, J. B., and L. Hong. 1994. Shanghai and the Lower Yangzi Valley. In *China Deconstructs*, ed. D. S. G. Goodman and G. Segal, pp. 224–252. London: Routledge.

Lary, D. 1997. Regions and Nation: The Present Situation in China in Historical Context. *Public Affairs* 70: 181–194.

Long, S. 1994. Regionalism in Fujian. In *China Deconstructs*, ed. D. S. G. Goodman and G. Segal, pp. 202–223. London: Routledge.

Moser, L. J. 1985. *The Chinese Mosaic: The Peoples and Provinces of China*. Boulder: Westview.

Naquin, S., and E. S. Rawski. 1987. *Chinese Society in the Eighteenth Century*. New Haven: Yale University Press.

Ohmae, K. 1995. *The End of the Nation State: The Rise of Regional Economies.* New York: Free Press.

Pye, L. W. 1990. China: Erratic State, Frustrated Society. *Foreign Affairs* 69: 56–74.

Shepherd, W. R. 1964. *Historical Atlas.* New York: Barnes and Noble.

Skinner, G. W., ed. 1977. *The City in Late Imperial China.* Stanford: Stanford University Press.

Tilley, C., ed. 1975. *The Formation of National States in Western Europe.* Princeton: Princeton University Press.

Whitney, J. B. R. 1970. *China: Area, Administration, and Nation Building.* Chicago: University of Chicago Department of Geography Research Paper no. 123.

Womack, B., and G. Zhao. 1994. The Many Worlds of China's Provinces: Foreign Trade and Diversification. In *China Deconstructs,* ed. D. S. G. Goodman and G. Segal, pp. 131–176. London: Routledge.

Wong, R. B. 1997. *China Transformed: Historical Change and the Limits of European Experience.* Ithaca: Cornell University Press.

Zheng, Y.-N. 1994. Perforated Sovereignty: Provincial Dynamism and China's Foreign Trade. *Pacific Review* 7: 309–321.

26

Afterword: China Enters the Twenty-First Century

Chiao-min Hsieh

The Historical-Geographical Roots of China's Future

In concluding this book, it is useful to speculate about the future—to ask what China might be like in the coming century and what implications this might have for other people and countries. Fifty years from now, China will be the world's second most populous country with 1.4 billion people (second only to India). Her economy should have successfully completed the industrial transition and will have developed an indigenous technological capacity.

The fuel for China's rise to economic pre-eminence will be provided by continual modernization and growth along the coastal regions stretching from Hong Kong through Xiamen, Shanghai, Tianjin, and Dalian. Three great regions of industrial strength will emerge: (1) Shanghai, Nanjiang, Hangzhou; (2) Guangzhou, Shenzhen, Zhuhai; and (3) Beijing, Tianjin, Tangshan. These areas are likely to have strikingly similar levels of prosperity as those found in North America and Western Europe such as "Bos-Wash" (from Boston to Washington, D.C.), San-San (from San Francisco to San Diego), Chi-Pitts (from Chicago to Pittsburgh), London and its immediate environs, the Ruhr-Rhine Zone in Germany, and Amsterdam-Rotterdam and the Hague in the Netherlands.

Beijing will become a cultural center akin to Paris or Washington, D.C. Shanghai is likely to join Hong Kong as a great global financial center as cosmopolitan as New York or London, offering a wealth of investment, insurance, and banking resources. China will further its proud record in aerospace, engineering, medicine, semiconductor, and laser technologies and emerge as a power in computer science, telecommunications, and pure sciences. China's food, films, music, and art will have more followers the world over. Even tourism will be increased, with the expectation that by 2020, 130 million people will visit China each year—a dramatic leap from today's modest 24 million annual visitors.

If the preceding is correct, China will become a world power in the twenty-first century, similar to Great Britain in the nineteenth century and the United States in the twentieth century. Competition among rising China, Europe, and the United States may be intense as they contend for center stage in the theater of world politics.

About a century ago, every region in the world except China seemed to be undergoing change. Although China has never been stagnant, today it appears that everything in China is changing. No other country has been so affected by the past as China. But today no other country is led by a regime so determined to ignore the past. History's most conservative nation has now seemed to become its most radical. To comprehend this situation, one needs to understand China's geographical background as well as its history.

China is an isolated country, surrounded by formidable land barriers in the north and west and by the sea in the east and south. Unlike the ancient empires of the Western world, China had no Athens or Alexandria to serve as a foundation culture. China developed its own language, customs, philosophies, and values and has considered its neighbors barbarians. The Chinese call their country *Zhongguo,* the Middle Kingdom or the realm between heaven and earth. They considered their emperors to be not just rulers of China but sons of heaven, who reigned over all civilization.

China occupies a unique position in space and time. Only the United States, Europe, and China occupy large territories in the middle latitudes. Whereas the history of the United States spans only two hundred years, China's spans several thousand. And whereas Europe is composed of many nations with no central authority, language, or customs, China has been able to centralize its power despite its imposing size.

Its modern history began in the eighteenth century, as European nations seeking raw materials and markets in the Far East sought trade with China. The Chinese, self-centered and disinterested in trading with the West, regarded the Europeans as barbarians. Failing in their efforts to establish trade, Western powers forced their way into China, touching off the Opium War between China and Great Britain. With guns and steam power, the armies of the West subdued the Chinese, who were helpless to resist the intruders. Totally defeated by foreigners for the first time, the Chinese lost confidence in their time-honored culture. They began to admire the technological material achievement of the West. They preferred Western machine guns and steam-powered warships to the traditional Chinese culture such as poems and Confucian philosophy.

China's contact with the Western world inspired two geopolitical changes. The first was the development of coastal areas; the second was basic social change, which led to reform movements and eventually to political revolutions. Before the eighteenth century, the front entrance to China lay in its interior northwest, with the southeastern coastal region as its back door. After contact with the West, the coastal regions replaced the interior as the front door, and the ports of Shanghai, Guangzhou, Tianjin, and Qingdao became metropolises that served as gateways to trade and new ideas.

In the early phases of political reform, the Chinese people tried to learn about modernization from Japan, Britain, and the United States. But it was the Russian revolution that ultimately drew their attention; the Chinese were impressed by the speed at which Russia was transformed from an agrarian to an industrial country. The Chinese set out to learn from Russia.

Dazzled by Western technology and power, China's early leaders, Hong Xiucuan, Kang You-wei, and Sun Yat-sen began the imposing task of transforming an agrarian, essentially medieval, society into a bustling, highly industrialized one. Progress was slow at first, but the success of the Russian economic transformation in the 1920s inspired the Chinese to emulate their neighbors.

Mao's Soviet-Style Planning and Deng's Second Revolution

No sooner had the Japanese been defeated in 1945, after eight years of war in China, than the "People's Revolution," under the leadership of Mao Zedong, succeeded in 1949. With the establishment of a new regime, the communist leadership embraced an ambitious plan of industrialization and a Soviet-style economic system, including central planning, collective agriculture, and land reform, with the goal of developing the nation's heavy industries. The entire nation was ready to learn from Russia's economy, science, engineering, art, and literature. Many Chinese students traveled to the Soviet Union to study, and many Soviet technicians came to China as advisors.

Convinced by the material achievements of the Soviets, Chairman Mao and other leaders believed that China should develop heavy industry based on steel and machinery, even though they lacked sufficient investment capital, engineers, and managers. China's industries relied heavily on Soviet engineers and technicians, so much so that when border conflicts erupted in 1960 and Soviet personnel were withdrawn, the industries they supported nearly collapsed. Had China built its development efforts upon light industry and agriculture, it might not have suffered economically from its adversarial relationship with Russia. Changing and often conflicting political winds periodically swept over China during the 1960s and 1970s. The "Hundred Flowers Campaign," the great proletarian "Cultural Revolution," and the "Great Leap Forward" frequently diverted attention and resources away from the goal of modernizing China's economy.

After Mao Zedong's death in 1976, Deng Xiaoping spectacularly changed the direction of Chinese communism and saved China from oblivion. In 1978, China initiated a program of economic reform and an open-door foreign policy so radical that it has been called the "Second Revolution." Deng sincerely believed in one of the fundamental principles of Western capitalism: A society consists of individuals and, therefore, no society will prosper unless each individual is encouraged to do his best. Deng realized that China's poor and unproductive condition was due to a lack of personal initiative, a condition summed up by the phrase *da guofan,* or "everyone has rice to eat, whether he works hard or not."

Deng decided to reform Chinese society with determination as well as direction. He designed special economic zones, based on his idea that "Hong Kong is particularly special, so we should have one hundred Hong Kongs inside China." Businessmen across Southeast Asia and the West have been encouraged to invest in China's economic future, thanks, in part, to the rapid growth of China's privately owned economy. Accompanying the move toward private investment was a push to reform state-owned enterprises through decentralization, contract responsibility, and shareholding. Indirect economic control through negotiation and legal processes has reduced rigid state controls on the market. Rural economies were consolidated and improved to absorb surplus rural labor and promote the integration of rural and urban areas.

Not wanting to repeat mistakes by the Soviets in their rush to a market economy, China has gradually integrated an economic policy of both public and private ownership. Out of this mixed communism and capitalism has come what is known as the "the Chinese style of communism," which offers the promise of continued success as China faces the opportunities of the next century.

Three Environmental Change Projects

China has not only changed the direction of national economic policy but also tried to modify the physical environment of the country, for some aspects of the physical environment present handicaps for economic development. The most difficult one is water resources.

China is well watered; no other country has been so troubled by water problems or paid so much attention to water control as China. The water problem in China, however, is due not to the total amount of water available but to the lack of balance in supply between regions. For instance, the northwestern part of China has 51 percent of the cultivated land of the country but accounts for only 7 percent of the surface flow of water; whereas southeastern China has only 33 percent of the cultivated land but accounts for 76 percent of the surface flow. In order to balance regional water supply, the Chinese government is making efforts to balance water supplies in different regions.

At present, the Chinese government has launched three large environmental change projects: the Three Gorges Dam Project, the South-North Water Diversion

Project or Nan Sui Bei Diao, and the West-East Gas Project or Xi Qi Dong Song. While the former two deal with water balance, the last one concerns energy resources.

The costliest and riskiest change to China's natural environment will be the Three Gorges Dam Project on the Yangtze River, which has caused much debate and now is near completion. A dam with a height of 593 feet is being constructed in Yichang in Hubei province with an investment of $24.65 billion. When totally completed in 2009, it will provide the benefits of flood control, improved navigation, and power generation with a capacity of 18.2 million kilowatts.

A major justification for the dam is the electrical power, which will be the equivalent of burning 40 million tons of coal a year. The power is to be sent through transmission lines in central and eastern China to reduce energy shortages there where the economy is still developing. Because Chinese per capita energy consumption is only one-third of the world average, the market potential of the project is vast.

However, opponents argue that the project will eventually displace more than a million people, inundate some 1,300 archeological sites, and destroy the legendary beauty of the Three Gorges and thereby substantially reduce tourism revenue.

The second project is the South-to-North Water Diversion Project. Water distribution has been uneven in China. The cities of Northern China are parched and constrained by a growing shortage of water. Yet in China's wet south, the Yangtze River carries large amounts of water into the sea. Therefore, the government has tried to transfer water from the rainy south to the dry north by building canals, a great Chinese water-shifting technology.

This project consists of the east route, the central route, and the west route. The west route will divert water from the Dadu River, the Yatong River, and the upper reach of the Yellow River in order to increase the water supply in the Ningxia Hui Autonomous Region, the Inner Mongolia Autonomous Region, and Shanxi province. This route will be difficult to build due to the physical environment of the area. The central route will divert water from the Han River, a tributary of the Yangtze River, to Beijing and Tianjin. The east route is relatively easy to construct for it follows much of the same path as the Grand Canal. The Grand Canal, hundreds of miles long, is still in use by thousands of inhabitants who depend on it for work and living space, for the canal is used as a trade route.

The North-South Project has been compared with the Tennessee Valley Authority Project in America in the 1930s, which spurred the rural development of the region. The Chinese project is an even greater project, because it will affect not only local regions but the whole of China.

The third is the West-to-East Gas Project. In September 2002, China launched a $17 billion natural gas transmission project. It is designed to supply billions of cubic feet of natural gas each year to eastern China from the northwest through a pipeline up to 2,610 miles long. The pipeline starts from western Xinjiang Uygur Autonomous Region, runs through Central China, and ends at Shanghai.

When the project is completed in 2005, it is expected to transfer 4 billion cubic feet of gas annually for industrial and domestic use in Shanghai and other parts of the Yangtze River Delta.

Transmission of large amounts of gas to eastern energy-deficient China will help improve the energy structure and greatly reduce pollution caused by industrial waste there. As construction of the pipeline requires 1.74 million tons of steel, large amounts of cement, and other building material, it will stimulate the cement, timber, machinery, and steel industries in areas along the pipeline. Operation of the pipeline will also make it possible for 85 million urban households in eastern China to use gas for cooking. In Shanghai and Jiangsu province alone, 17 million urban households will have access to natural gas.

Current Internal and External Challenges

At the present time, China faces two important challenges; one is internal and the other external. There are significant regional disparities in China's population and resource distribution. The country has become divided into two economic regions. The eastern coastal region is rich, densely populated, and technologically advanced, while the interior west is poor, less populated, and technologically backward. For example, from 1982 to 1992 the total output value of industry and agriculture in Guangdong province, representing the eastern coastal region, increased from 41.5 billion to 492 billion yuan; per capita farmer's annual income increased from 182 yuan to 1,370 yuan. Meanwhile, total output value in Guizhou province, representing the western interior region, increased from 10.2 million to 44.6 million yuan and farmer's per capita yearly income increased from 108 yuan to 506 yuan. Thus, in 1982, the total output value of industry and agriculture for Guangdong was 3.07 times that of Guizhou province and increased to 11.03 times that for Guizhou in 1992. In 1982, farmer's per capita yearly income in Guangdong was 1.69 times that in Guizhou and had increased by more than 250 percent in those ten years. This regional disparity will continue to increase. The individual income gaps between different areas may create social instability.

China's immediate problem is how to improve the internal markets. The open-door policy to the world has played a decisive role in China's economic growth, and labor-intensive manufactured exports have constituted the most important source of China's wealth for the past several decades. In modern economic history, China's industrial development and trade expansion have been unique and unparalleled. Through adapting a policy of capitalism, absorbing massive foreign investment, and encouraging international trade, China's economy, with its abundant cheap labor, has experienced explosive growth. In 2000, China was reported to be the sixth largest trading nation in the world. Some economists predict that China will soon be second to the United States and ahead of Germany and Japan. As China's living standards improve, Western economies will benefit, for prosperous China will buy Western airplanes, foodstuffs, and pharmaceuticals.

Externally, China's most important issue is its relationship with the Western powers, especially the United States. China does not want isolation and needs friendship with the United States, for it is the only country capable of providing the capital and industrial technology that the Chinese most need. At present, the Chinese are in the mood to make money, not enemies.

Unlike Japan, China has a vast landmass with rich resources of uranium, oil, coal, copper, and bauxite and a fifth of the world's population, ensuring tremendous demand for all kinds of goods and much cheap labor. China has had 8–10 percent growth for more than a dozen years. She is still a poor country, but as her middle class grows, China will have an appetite for all sorts of American products. America buys from China clothing, shoes, toys, and electronics and sells grain, aircraft, and industrial equipment.

China will not be a threat to the West. China is not intent on world domination or global revolution. With 20 percent of the world's people and less than 7 percent of the world's arable land, China will not be self-sufficient in food, or many other commodities. For China, friendly relations and cooperative development with the rest of the world are in its best interests.

In 1990 the total value of China's trade with the other nations of the world was around $34 billion. In the first year of the twenty-first century, this trading value leaped to over half a trillion dollars—an increase of nearly sixteen-fold in one decade. General economic growth during this period was about nine percent. New highways and airports and efficient ports are all supporting the rapidly growing manufacturing sector of China's economy. Significantly, the United States, with its forty-one percent of China's exports, had the largest trade surplus with China.

Universities are also modernizing their programs of study for their students so that they will be prepared for positions of leadership in all phases of life in this new China. Other schools are also busy offering youth a curriculum that would enable them to find gainful employment in an economy that is demanding ever-higher levels of education.

As China searches for other ways to expand its economy, it increasingly looks to foreign investors and especially to the United States for essential capital, industrial technology, and managerial expertise. Such assistance is urgent as China searches for ways to develop its rich natural resources in oil, coal, copper, bauxite, and uranium, and to build the superstructure that will efficiently utilize these materials.

Among the most important of China's international relationships is that with the United States. Cooperation between the United States, the sole remaining world superpower, and China, the emerging world power, could bring stability not only to Asia but to the entire world. China's growth as an economic power, fueled by its ample resources, cheap labor, and increasing consumption of all sorts of goods, is also of interest to the United States. Although predicting how the relationship between these two world powers will unfold is difficult, the story of two similar empires of the past may provide some insight.

About two thousand years ago, two great states dominated the globe: the Roman Empire in the West and the Han Dynasty of the Central Kingdom in the East. Both were militarily strong, politically powerful, economically prosperous, and socially stable. Despite physical barriers such as the Tibetan Plateau and the Gobi Desert, the so-called "Silk Road" economically linked the two nations. Unfortunately, after four hundred years, the Han Dynasty and the Roman Empire both collapsed. Europe was overrun and entered a cultural Dark Age. Similarly, northern nomads marched into China, dividing it into sixteen states. But European civilization was kept alive by the Christian religion during the Dark Ages, just as Buddhism enriched Chinese religious life after the Han Dynasty. Twenty centuries later, it is vital to realize that knowledge and understanding are still more important to civilization than military power.

For people who wonder just what the twenty-first century has in store for us, some might already anticipate that the rise of a modern, technologically advanced China could well be the single most important event to happen in our community of nation-states. In a world whose nations are becoming increasingly interdependent, the ripple effect of a new economically and politically powerful China would obviously be felt everywhere. The consequences for China herself, in both her internal and external worlds, would likewise be profound.

Whatever happens in the future, we can be sure that geographers, such as those who shared their most informative research for *Changing China*, will continue to monitor and analyze these events and, by so doing, help people everywhere to understand this latest chapter in the lives of the Chinese whose civilization stretches back to antiquity.

Chronology of Events

China's Dynasties

Period/Dynasties	Dates
Yangshao Culture	ca. 5000–3200 B.C.
Longshanoid Culture	ca. 3200–1850 B.C.
Xia (mythological?)	ca. 1994–1766 B.C.
Shang (Yin)	1766–1122 B.C.
Zhou	1122–221 B.C.
Qin	221–207 B.C.
Han	207 B.C.–A.D. 220
Sui	A.D. 581–618
Tang	A.D. 618–907
Song	A.D. 960–1279
Yuan (Mongol)	A.D. 1279–1368
Ming	A.D. 1368–1644
Qing (Manchus)	A.D. 1644–1911
Republic	A.D. 1912
People's Republic	1949–present

Key Events Before the Founding of the People's Republic

1839–1842	First Sino-British Opium War
1942	Treaty of Nanjing is signed; Hong Kong is ceded to Britain.
1851–1864	Taiping Rebellion
1856–1860	Second Opium War
1894–1895	First Sino-Japanese War
1900	Boxer Rebellion

October 10, 1911	Imperial rule collapses when provinces declare independence from the Qing dynasty.
January 1, 1912	Sun Yat-sen, regarded as the father of modern China, is named provisional president of the Republic of China.
February 1912	Sun resigns and Yuan Shih-kai, a reformist official and chief trainer of the army, becomes first president.
1913	Yuan dissolves parliament and takes dictatorial power.
1916	Yuan dies; China disintegrates into regionalism ruled by feudal warlords.
April 1917	Sun declares himself generalissimo of his own regime.
1919	May 4th Movement
July 1921	Chinese Communist Party is founded in Shanghai.
1923	Communists and Sun's Nationalists ally to drive out the warlords.
1925	Sun dies and is succeeded by Chiang Kai-shek.
1927	Communist-Nationalist alliance breaks; civil war starts.
1928	Chiang establishes Republic of China in Nanjing, but most of China still ruled by warlords.
1931	Japan invades Northeast China (Manchuria) and sets up former Qing emperor as puppet-emperor of Manchukuo.
1934–1935	The Long March.
1937	Japan attacks near Beijing in July; Nanjing falls to the Japanese in December. Communist-Nationalist civil war suspended for Anti-Japanese War, which merges into World War II.
1945	World War II ends; civil war resumes.
1949	Chiang Kai-shek and his supporters flee to Taiwan.
February 3, 1949	Communist troops enter Beiping (Beijing).
October 1, 1949	Communist leader Mao Zedong (Mao Tse-tung) declares founding of the People's Republic of China.

Key Events Between the Founding of the PRC and the Inception of Reform

1950	Mao quickly brings China under his control and allies the country with the Soviet Union.
June 25, 1950	The Korean War breaks out. Over 700,000 Chinese troops ("volunteers") are involved in the Korean War, with heavy casualties.
October 25, 1950	Troops from the United States and China clash near Pukchin, Korea.
November 26–27, 1950	China sends a huge army against United Nations troops in Korea.
1952	Land reform is completed.
1953	First Five-Year Plan (FYP, 1953–1957) for economic development is announced with large-scale support from the Soviet Union for industrialization. Many Chinese technicians go to study in the Soviet Union, and 11,000 Soviet specialists come to China.

First National Population Census reveals the total population as 582,600,000.

Stalin dies.

The Korean War ends as the principal combatants sign an armistice.

1955   Agriculture is collectivized; the process completed in 1957.

1957   The Hundred Flowers Campaign begins, during which many intellectuals voice grievances against the state. Mao immediately begins an "anti-rightist" campaign to punish those who speak out. Numerous intellectuals disappear into labor camps or the countryside.

1958–1960   The Great Leap Forward is implemented, establishing People's Communes (huge rural collectives) in the countryside. By the end of 1958, about 99% of the rural population is reorganized into 26,000 communes. "Backyard furnaces" are built to produce iron and steel locally, which is a disastrous failure.

1959   Armed revolt erupts in Tibet and is suppressed by the Chinese army; the Dalai Lama goes into exile in India.

1960   The Soviet Union withdraws all of its technicians and specialists from China.

1960–1962   Three years of natural calamities, also called "Three Difficult Years," occur, during which mass starvation takes place due to poor planning, bad harvests, and social dislocation; millions of people perish.

1962   Border clashes erupt between China and India over disputed boundaries ("the McMahon Line") imposed during British rule in India.

1963   The Socialist Education Campaign commences, with the hope of reviving interest in socialist ideology.

1964   China tests its first atomic bomb.

Second National Population Census.

1965   China achieves oil self-sufficiency.

Border conflicts erupt with India.

Tibet becomes an autonomous region of China.

1966–1976   The Cultural Revolution

1967   China tests its first hydrogen bomb.

1968   Urban youth are sent to the countryside during the rustication campaign.

1969   Chinese and Soviet border patrol troops clash.

1970   U.S. Secretary of State Henry Kissinger secretly visits China in July, following the goodwill invitation of the U.S. table tennis team in April; China's first satellite is put into orbit.

1971   China is admitted to the United Nations; the seat for "the Republic of China" is suspended.

1972   President Nixon visits China, ending decades of isolation. China and the United States jointly issue the Shanghai Communiqué.

1975	Chiang Kai-shek dies. His son, Chiang Ching-kuo, succeeds him.
1976	Premier Zhou Enlai dies on January 8, and a mass demonstration, paying tribute to him, occurs in Beijing on April 4–5. After an extended illness, Mao Zedong dies on September 9. The "Gang of Four," scapegoats for the chaotic, ultra-leftist 1966–1976 Cultural Revolution, led by Mao's widow, Jiang Qing, are removed, and Hua Guofeng assumes power.
July 28, 1976	China's worst earthquake in many centuries occurs in Tangshan, with a death toll of close to one-quarter million.
1977	Urban youth start to return to cities.
July 22, 1977	Deng Xiaoping makes his first appearance after Mao's death and is named vice premier.
August 12, 1978	China and Japan establish diplomatic relations.

Key Events Since the Reform (Deng Xiaoping Era)

December 1978	The Third Plenum of the 11th Central Committee of the Communist Party adopts Deng's program of "reform and opening up" to the outside world; this marks the start of China's transformation from a planned economy to a market-oriented economy.
Winter 1978–1979	Political activists post pro-democracy essays on a wall in central Beijing in the short-lived Democracy Wall movement.
1979	Implementation of the household responsibility system in agriculture begins, and the "one couple, one child" policy is initiated, with strong disincentives for those who bear more than one child.
January 1, 1979	China and the United States establish diplomatic relations. The U.S. breaks relations with the Republic of China (Taiwan) and recognizes Beijing as the capital of China.
January 29, 1979	Deng Xiaoping visits the United States.
February 17, 1979	Chinese troops invade Vietnam to retaliate for Vietnam's invasion of Cambodia; the troops retreat three weeks later after heavy losses.
March 29, 1979	Police arrest Wei Jinsheng, one of at least 40 dissidents involved with the Democracy Wall who are arrested or silenced. Wei is later tried and sentenced to a 15-year prison term.
1980	Shenzhen, Zhuhai, Shantou, and Xiamen are designated as special economic zones. Zhao Ziyang replaces Hua Guafeng as premier; Hu Yaobang is appointed party chief. The U.S. extends Most Favored Nation status to China.

November 1980	The Gang of Four are tried and sentenced in nationally televised court proceedings.
	Mao's wife, Jiang Qing, is sentenced to death, later commuted to life imprisonment.
1982	The Third National Population Census reveals that China's population has surpassed 1 billion.
	The commune system is abolished.
1983	Campaign against spiritual pollution, aimed at eradicating the influence of the West, is initiated.
1984	Fourteen coastal ports are designated "open cities."
October 1984	The Communist Party announces momentous economic reforms, including plans to lift government price subsidies and relinquish party control over state enterprises.
December 19, 1984	China and Britain sign an agreement to return the British colony of Hong Kong to Chinese rule on July 1, 1997.
September 1986	China's first stock market opens in Shanghai.
December 10–30, 1986	Thousands of students in Shanghai protest, calling for more democracy; demonstrations spread to Beijing.
January 16, 1987	Communist Party chief Hu Yaobang becomes the scapegoat for student protests and is forced to resign. He is succeeded by Zhao Ziyang.
April 1987	Portugal and China sign an agreement laying the ground for the return of Macao to China in 1999.
July 14, 1987	Martial law is lifted in Taiwan, allowing new political parties and ending military censorship.
October 25, 1987	The Communist Party convenes its 13th Congress; Deng steps down from all but the top military post. The Party vows to continue economic reform and open-door policies.
January 1988	Chiang Ching-kuo dies. Lee Teng-hui is named as his successor.
March 1988	Riots erupt in Tibet.
April 1988	Li Peng is named premier.
	Hainan province is designated a Special Economic Zone.
August 1988	Months of runaway inflation set off panic buying. By year's end, consumer prices rise as much as 36 percent over the previous year.
April 15, 1989	Former party chief Hu Yaobang dies of a heart attack; two days later, thousands of students in Beijing and Shanghai take to the streets to mourn his death. This is the start of the seven-week Tiananmen Square prodemocracy protests, which eventually spread to about 80 cities.
May 15–18, 1989	Soviet leader Mikhail Gorbachev visits Beijing to mend the 30-year Sino-Soviet rift.

May 18, 1989	One million protesters fill Tiananmen Square. Two days later, martial law is imposed in Beijing.
June 3–4, 1989	Troops enter Beijing to stifle pro-democracy student demonstrations, firing on the crowds and clearing Tiananmen Square. Many unarmed civilians are killed.
June 24, 1989	The Communist Party chief, Zhao Ziyang, is ousted for allegedly supporting the protests; Jiang Zemin becomes Deng's third hand-picked successor; Deng resigns from his last official post.
1990	The development of the Pudong district in Shanghai is approved. The first free trade zone is established in Pudong.
	Fourth National Population Census.
1991	A second official stock exchange is opened in Shenzhen.
1992	Deng tours the southern provinces in January and February, urging more bold economic reforms; his remarks jump-start reform measures that were stalled after the Tiananmen Square crackdown.
	In April, the National People's Congress approve the Three Gorges Project. The first phase is to be completed by June 2003 and the whole project by 2009.
	Japanese Emperor Akihito visits China in April and admits war-time guilt for the "great suffering" caused by Japan but stopped short of apologizing.
1993	China's highly publicized campaign to win the right to host the Year 2000 Summer Olympic Games fails. That right is awarded to Sydney, Australia.
1994	Deng makes his last public appearances at Spring Festival in February.
	Work begins on the Three Gorges Dam project.
	China establishes its first direct link to the Internet through the National Computing and Networking Facility of China, anchored by the Chinese Academy of Sciences.
January 30, 1995	Jiang Zemin proposes 8 points on unifying Taiwan.
1996	Lee Teng-hui wins the first democratic elections in Taiwan on March 23.
	The Beijing-Kowloon Railway is completed.
February 19, 1997	Deng dies from complications from Parkinson's disease and a lung infection at age 92.
March 14, 1997	Chongqing becomes China's fourth municipality, after Beijing, Shanghai, and Tianjin.
July 1, 1997	Hong Kong reverts to Chinese sovereignty after 156 years of British colonial rule. Tung Chee-hwa becomes the first chief executive of the Hong Kong Special Administrative Region.
October 1997	Jiang Zemin visits the United States.

November 8, 1997	The Yangtze River is diverted from its natural course, clearing the way for construction to begin on the Three Gorges Dam.
March 1998	The 9th National People's Congress elects Jiang Zemin as China's president and Zhu Rongji as premier.
June–September 1998	Devastating floods occur along the Yangtze River, Neng Jiang, and Song Hua Jiang in the Northeast.
June 25–July 3, 1998	U.S. President Clinton visits China, the first such visit by a U.S. president since 1989.
October 1998	Jiang Zemin orders Chinese military forces to relinquish their vast business operations in an attempt to contain smuggling, a trade dominated by the People's Liberation Army (PLA).
November 1998	Jiang Zemin visits Japan, the first visit to the country by a Chinese president.
April 25, 1999	About 10,000 believers of Falun Gong, a meditation sect, stage a silent demonstration around Zhong Nanhai, the central government compound in Beijing.
May 8, 1999	The United States bombs the Chinese Embassy in Belgrade during NATO's military campaign against Yugoslavia, killing three Chinese journalists and wounding more than twenty diplomats.
July 9, 1999	Taiwanese President Lee Teng-hui claims that the relationship between Taiwan and Mainland China should be considered a state-to-state one, which appears to abandon the long-standing "One China" policy.
July 22, 1999	Falun Gong is labeled an "Evil Cult" and banned in Mainland China.
October 1, 1999	China celebrates the fiftieth anniversary of the establishment of the People's Republic, with extravagant ceremonies and a military parade in Beijing's Tiananmen Square.
November 15, 1999	China and the United States sign a trade agreement, and the U.S. grants China normal trade status and backs China's application to join the World Trade Organization.
November 20, 1999	The Chinese unmanned space capsule Shenzhou is successfully launched and recovered.
November 1999	China starts a major new initiative to develop the western provinces and makes that a top national priority for the next decade.
December 20, 1999	Portugal hands the gambling enclave of Macao back to China, bringing an end to 442 years of Portuguese rule and closing the era of European colonialism in Asia.
February 2000	Jiang puts forth his "Three Represents" idea while visiting Guangdong, which states that the Communist Party must represent the need to develop advanced productive forces, the trend of advanced culture, and the fundamental interests of the Chinese people.

March 18, 2000	Chen Shui-bian, a Democratic Progressive Party leader, is elected president of Taiwan, marking the end of five decades of Nationalist Party rule on the island.
October 2000	New policy explicitly encourages individuals and families to purchase automobiles.
	The United States passes a law guaranteeing permanent trade relations with China.
December 2000	The State Statistics Bureau announces that China's GDP in 2000 exceeded $1,000 billion for the first time.
2000	Private banks open for business.
March 2001	The 4th Plenum of the 9th National People's Congress approves the essentials of the 10th economic and social development Five-Year Plan (2001–2005).
April 2001	A Chinese fighter jet and an American surveillance plane collide in midair off Hainan Island, leading to a tense standoff between China and the United States.
July 13, 2001	The International Olympic Committee selects Beijing as the host city for the 29th Summer Olympic Games in 2008, the first time ever for China.
November 10, 2001	China is accepted into the World Trade Organization (WTO), following nearly fifteen years of tortuous negotiations with the United States and other major trading partners; on December 11, China formally becomes the 143rd full-fledged member of the organization. On November 11, the WTO also extends membership to Taiwan.
March 2002	Tens of thousands of workers in Daqing and Liaoyang take to the street in protest of unpaid pensions, job losses, and inadequate severance packages provided to them by their state-owned factory employers.
July 4, 2002	West-east gas transfer project commences.
November 8–14, 2002	Chinese Communist Party's 16th Congress is held in Beijing. Jiang Zemin is replaced by Hu Jintao as the general secretary of the party, which marks the first peaceful transition of rule in modern China. Jiang, however, retains control of the all-powerful party Central Military Commission.
	The 16th Congress approves an amendment to the party constitution that would admit capitalist entrepreneurs into membership—a controversial measure first proposed by Jiang Zemin.
December 27, 2002	Work starts on the huge project that will move water from the Yangtze basin to the thirsty north over three pathways. With an official price tag of $58 billion, this is boasted as the world's largest hydrological project.
December 30, 2002	Successfully launched Shenzhou IV, the last unmanned space flight. China's first manned space flight is expected in late 2003.

2002   Hong Kong government enacts security legislation
                             (Article 23), which many believe will sharply curtail
                             the former British colony's freedoms by giving the
                             government a pretext to crack down on political activ-
                             ities and freedom of speech.

March 5–18, 2003   The 10th National People's Congress is held in
                             Beijing. Hu Jintao takes over from Jiang Zemin as the
                             President of China; Zhu Rongji is succeeded by Wen
                             Jiabao, a Deputy Premier under Zhu, as Premier.
                             The National People's Congress also approves the for-
                             mation of the State Asset Management Committee,
                             charged with managing state assets in the process of
                             state-owned enterprise reform.

June 24, 2003   The World Health Organization lifts its warning
                             against travel to Beijing because of concerns about
                             severe acute respiratory syndrome (SARS). SARS orig-
                             inated in South China's Guangdong province in
                             November 2002 and became an epidemic affecting
                             Beijing and many other places in the world. By June,
                             the flu-like illness infected nearly 8,500 people world-
                             wide and claimed 800 lives. Over 2,500 cases and
                             190 deaths were in Beijing.

(Compiled by Max Lu)

Glossary

autonomous regions
provincial-level administrative units established in China's minority areas, usually named after the dominant ethnic group. There are currently five autonomous regions: Nei Mongol, Ningxia Hui, Xinjiang Uygur, Guangxi Zhuang, and Xizang (Tibet).

Confucianism (Kongfuzi)
system of ethics based on the teachings of Kongfuzi (ca. 551–479 B.C.), who held that man would be in harmony with the universe if he behaved with righteousness and restraint and adhered properly to specific social roles. With its emphasis on the study of the classics, the worship of ancestors, and the submission to authority, Confucianism formed the dominant ethic of Chinese social units from the imperial government to the peasant family.

Cultural Revolution (1966–1976)
formally called the Great Proletarian Cultural Revolution. It was a complex social upheaval that began as a struggle between Mao Zedong and other top party leaders for dominance in the Chinese Communist Party and went on to affect all of China with its call for "continuing revolution." The excesses of the militant Red Guards intimidated the entire nation by trying ferociously to root out the "reactionary" elements in Chinese society. The revolution led to disastrous economic setbacks in the country.

danwei
place of work in post-1949 China; a company or an organization that functions as an employer and often provides housing and social services for its employees.

Five-Year Plan
the planning instrument used by China and other socialist countries for social and economic development. Production and other development goals are set by the government for every five-year period in its attempt to generate fast and desired development, but those goals are often too ambitious for the plan to be of much practical use.

floating population
the Chinese term for rural migrants. The successful rural reform has released a substantial amount of labor from agricultural production. Many rural people have migrated to cities in search of employment opportunities. The total number of such migrants was estimated at over 80 million in 1997.

Four Cardinal Principles
the Chinese Communist Party's instrument for maintaining a socialist system and the party's unchallenged political control. The Four Cardinal Principles state that China will stick to the leadership of the Chinese Communist Party; socialism; the dictatorship of the proletariat; and the ideologies of Marx, Lenin, and Mao Zedong.

Four Modernizations
the goal of Chinese domestic policy, announced in 1978, to develop the four areas of agriculture, industry, national defense, and science and technology. In pursuit of this goal, China under Deng Xiaoping implemented an open-door policy toward the West, developing special economic zones and sending students abroad to study.

Gang of Four
the label applied to the four "radicals" or "leftists" who dominated first the cultural and then the political events during the Cultural Revolution. The four members were removed from power shortly after Mao's death in 1976 and put on trial for their crimes.

Great Leap Forward (1958–1962)
the attempt launched by Mao Zedong to heighten economic productivity dramatically in China through mass organization and the inspiration of revolutionary fervor among the people. Exaggerated reports of the success of policies such as the radical collectivization of peasants into large "People's Communes" and the decentralization of industrial production temporarily masked the actual economic disaster and widespread famine brought by the Great Leap.

household responsibility system
a system of rural production introduced in China after economic reform to replace the People's Communes and to improve the dismal agricultural productivity. Under this system, the land owned by rural collectives was divided and contracted by peasant households, who manage the production on their land. Farmers are required to fulfill their contractual quota by selling a certain portion of their agricultural produce to the government but may retain however much remains or sell it in the market for cash. The household responsibility system restored the incentive system in Chinese agriculture and resulted in a huge increase in agricultural efficiency and production.

hukou
the household registration system adopted by the Chinese government in 1958 for political and social control. The system assigns either urban or rural status to every resident of the country. The status is inheritable and change from one status to

another is strictly controlled. Before the reform, the rationing of food and other life necessities was based on the *hukou* system, and only those with urban *hukou* were entitled to official residence and jobs in cities as well as government welfare benefits. The system has been slightly loosened since the reform.

Kuomintang (KMT)
the Chinese Nationalist Party, founded by Sun Yat-sen in 1912. It is currently the ruling party in Taiwan.

Long March
the retreat of the Communist troops in 1934–1935 from the eastern core areas of China to the interior refugee camps to escape annihilation by the Nationalist government. Thousands died during the journey, but the Communists gathered sufficient strength in the remote interior to challenge the Nationalists and emerge victorious in the late 1940s.

Mandarin *(Guo Yu)*
the standard Chinese language based on the northern dialect. It is the official language of China and Taiwan.

one country, two systems
a policy devised by the Chinese government to unify the country. Under this system, capitalism is allowed to exist in the former British colony of Hong Kong, which reverted to Chinese control on July 1, 1997. China hopes to reunify with Taiwan, which Mainland China considers a renegade province, using the same formula.

one couple, one child
The Chinese family planning policy adopted in 1979 to curb rapid population growth. Each couple is allowed to have only one child, and those who sign a pledge receive certain benefits from the government in child care, health care, and education, whereas couples having more than one child are severely punished. The policy has been widely accepted in urban areas but less successful in rural areas where the traditional preference for boys and the need to have male laborers work against the policy. In recent years, China has loosened the one child policy in order to remedy its drawbacks such as population aging and female infanticide. Ethnic minorities are generally exempted from this policy.

Pacific Rim
refers to the group of countries and parts of countries along the western arc of the Pacific Ocean, from Japan to New Zealand. This region has exhibited high levels of economic growth, industrialization, and urbanization since the 1960s and contains a number of the world's economic giants and fast-rising economic tigers.

People's Commune
central unit of economic and political organization in the countryside, some consisting of tens of thousands of families, introduced in the Great Leap Forward and

popularized again in the Cultural Revolution. Communes were further divided into brigades and production teams that directed labor and divided work points.

Republic of China (R.O.C.)
the government established by Sun Yat-sen following the 1911 revolution. It was ousted by the Chinese Communists in 1949 and maintained in Taiwan.

Shanxian **(Third Front) construction**
the campaign taking place in China during the 1960s and 1970s. Its purpose was to promote industrial development in dispersed, mountainous, and remote locations so as to increase the country's readiness for potential foreign attack. Many manufacturing plants were relocated from coastal cities to inland areas. Bad location and mismanagement of the projects resulted in huge losses and inefficiency.

Special Administrative Region (SAR)
the political status of Hong Kong after it returned to Chinese sovereignty in 1997. The SAR has much greater political, economic, and cultural autonomy from the central government in Beijing than other political subdivisions of China.

special economic zone (SEZ)
cities targeted by the Chinese Communist Party to accept direct foreign investment. They were designed to increase Chinese exports and act as a bridge for the adoption of foreign technology. The first four special economic zones—Shenzhen, Zhuhai, Shantou, and Xiamen—were established in 1979 and were followed by fourteen other cities plus the island of Hainan in 1986.

township and village enterprise (TVE)
nonagricultural enterprises set up in the Chinese countryside to absorb rural surplus labor and increase rural income. They may be owned collectively by a township or a village, or they may be privately owned. The Chinese government promoted TVEs to prevent rural people from flocking to already crowded cities.

xian
county, or the unit of government administration below the level of province.

xiang
township, or the unit of local government below the level of *xian*.

Index

Printed in the United States
by Baker & Taylor Publisher Services